Biomass—a Renewable Resource for Carbon Materials (2nd Edition)

Biomass—a Renewable Resource for Carbon Materials (2nd Edition)

Editors

Indra Neel Pulidindi
Pankaj Sharma
Aharon Gedanken

Basel • Beijing • Wuhan • Barcelona • Belgrade • Novi Sad • Cluj • Manchester

Editors

Indra Neel Pulidindi
Department of Chemical Sciences
G S F C Univeristy
Vadodara
India

Pankaj Sharma
Department of Applied Chemistry
University of Baroda
Vadodara
India

Aharon Gedanken
Department of Chemistry
Bar Ilan University
Raman Gan
Israel

Editorial Office
MDPI
St. Alban-Anlage 66
4052 Basel, Switzerland

This is a reprint of articles from the Special Issue published online in the open access journal *C* (ISSN 2311-5629) (available at: www.mdpi.com/journal/carbon/special_issues/5A8R85W45H).

For citation purposes, cite each article independently as indicated on the article page online and as indicated below:

Lastname, A.A.; Lastname, B.B. Article Title. *Journal Name* **Year**, *Volume Number*, Page Range.

ISBN 978-3-0365-9862-8 (Hbk)
ISBN 978-3-0365-9861-1 (PDF)
doi.org/10.3390/books978-3-0365-9861-1

© 2023 by the authors. Articles in this book are Open Access and distributed under the Creative Commons Attribution (CC BY) license. The book as a whole is distributed by MDPI under the terms and conditions of the Creative Commons Attribution-NonCommercial-NoDerivs (CC BY-NC-ND) license.

Contents

About the Editors vii

Preface ix

Farhan Chowdhury Asif and Gobinda C. Saha
Graphene-like Carbon Structure Synthesis from Biomass Pyrolysis: A Critical Review on Feedstock–Process–Properties Relationship
Reprinted from: C **2023**, 9, 31, doi:10.3390/c9010031 1

Tomy Muringayil Joseph, Aparna Beena Unni, K. S. Joshy, Debarshi Kar Mahapatra, Józef Haponiuk and Sabu Thomas
Emerging Bio-Based Polymers from Lab to Market: Current Strategies, Market Dynamics and Research Trends
Reprinted from: C **2023**, 9, 30, doi:10.3390/c9010030 30

Raúl Llamas-Unzueta, Luis A. Ramírez-Montoya, J. Angel Menéndez and Miguel A. Montes-Morán
Customised Microporous Carbon 3D Structures with Good Mechanical Properties and High Nitrogen Content Obtained from Whey Powders
Reprinted from: C **2023**, 9, 100, doi:10.3390/c9040100 53

Roberta Mota-Panizio, Ana Assis, Luís Carmo-Calado, Catarina Nobre, Andrei Longo and José Silveira et al.
Production, Characterization, and Activation of Biochars from a Mixture of Waste Insulation Electric Cables (WIEC) and Waste Lignocellulosic Biomass (WLB)
Reprinted from: C **2023**, 9, 49, doi:10.3390/c9020049 70

Silvia Izquierdo, Nazaret Pacheco, Carlos J. Durán-Valle and Ignacio M. López-Coca
From Waste to Resource: Utilizing Sweet Chestnut Waste to Produce Hydrothermal Carbon for Water Decontamination
Reprinted from: C **2023**, 9, 57, doi:10.3390/c9020057 87

Anawat Pinisakul, Nattakarn Kruatong, Soydoa Vinitnantharat, Ponwarin Wilamas, Rattikan Neamchan and Nareerat Sukkhee et al.
Arsenic, Iron, and Manganese Adsorption in Single and Trinary Heavy Metal Solution Systems by Bamboo-Derived Biochars
Reprinted from: C **2023**, 9, 40, doi:10.3390/c9020040 103

Raja Mistry, Tapash Kumar Roy, Sand Aldagari and Elham H. Fini
Replacing Lime with Rice Husk Ash to Reduce Carbon Footprint of Bituminous Mixtures
Reprinted from: C **2023**, 9, 37, doi:10.3390/c9020037 117

Michael Egyir, Innocent Yao Dotse Lawson, Daniel Etsey Dodor and Deogratius Luyima
Agro-Industrial Waste Biochar Abated Nitrogen Leaching from Tropical Sandy Soils and Boosted Dry Matter Accumulation in Maize
Reprinted from: C **2023**, 9, 34, doi:10.3390/c9010034 134

Otmane Sarti, Fouad El Mansouri, El Habib Yahia, Emilia Otal, José Morillo and Mohamed Saidi
Efficient Removal of Tannic Acid from Olive Mill Wastewater Using Carbon Steel Slag
Reprinted from: C **2023**, 9, 32, doi:10.3390/c9010032 148

Zhihong Liu, Weitao Cao, Man Zhang, Wenke Zhao and Yaning Zhang
Changes of C, H, and N Elements of Corn Straw during the Microwave Heating Process
Reprinted from: *C* **2023**, *9*, 117, doi:10.3390/c9040117 . . . . **172**

About the Editors

Indra Neel Pulidindi

Dr Indra Neel Pulidindi is currently working as a scientific consultant at JSCIAR, India. Dr Neel has worked as an assistant professor at GSFC University, Vadodara (19/9/2022–18/10/2021). He received his Ph.D degree from the Indian Institute of Technology Madras under the supervision of Professor T K Varadarajan and Professor (Em) B Viswanathan (2002–2010). He has worked in the laboratory of Professor Aharon Gedanken Israel on the biomass conversion to biofuels and biochemicals (2/9/2010–30/10/2016). He has worked in the laboratory of Professor Tae Hyun Kim at Hanyang University on biomass composition analysis, pretreatment, and conversion (1/11/2016–9/12/2017). Subsequently, he then worked in the laboratory of Professor Xinling Wang at Shanghai Jiao Tong University on carbon-fiber-reinforced plastics (14/4/2019–24/4/2021). Dr Neel has published 48 research papers and has 3 Indian patents, 4 US patent applications, 5 books, 1 ebook, and 7 book chapters and 5 ebook chapters to his credit. Dr Neel has guided several Ph.D, masters, undergraduate, and freshmen in their academic research curriculum and helped them to earn their degrees. His research metrix, namely H index and number of citations, are 23 and 1500, respectively. His researcher interests include CO_2 conversion, biomass conversion, and energy conversion and storage.

Pankaj Sharma

Pankaj Sharma (Varanasi, INDIA, 1976) is currently working as an Assistant Professor in the Department of Applied Chemistry, Faculty of Technology and Engineering, the Maharaja Sayajirao University of Baroda, Vadodara, Gujarat, INDIA. He earned his B.Sc. (Chemistry Hons.) and M.Sc. (Organic Chemistry) from Banaras Hindu University, Varanasi, India, and Ph.D (Chemistry) from the Maharaja Sayajirao University of Baroda, Vadodara, Gujarat, India. He entered into the profession of academics by choice. The prime goal of his life is to encourage, inspire, and motivate young talents to prepare them to achieve their aim and contribute their best to society. He has carried out research in the area of pharmaceutical chemistry, nanotechnoogy, energy storage materials, bioplastics, heterogeneous catalysis, mesoporous materials, task-specific ionic liquids (TSILs), green chemistry, and carbon capture and sequestration (CCS). Up until now, he has published 18 peer-reviewed papers in reputed journals, including SCI and SCIE. In addition to this, he has published one book chapter with Elsevier. He has edited two Books published by MDPI. He is also a Guest Editor of four MDPI Journals. He has been granted two patents, one Korean and one US. He has also published two Indian patents. His H index is 12 and his I index is 11. Before obtaining his Ph.D., he acquired the skills of a researcher by working in India's most advanced CSIR group of laboratories, like the Central Drug Research Institute (CDRI) and Central Salt and Marine Chemicals Research Institute (CSMCRI). But his research skills reached a new height by working with his Korean mentors, Dr. Il Hyun Baek at the Korea Institute of Energy Research (KIER), Prof. Chang Sik Ha at Pusan National University (PNU), and Prof. Jae Won Lee at Dankook University (DU) in the Republic of Korea, respectively, in their well-established and highly advanced laboratories during his postdoctoral experience.

Aharon Gedanken

Professor Aharon Gedanken is an outstanding and legendary scientist with over 900 publications in peer-reviewed journals of international repute and with a high impact factor. He has 37 patent applications, 5 books, and over 10 book chapters to his credit. Professor Gedanken has made remarkable contributions to the fields of sonochemistry and microwave technology and their applications to nanomaterials, biomaterials, and biofuels. He served as a faculty member in the department of Chemistry at Bar Ilan University for over 34 years (1975–2009) and has been an Emeritus professor in the same institute for over 13 years (2009–present). Professor Gedanken's research metrics, namely H-index and citations, are 117 and 51510, respectively. His research interests include solid-state chemistry, catalysis, energy, materials science, and biochemistry.

Preface

Carbon materials are the materials of the past, the present, and the future, catering to the needs of mankind. As with chemicals and fuels, a paradigm shift is happening in the production of carbon materials as well. A rapid transition is occuring in the nature of feedstock used for the production of carbon materials, namely from fossil-based to biomass-based. Biomass is the single largest carbon resource on the surface of planet Earth. The use of biomass as a starting material for the production of carbon materials has made the material sustainable, environmentally benign, and economically viable. Most often, the properties and the resulting applications of such carbon materials are either comparable or superior to the fossil-based counterparts. As is the diversity of the nature and the biomass contained therein, so is the diversity in the synthetic strategies, properties, and applications of the resulting carbon materials. The first volume of this Special Issue in C—*Journal of Carbon Research* on "Biomass – A renewable resource for carbon materials" has been successful, and the second volume in this series (second edition) has seen similar success, attracting 10 research groups from 14 different nations, showing the intensity of the research and development activity in the field globally. The Editors wish to acknowledge the 10 research groups that have contributed 10 scholarly papers (2 state-of-the-art reviews and 8 research articles) that have formed the 10 chapters of this Book. The content of this reprint deals with the use of novel biomass feedstock for carbon materials production, exotic synthetic strategies, new insights into characterization techniques, and diverse and unconventional applications, all centered around environmental remediation, a problem requiring a quick remedy for the wellbeing of mankind. Indebtedness is due to Dr William Wang, the Managing Editor of C—*Journal of Carbon Research* for his steadfast and unconditional support, without which this endeavor could not have been possible. Grateful thanks are also expressed to Professor B Viswanathan for introducing the astounding field of carbon materials to Dr Indra Neel.

Dedicated to "My LORD, my GOD, Jesus Christ. John 20:28". "My grace is sufficient for thee. 2 Corinthians 12:9".

Indra Neel Pulidindi, Pankaj Sharma, and Aharon Gedanken
Editors

and nitrogen-doped graphene quantum dots (NGQDs) for pH-responsive drug delivery have been developed by Darwin et al. [26]. Using microplasma processing, it has been found that only 4.5% of the NGQD ratio is necessary for the SDNCs to become tough, thereby avoiding any exposure to high temperatures or hazardous chemical cross-linking agents. These composites can be used in biomedical applications and offer advantages such as strong drug-loading efficiency, pH-controlled sustained release of drugs, and stable solid-state PL properties for monitoring and therapeutic treatment. This study has opened new possibilities in the development of environmentally friendly and biocompatible nanographene hydrogels with potential biomedical applications. Meanwhile, nano-graphene oxide (GO) is showing successful outcomes in orthopedic fields. According to the investigation conducted by Yitian et al. [27], 3D-printed biphasic calcium phosphate (BCP) scaffolds containing nano-GO have been identified to significantly enhance angiogenic effects as well as raise bone volume.

The food industry has also taken advantage of graphene-based materials for a variety of purposes, including aiding in plant growth, removing and detecting contaminants, and detecting patulin and quinolone [28]. Additionally, there are several articles discussing the application of graphene-based materials in medicine and biology [29], self-healing/protective coatings [30,31], smart drug/gene delivery [32], antimicrobial and coating applications in medicine and dentistry [33], strain sensors [34], catalysis [35], cryptography [36], electrosorption [37], desalination [38], electric vehicles, [39] as well as space technology [40].

The remarkable qualities of graphene have been well documented, but its utility is limited by the fact that its improved properties are only available with high purity. Moreover, the performance of graphene samples is affected by both the purity of the sample and the number of layers [15]. Thus, in the arena of graphene synthesis, the optimum outcome would be the mass production of graphene with purity levels equivalent to those produced by laboratory-scale synthesis.

The production of graphene and GLC materials can be performed using top-down and bottom-up methods. Top-down methods, such as mechanical exfoliation, chemical exfoliation, and chemical synthesis, are easy to apply for large-scale graphene production, but the quality of the graphene produced is often low. Bottom-up methods, such as chemical vapor deposition (CVD), epitaxial growth, and pyrolysis, are better for producing high-quality graphene with some structural defects and good electronic properties, though the amount produced is small. Defect-free, adjustable layer graphene can also be produced using bottom-up methods for special applications [41].

The CVD approach is highly cost-effective, but its yield is lower than other techniques and the removal of graphene from the metallic substrate is a complex operation. Moreover, this procedure can produce a considerable amount of hydrogen, which is a disadvantage. Mechanical exfoliation, on the other hand, presents a low yield output in overcoming the van der Waals force between the first and second layers without affecting the subsequent layers. This can have a detrimental effect on the performance of the devices due to the alteration of the 2D crystal's lattice structure. Additionally, the insolubility of macromolecules in organic syntheses, the emergence of unpredictable side effects with increasing molecular weight, and the inability of mechanical cleavage techniques to undergo mass production are all issues that must be addressed. Furthermore, liquid exfoliation techniques can yield low-conductivity graphene. Therefore, the development of a novel production process that can fulfill mass production and yield a superior end product is a subject of ongoing research [15].

Pyrolysis is a straightforward and well-liked thermochemical technique for synthesizing nanostructured carbon, which breaks down carbon sources into tiny pieces without oxygen. Historically, it has been employed in separate petroleum products but it is now utilized to convert waste agricultural materials into commercially and environmentally beneficial products. In recent years, microwave-assisted pyrolysis (MAP) of biomass has emerged as a promising pyrolysis method for minimizing time and energy, achieving

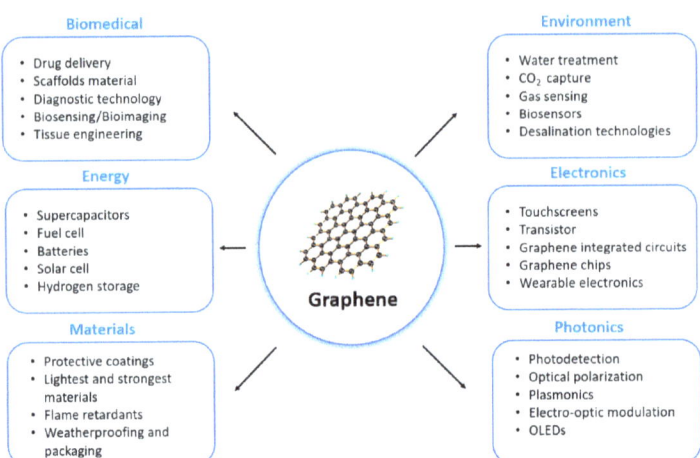

Figure 3. A diagram illustrating the various applications of graphene.

Graphene has been found to be an ideal material to improve the corrosion resistance of self-healing coatings due to its ability to enhance the mechanical properties, physical shielding ability, and energy conversion efficiency of the coatings [17]. For example, incorporating graphene into DA-epoxy resin composites has been shown to increase their self-repairing capacity and anti-corrosion performance [18]. Furthermore, graphene can also be used to tailor the flammability and flame retardancy of composite materials for safety applications. Taj et al. [19] investigated the flame properties of polymer-reinforced composites, with fillers consisting of nano-aluminum oxide and nano-graphene. It was determined that these fillers reduced flammability and improved flame retardancy, making them suitable for use in building materials. Tests were performed to analyze the burning rate, mass loss, and length loss of the composites, revealing that increasing the nano-graphene content up to 3% prevented flame propagation and resulted in a burning rate of zero.

Agrawal et al. [20] have developed a graphene-based filter to absorb CO_2 that met not only performance expectations for CO_2 collection but also recorded the highest CO_2 permeance. For instance, the filter exhibited a CO_2 permeance of 6180 GPU with a remarkable CO_2/N_2 separation factor of 22.5. This highlights the vital role that graphene can play in environmental applications. Additionally, a special form of graphene known as "Graph Air" is being employed in Australia to produce clean drinking water from highly contaminated water, with remarkable success compared to traditional filtration systems [21], demonstrating the versatility of graphene.

Graphene can have a significant positive impact on the energy sector too. Graphene Nanosheets derived from recycled plastics are being utilized in dye-sensitized solar cells (DSSCs) and supercapacitors, with DSSC yielding an impressive fill factor of 86.4% and a V_{oc} of 0.77 V, and supercapacitors achieving a remarkable specific capacitance of 398 Fg^{-1} [21,22]. In addition, the incorporation of highly conductive additives, such as graphene nanoparticles (GNPs), into phase change materials (PCMs) has been found to boost the thermal conductivity of these heat storage systems by as much as 220% when 3 wt.% GNPs are added to the PCM [23]. Graphene is also being investigated for potential use in smart wearable technology; for instance, researchers from Queen Mary University of London have developed a communicative piece of cloth utilizing graphene [24].

Graphene is also demonstrating its potential in the medical sector, with its antibacterial efficacy against both gram-positive and gram-negative bacteria being remarkable, and its high electrical and thermal conductivity making it a promising candidate in combating coronavirus [25]. In addition, smart drug nanocarriers (SDNCs) composed of chitosan

In addition, this versatile material is also lightweight, impermeable to all gases, highly resilient to high current density, and easily amenable to chemical functionalization. The potential of graphene has yet to be fully explored, and its versatile characteristics present numerous opportunities for further research. Figure 1 shows a schematic representation of graphene properties.

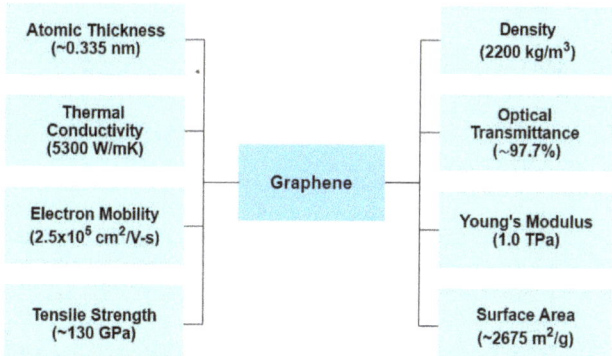

Figure 1. Schematic representation of graphene properties.

Graphene's honeycomb structure is the fundamental building block for the formation of other carbon allotropes, such as graphene oxide (GO), carbon nanotubes (CNTs), carbon nanodots, carbon nanoparticles, and fullerenes [14]. These allotropes differ structurally, with stacked honeycomb structures forming graphite, and rolled or wrapped honeycomb structures resulting in one-dimensional nanotubes and zero-dimensional fullerenes, respectively [15], as depicted in Figure 2a. Figure 2b–e represents the molecular structure of graphite, graphene, GO, and rGO.

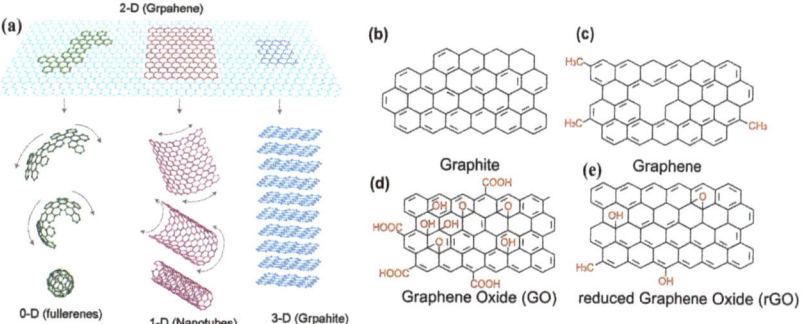

Figure 2. (**a**) Various carbon allotropes, such as 0D, 1D, and 3D, using 2-D graphene nanosheets as their basic building units (with permission from Ref [6]); the molecular structure of (**b**) graphite, (**c**) graphene, (**d**) GO, and (**e**) rGO [16].

The potential applications of graphene are far-reaching, covering areas such as photonics, composite materials and coating, energy generation and storage, environmental protection, biomedicine, and so on. Figure 3 depicts some of the important applications of graphene in research and industry.

Review

Graphene-like Carbon Structure Synthesis from Biomass Pyrolysis: A Critical Review on Feedstock–Process–Properties Relationship

Farhan Chowdhury Asif and Gobinda C. Saha *

Nanocomposites and Mechanics Laboratory (NCM Lab), University of New Brunswick, Fredericton, NB E3B 5A3, Canada
* Correspondence: gsaha@unb.ca; Tel.: +1-506-4587784

Abstract: Biomass pyrolysis is a promising route for synthesizing graphene-like carbon (GLC) structures, potentially offering a cost-effective and renewable alternative to graphene. This review paper responds to the call for highlighting the state of the art in GLC materials design and synthesis from renewable biomass microwave pyrolysis. This paper includes an introduction of the microwave pyrolysis technology, information on feedstock variability and selection, discussion on the correlation between microwave pyrolysis process conditions and pyrolyzed product characteristics, and, more importantly, a section identifying any differences between pyrolyzing feedstock using the microwave pyrolysis method vs. conventional pyrolysis method. Furthermore, this work concludes by detailing the knowledge currently missing with the recommendation for future research/innovation directions.

Keywords: graphene; graphene-like materials; carbon; biomass pyrolysis; microwave pyrolysis

Citation: Asif, F.C.; Saha, G.C. Graphene-like Carbon Structure Synthesis from Biomass Pyrolysis: A Critical Review on Feedstock–Process–Properties Relationship. C 2023, 9, 31. https://doi.org/10.3390/c9010031

Academic Editors: Indra Pulidindi, Pankaj Sharma and Aharon Gedanken

Received: 21 January 2023
Revised: 4 March 2023
Accepted: 8 March 2023
Published: 11 March 2023

Copyright: © 2023 by the authors. Licensee MDPI, Basel, Switzerland. This article is an open access article distributed under the terms and conditions of the Creative Commons Attribution (CC BY) license (https:// creativecommons.org/licenses/by/ 4.0/).

1. Introduction

In recent times, there has been a surge of enthusiasm for the utilization of renewable energy sources and the implementation of sustainable production approaches in order to overcome the forthcoming energy shortage and reach carbon neutrality [1–3]. As humanity's technological advances outpace the availability of resources, research endeavors are shifting towards the development of sustainable technologies that exploit novel materials with enhanced properties. Carbon, one of the most abundant elements on earth, has the potential to address this challenge. Traditional carbon materials, such as diamond, graphite, and carbon fiber, have been utilized for decades and have had a major role in economic and social progress [4]. The research in carbonaceous materials has evolved further since the discovery of carbon nanomaterials (CNMs) in the 1990s [5,6]. The distinctive and adaptable surface of carbon materials, along with their simplicity of production, makes them versatile targets and they are becoming increasingly important for use in the energy, biotechnology, biomedicine, and environment sectors.

Graphene, a single sheet of carbon atoms connected via sp^2 hybridization and arranged in a honeycomb lattice, has been widely regarded as one of the most revolutionary substances of the 21st century. Graphene has generated widespread interest in the scientific community since its discovery by Konstantin Novoselov and Andre Geim in 2004 [5], owing to its unparalleled characteristics [6]. Graphene is the world's first two-dimensional atomic crystal that has extraordinary characteristics, such as an electron mobility of $2.5 \times 10^5 \text{cm}^2\text{V}^{-1}\text{S}^{-1}$ at room temperature, extremely high thermal conductivity above 5300 W/mK, a Young's modulus of up to 1 TPa, a tunable surface area of about 2675 m^2/g, an atomic thickness of ~0.335 nm, a density of 2200 Kg/m^3, ~97.7% optical transmittance, and an intrinsic strength of 130 GPa [7–12]. Theoretically, graphene has a higher electric double layer capacitance and specific capacitance than activated carbon, with 550 Fg^{-1} and 268 Fg^{-1}, respectively, compared to activated carbon's 210 Fg^{-1} [13].

better heating efficiency, gaining greater control over the process, and producing more desired products than conventional pyrolysis [42]. However, the major challenge with this approach is the difficulty in regulating particle size, which leads to a wide range of diameters from 1 to 5 nm [43].

Biomass, composed of a high concentration of carbon and being a renewable resource, has been identified as a new potential source from which graphene and GLC materials can be derived. The amount of biomass waste produced around the world each year is estimated to be around 10 billion metric tons, and this number is anticipated to grow [44]. The advantages of biomass resources are their low cost and wide accessibility, which can potentially decrease the cost of graphene and other carbon-derived compounds. So far, several biomass materials had been effectively transformed into graphene and carbon-derived compounds, including wheat straw, sawdust, gumwood, bamboo, peanut shell, rice husks, sugarcane bagasse, orange peels, ginger, cotton, corncobs, and camphor leaves [45,46]. Biomass-derived graphene is frequently made up of aligned nanographene domains, which differ from ideal two-dimensional (2D) stacked graphene sheets and result in a variety of shapes, special functional groups, and amazing capabilities [47]. These carbons generated from biomass are frequently referred to as graphene-like materials.

In this review paper, we summarize and discuss some recent strategies to synthesize GLC material using solely the biomass pyrolysis process. We discuss the pyrolysis process and the correlation between different process parameters and pyrolyzed products, as well as the mechanisms of GLC material formation via biomass pyrolysis. Additionally, we provide a brief discussion on the special effect of microwave irradiation during pyrolysis on the morphology and microstructure of the pyrolyzed product.

2. Pyrolysis Process

Pyrolysis is a type of thermochemical conversion process that takes place without the presence of oxygen and is designed to break down the chemical bonds in a particular feedstock to decompose organic materials. This yields biochar, bio-oil, syngas, and other value-added products. The process is conducted in an oxygen-free environment with temperatures ranging between 400 °C and 1200 °C or even higher [48,49]. In this environment, biomass can be heated beyond its thermal stability limit without initiating combustion. Pyrolysis is a complex process involving various reactions and pathways such as depolymerization, dehydration, decarboxylation, intramolecular condensation, and aromatization, which take place at different temperatures and yield diverse product states for lignocellulosic components [49].

Depending on the heating mechanism used, pyrolysis can be classified into two categories: conventional pyrolysis (CP) and microwave-assisted pyrolysis (MAP). CP usually relies on an electric heating mechanism, which is often inefficient and energy intensive. MAP, on the other hand, has gained considerable attention from the research community due to its advantages over CP. MAP is faster, more energy-efficient, and offers greater precision over the process. Furthermore, MAP also results in higher heating rates and yields of desired products compared to CP [50–52]. Microwaves are a type of electromagnetic wave that falls between infrared and radio frequencies, with a frequency range of 300 MHz to 300 GHz and a wavelength that varies between 0.001 and 1 m. The majority of microwave reactors used in chemical synthesis, including those found in household kitchens, have a wavelength of 12.25 cm and a frequency of 2.45 GHz [53].

3. Microwave Pyrolysis Reaction Mechanism

In conventional pyrolysis, the heat is transmitted from an external source to the material's exterior and then to its core through conduction, convection, and radiation. Therefore, CP is inefficient, energy-consuming, and relies on convection and the thermal conductivity of the material being processed. In contrast, electromagnetic energy is transformed into heat energy in MAP. This occurs through microwaves entering the feedstock and then being stored as energy, which is then converted into heat inside the feedstock's core. This method is advantageous because it avoids heat losses due to volumetric heating of the feedstock [53].

In MAP, the temperature of the biomass particle increases from the interior to the exterior, but in CP, it is the opposite. Furthermore, for both MAP and CP, the diffusion of volatile materials (mass flow) is always outward. Thus, heat flow and mass flow are concurrent for MAP and countercurrent for CP. Figure 4 shows the schematic of microwave and conventional heating methods. While the volatile elements diffuse from the interior core of the feedstock to its exterior surface, the surrounding of the feedstock is extremely hot during the CP process and relatively cooler for the MAP process. As a result of the improved heating mechanisms described above, MAP's heating and response mechanisms have significant advantages over CP. MAP offers many advantages, such as quick, precise, and even heating, saving time and energy, eliminating direct contact between the heat source and the material, transferring energy instead of heat, low thermal inertia, no need for prior treatment of the feedstock, quicker response time, better control, improved safety, and so on. Because MAP speeds up thermochemical processes and shortens reaction times, it also has the added benefit of reducing energy consumption [53,54].

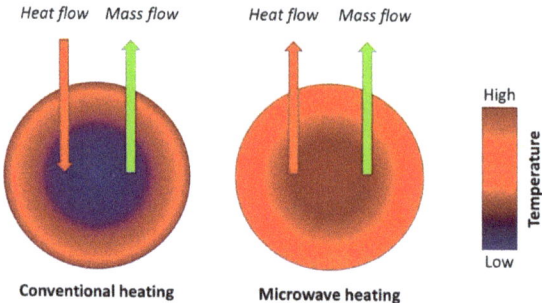

Figure 4. Illustration of the differences between microwave and conventional heating techniques. Reprinted from [54], with permission from Elsevier, Amsterdam, The Netherlands.

Also, it is worth noting that biomass does not absorb microwaves well, and therefore does not reach the necessary temperatures for pyrolysis when exposed to microwaves. Thus, pyrolysis systems employ external microwave absorption materials [55].

4. Key Distinction between MAP and CP

Robinson et al. [56] conducted a study that combined microwave pyrolysis, dielectric measurement, and fluid flow modeling to better understand the differences between microwave and conventional pyrolysis. Through their research, they were able to analyze and contrast the mechanisms of both processes. Their key finding was that the distinction is not between microwave and conventional heating, but rather between low and high heating rates. With low microwave power or domestic ovens, the heating rate can be comparable to conventional methods. When heating rates are low, vaporization of water within biomass structures is slow and pressure remains close to atmospheric. Pyrolysis in this case would proceed similarly to conventional methods with hemicellulose depolymerizing at temperatures over 200 °C, cellulose at 300 °C, and lignin in the 220–400 °C range with a diverse chemical composition produced. When heating rates are high (microwave heating), the vaporization rate increases, leading to pressure build-up, which elevates the boiling point of water remaining within biomass at temperatures well over 100 °C. This results in hydrolysis of hemicellulose at ~130 °C, producing furfural as the primary product; hydrolysis of cellulose occurs at ~175 °C, producing levoglucosan as the primary product; and lignin follows the same reaction scheme as conventional pyrolysis due to its lack of hydrolyzable linkages upon further heating, as illustrated in Figure 5.

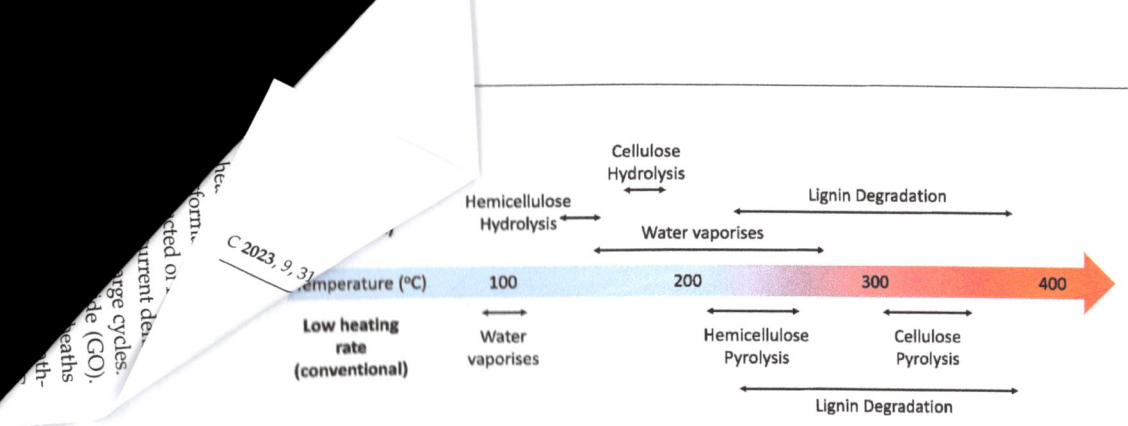

Figure 5. Scheme for MAP and CP. Reprinted from [56], with permission from Elsevier.

It was hypothesized that different behaviors with different biomasses occur not because of inherent chemistries but due to their micro- and macro-structures, which are characterized by permeability; high permeability would not sustain high pressures, so no difference between microwave vs. convectional pyrolysis would be expected. In contrast, low permeability does sustain those pressures, allowing for a difference in the mechanistic pathway based on the heating rate. The key finding of Robinson is also evident from the study of Zoraida et al. [57] who used anthracene oil as raw material for the production of carbon precursors via conventional and microwave technologies. In addition, they observed that the energy usage of microwave-assisted technology for producing these materials was up to 60% lower than traditional heating. The graphite obtained from microwave-based precursors had an excellent degree of graphitization, with smaller crystallite sizes than those from an exclusively microwaved coke. Graphene materials synthesized from these microwaved precursors showed improved lattice recovery, structure, and reduced oxygenated surface functional groups, especially when derived from microwave-derived coke.

5. Current Trends on Synthesis of GLC Materials via Biomass Microwave Pyrolysis Process

The process of creating GLC materials via microwave pyrolysis of biomasses has been explored by researchers in recent years. Even though they used various feedstocks and varied process settings, the basic process is nearly identical. This process involves three distinct steps: sample pre-treatment, pyrolysis, and post-treatment. During pre-treatment, the biomass sample is washed with deionized water to remove any contaminants, dried to eliminate moisture, and ground into a fine powder. Pyrolysis then takes place, followed by a post-treatment process that involves filtering, washing with deionized water, and drying. In this section, we have highlighted some of the recent studies on GLC materials synthesis through the biomass pyrolysis process.

Zhang et al. [58] investigated the synthesis of hollow carbon nanofibers (HCNFs) on the surface of biochar, using pine nutshell (PNS) as feedstock, without the use of a catalyst. The PNS was crushed until it had a particle size of 65–200 μm and then put into a vacuum oven and heated to 105 °C for 10 h to take out the moisture. Afterward, it was blended with commercially accessible biomass-based activated carbon (AC) in an 8:2 mass ratio. The samples were then pyrolyzed for 20 min at 400 °C, 500 °C, 600 °C, and 700 °C in a N_2 atmosphere. The results indicated that 600 °C was the optimal temperature for the synthesis of the HCNFs, resulting in the formation of fewer organic matrixes, functional groups, structural defects, and imperfections. It was proposed that pyrolysis volatiles forced their way out of surface pores, solidified, and graphitized the vapor on the biochar exterior, leading to the formation and development of HCNFs. On the other hand, Gopalakrishnan et al. [59] developed a simple one-step method to synthesize few-layer graphene-like porous carbon nanosheets (FLG-CNs) using ginger as a feedstock. The initial step of the process was to slice up the fresh ginger and rinse it with deionized water to eliminate any external impurities. This was followed by oven drying, and then pyrolyzing the ginger at temperatures of 600 °C, 800 °C, and 900 °C for an hour under an argon gas

atmosphere. Since ginger contains a variety of minerals, when it is
become porous and increase the surface area, which is ideal for the pe
double-layer capacitors (EDLCs). It was validated by the electrode constr
at 800 °C exhibiting an excellent specific capacitance of 390 Fg^{-1} when the
was 1 Ag^{-1}, and the capacity remained at 93.3% even after 3500 charge/disch

Researchers have utilized different biomasses to synthesize graphene ox
Danafar et al. [60] demonstrated the synthesis of nano-sized GO flakes from onion s
through pyrolysis coupled with sonochemistry. To remove the surface dust, onion she
ings were washed with deionized water and air dried. Then, they were pyrolyzed at 700
under N$_2$ gas flow. The obtained GO-like carbon flakes were washed with a combination
of water and ethanol in order to eliminate any remaining byproducts, then sonicated in
deionized water to convert them into nano-sized flakes (6.6 ± 2.4 nm) of a uniform size.
Figure 6 illustrates the X-ray diffraction (XRD) patterns of both nano-sized GO-like carbon
flakes and pyrolyzed GO-like flakes, where an intensification in the intensity of the carbon
flakes following ultrasound treatment was noticed, implying the reinforcement of the
stacking arrangement of the aromatic layers in the nanoflakes.

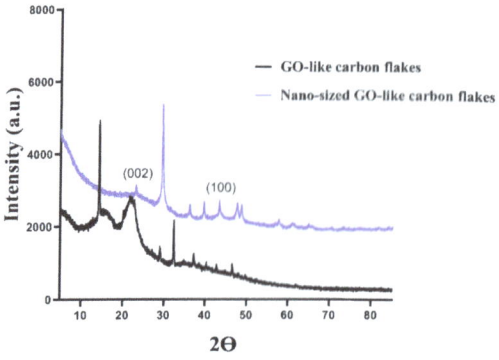

Figure 6. The XRD patterns of both pyrolyzed and nano-sized GO-like carbon flakes. Reprinted from [60], with permission from Elsevier.

A straightforward and economic procedure was developed by Somanathan et al. [61] to synthesize GO from agricultural waste. For this study, sugarcane bagasse was first crushed and grounded to produce a powder. A muffle furnace was used to heat 0.5 g of powder and 0.1 g of ferrocene for 10 min at 300 °C under atmospheric conditions. They found that the produced GO presented a well-graphitized structure. Another study was carried out by Hashmi et al. [62] using orange peel (OP), sugarcane bagasse (SB), and rice bran (RB) both individually and as tri-composite agro-waste (TAW) mixtures to produce GO. They prepared the feedstock by washing it with water and then drying it in the sun. The dry feedstocks were then processed using a mortar, pestle, and mixer to create fine powder. For preparing GO from the individual agro-waste, they mixed 0.3 g of each feedstock powder with 0.1 g of ferrocene. The mixture was then heated in a muffle kiln at 300 °C for 15 min. To prepare GO from the tri-composite mixture, they mixed half a gram of each feedstock powder with 0.3 g of ferrocene and then heated the mixture in a muffle kiln at 400 °C for 15 min.

Figure 7 is the illustration of the XRD spectrum of their study. From the XRD patterns, they observed that GO was successfully prepared only by TAW with the main diffraction peak located at 2θ = 12.705. They also observed excellent crystallinity that was indicated by the prominent and strong peak of GO (Figure 7). Debbarma et al. [63] also synthesized graphene oxide from sugarcane bagasse using pyrolysis at a low temperature ranging from 250 °C to 450 °C. The first step in the process was to chop the sugarcane bagasse into tiny pieces then wash it with deionized water to get rid of any contaminants. The samples were left to dry in the sun for a few days before being heated in an oven at 70 °C for a period of 24 h. Afterwards, the samples were ground into a fine powder and subjected to pyrolysis at 250 °C, 350 °C, and 450 °C for durations of 1 h, 30 min, and 10 min, respectively. After this, the material that resulted from pyrolysis was filtered, rinsed with warm, deionized water, and left to dry for 24 h. In this study, it was observed that 350 °C was suitable for condensation and aromatization of the glucose monomers to form graphene oxide nanosheets at a large scale. In another study, this same research group synthesized graphene nanosheets following the same process mentioned above in the presence of sodium hydroxide [64] and they found that mixing sodium hydroxide with it in a 1:1 ratio prevented oxygen from attacking the sample during the pyrolysis process.

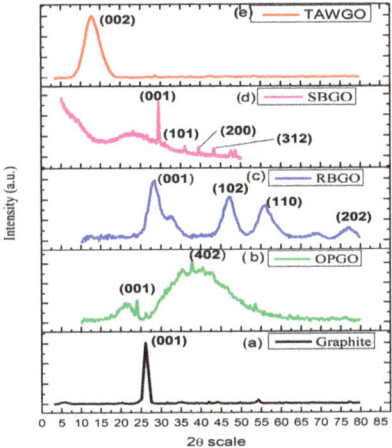

Figure 7. XRD spectrum of (**a**) graphite; and GO synthesized from (**b**) OP, (**c**) RB, (**d**) SB, and (**e**) TAW [62], reprinted with permission from the publisher (Taylor & Francis Ltd., Abingdon, Oxfordshire, England, http://www.tandfonline.com (accessed on 20 January 2023)).

Liu et al. [65] and Wang et al. [66] synthesized biobased graphene by pyrolyzing kraft lignin (KL) and bamboo biomass, respectively. Liu prepared biobased graphene in the presence of iron catalyst using commercial KL as feedstock. In this experiment, five grams of KL was mixed with Fe powder of varying ratios (1:2, 1:3, 1:5, and 1:7) before being heated in a quartz tube at 1000 °C under an argon flow. The samples were kept in the quartz tube for periods of 60, 75, 90, 105, and 120 min. Once the thermal treatment was finished, the sample was cooled back to room temperature and rinsed multiple times with deionized water. The iron particles were isolated by means of magnetic separation and any remaining iron was eliminated by washing the specimens with 10% hydrochloric acid. It was observed that the thermal treatment process lasting 90 min, with a ratio of 3:1 of carbon source to iron, resulted in graphene of superior quality. Moreover, carbon nanotubes (CNTs) were seen when the thermal treatment lasted 105 min. In contrast, Wang et al. [66] synthesized graphene-containing biochar from waste bamboo biomass with the activating agent K_2CO_3 and the help of microwave-assisted catalytic graphitization. The process of this study began by washing the raw bamboo with deionized water to clear away any contaminants, followed by drying and reducing it to particles that were smaller than 0.1 mm. It was then carbonized at 400 °C while under a flow of nitrogen gas for a period of three hours. The carbonized material was combined with potassium carbonate at a ratio of 1:3, and pyrolyzed at 900 °C under a nitrogen atmosphere for 25 min. For the post-treatment, the sample was given multiple rinses with deionized water and had its pH adjusted by adding a weak solution of hydrochloric acid. Lastly, the sample was dried at 105 °C for 12 h. The resulting biochar exhibited a typical graphene structure, plenty of micropores, and a huge surface area of up to 1565 m²g⁻¹. The activating agent K_2CO_3 played a significant role in facilitating transformation of amorphous carbon into graphene-like carbon.

CNTs were also synthesized by different researchers using different biomasses and varied pyrolysis processes. Yu et al. [67] developed a new approach that uses microwave pyrolysis for producing super-long carbon nanotubes (SL−CNTs) without the need for an outside catalyst. By pyrolyzing cellulose at temperatures ranging from 1200 to 1400 °C, they were able to generate CNTs with lengths between 0.7–2 mm. For this study, cellulose was derived from a palm kernel shell (PKS) and AC was adopted as the microwave absorbing material. The cellulose sample was first oven dried for 12 h at 80 °C and then mixed with AC at the ratio of 10:2. Next, the specimen was placed inside a quartz tube and pyrolyzed at 600 °C for a period of 30 min under N_2 environment. In the second step, 5 g of the produced char from the first step pyrolysis was again pyrolyzed at 1200 °C, 1300 °C, and 1400 °C for 30 min under N_2 environment. As the pyrolyzing temperature increased, they observed that the average length of CNTs increased. They also observed that CNTs' shapes changed from twisted, coiled, and threadlike to straight structures as the temperature varied. Furthermore, they noticed that the carbon order in the SL−CNTs increased after microwave treatment at 1400 °C, which was evidenced by the low Raman I_D/I_G ratio of 0.84. Additionally, the inorganic components found in the biomass were thought to act as a catalyst, accelerating the growth of the SL−CNTs. These observations are depicted in Figure 8.

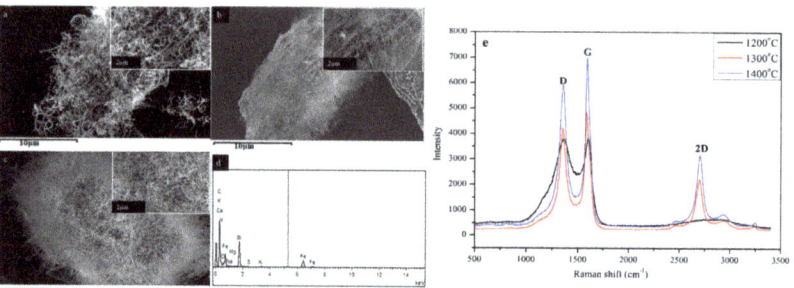

Figure 8. SEM images of SL-CNTs following microwave treatment at temperatures of: (**a**) 1200 °C, (**b**) 1300 °C, and (**c**) 1400 °C; (**d**) EDS spectra of SL−CNTs synthesized at 1300 °C; (**e**) Raman spectra of SL−CNTs at different temperatures. Reprinted from [67], Copyright 2022, Esohe Omoriyekomwan J, Tahmasebi A, Zhang J, Yu J., used under Creative Commons Attribution License (CC BY) (https://creativecommons.org/licenses/by/4.0/ (accessed on 20 January 2023)).

Hidalgo et al. [68] developed another method for synthesizing CNTs from biochar using microwave irradiation and ferrocene as a catalyst. The first step in this process was to pyrolyze 100 g of agro-industrial residual biomass containing wheat straw, rapeseed cake, oat hulls, and hazelnut hulls at temperatures of 400 °C and 600 °C under N_2 gas flow for 3 hrs. The produced biochar was then separated, milled, and fractioned using a sieve (size < 75 μm). In the second step, the biochar was mixed with ferrocene and pyrolyzed at 80 °C for 5 min. Researchers observed that CNTs were present in higher concentrations when biochar was pyrolyzed at 600 °C. Furthermore, a superior quality of CNTs with a higher rate of graphitization was observed when biochar created from hazelnut hull and wheat straw was employed.

Researchers have also synthesized graphene sheets using a biomass pyrolysis process. Xia et al. [69] synthesized three-dimensional porous graphene-like sheets (3DPGLS) with an impressive level of purity, negligible defect rate, great electrical conductivity, and a large specific surface area (1506.19 m²g⁻¹). For this experiment, coconut shells were ground to a particle size of less than 100 μm and then carbonized at a temperature of 400 °C for three hours in a nitrogen atmosphere. The biocarbon powder created from the carbonization process was blended with K_2CO_3 in a ratio of 1:2 and pyrolyzed at 900 °C in a nitrogen atmosphere for 2 h. Lastly, the samples were cooled down, followed by a treatment of dilute hydrochloric acid and de-ionized water washes, and dried for 12 h at 60 °C. Widiatmoko et al. [70] also synthesized graphene sheets through a two-step pyrolysis process using oil palm empty fruit bunch (EFB). In the first step, they ground the EFB into powder using a ball mill and blended it with 3M $FeCl_2$ as a catalyst, with urea as a nitrogen source (1:1) and with $ZnCl_2$ as an activator (2:1). Then the sample was heated on a hot plate at 80 °C for 2 h, followed by 2 h in the oven for drying. For the first stage pyrolysis, the samples were pyrolyzed at 250 °C, 350 °C, and 450 °C for a duration of 60 min. The second stage involved raising the temperature to 900 °C and maintaining it for an additional 90 min. It was observed that the yield of the pyrolysis product was influenced by the temperature used in the initial pyrolysis step. Furthermore, it is a well-known fact that lignin yields more aromatics and char than cellulose when pyrolyzed, and that decomposition of lignin is most effective in the 350–450 °C range. Widiatmoko's study showed that, when pyrolysis began at a temperature that corresponded to the decomposition temperature of lignin, it resulted in a great deal of graphene with a very high carbon concentration. There are also other studies on synthesizing GLC materials via pyrolysis. A brief summary is given in Table 1.

Table 1. GLC materials synthesis via pyrolysis from different bio sources.

Biomass Sources	Reaction Temperature	Pyrolysis Environment	Retention Time	Catalyst	Pyrolyzed Product	Ref.
Waste Tea	800 °C	N_2 gas	1 h	Potassium Ferrate	Multi-hierarchical porous carbon	[71]
Peanut Shell	800 °C	N_2 gas	1 h			
Pomelo Peel	800 °C	N_2 gas	1 h			
Spent Tea	1st stage: 1000 °C 2nd stage: 100–900 W	Inert	3 h 15–180 min	HNO_3	Graphene quantum dots	[72]
Quercus ilex leaves	820 °C	-	3 h	ZSM-5 and bentonite clay	Metal-doped graphene sheets (MDGs)	[73]
Waste biomass-derived cellulose	800 °C	N_2 gas	2 h	KOH	Multilayered graphene	[74]
walnut shell	850 °C	Ar gas	90 min	KOH	Graphene-like (GL) porous carbon	[75]
Dried green tea leaves	900 °C, 1100 °C	N_2 gas	3 h	-	Few-Layer Multifunctional Graphene	[76]
Chitosan	600 °C–800 °C	Ar gas	-	-	N-doped graphene	[77]
Biomass guanine	1000 °C	N_2 gas	4 h	-	GL 2D carbon	[78]
Gumwood	500 °C	N_2 gas	30 min	-	CNTs	[79]
Okara	800 °C	N_2 gas	2 h	-	N-doped GL mesoporous nanosheets	[80]

6. Effect of Microwave on GLC Materials Synthesis via Pyrolysis

In the process of GLC materials synthesis via pyrolysis, the choice of heating method plays a key role in determining the amorphous phase transition of graphene. Direct pyrolysis via traditional heating methods typically results in an amorphous phase and small graphite clusters. On the other hand, microwave heating can be used to convert electromagnetic energy into heat energy at the molecular level, leading to the creation of localized hotspots with a much higher temperature than the bulk material. These hotspots act as nucleation sites, which promote the rearrangement of molecules from an unordered phase to a crystalline phase. Furthermore, microwave radiation treatment can be used to transform sp^3 bonds to sp^2 bonds, in preparation for graphene formation [81]. Additionally, this method can improve the degree of graphitization of carbon material at a lower temperature in a shorter period without the need for a catalyst, while also producing a higher yield of few-layer graphene compared to conventional heating techniques [82]. A better understanding of the effect of microwave on the synthesis of carbon-based materials could be gained from the studies of Omoriyekomwan et al. [83] and Kaiqi et al. [79].

Omoriyekomwan et al. [83] conducted a study where they compared the results of forming hollow carbon nanofibers (HCNFs) via microwave pyrolysis of palm kernel shells at 500 °C and 600 °C to those synthesized with fixed-bed pyrolysis. They observed that the development of HCNFs could only be detected during microwave pyrolysis, implying that microwave radiation played an important role in the production of these nanostructures. This growth was believed to be due to the microwave radiation being absorbed by the biomass, causing an electric arc formation and devolatilization. The heavy components of the volatile matter then resolidified on the surface as a result of lower temperatures, forming carbon nanospheres. These nanospheres then self-extrude outward from the biomass particle through nano-sized channels, initiating HCNF growth, which is known as the "self-extrusion model growth". In comparison, when conventional heating was used in fixed-bed pyrolysis, the surrounding temperature was higher than the particle core, preventing volatiles from solidifying and undergoing secondary cracking instead.

Kaiqi et al. [79] synthesized multi-walled CNTs via microwave-induced pyrolysis of gumwood. In this experiment, gumwood was pyrolyzed at 500 °C and maintained for 30 min under an oxygen-free atmosphere with nitrogen gas flowing at a rate of 100 mL/min. The gumwood was then combined with SiC in a 20:1 mass ratio, and the resultant compounds were separated for further analysis. For comparison, conventional pyrolysis was also performed at the same temperature and nitrogen flow rate. In terms of morphology and microstructure, the researchers observed that the chars formed by microwave-induced pyrolysis differed from those produced by conventional pyrolysis, with the latter having no CNTs on their surfaces. They attributed the formation of CNTs under microwave-induced pyrolysis to the special effect of microwave radiation on the thermochemical processing of biomass. They proposed a mechanism for CNT development under microwave irradiation, in which volatiles released from biomass formed char particles, which then served as substrates. Mineral matter in char particles served as a catalyst, while released volatiles served as a carbon source gas, undergoing thermal and/or catalytic breaking on char particle surfaces. As a result of the impacts of microwave irradiation, amorphous carbon nanospheres formed, which then self-assembled into multi-walled CNTs. This method had the benefit of producing localized hot spots that could graphitize CNTs at far lower temperatures than conventional heating. The advantage of this approach was that it efficiently produced localized hot spots that graphitized CNTs at much lower temperatures than conventional heating requires.

7. Suitable Biomass Feedstock for GLC Materials Synthesis via Pyrolysis

Recently, there has been an increased focus on the synthesis of graphene or GLC materials from various biomass sources due to their sustainability, non-toxicity, environmental friendliness, cost-effectiveness, and ease of acquisition. Lignocellulosic biomass, composed primarily of lignin, cellulose, and hemicellulose, is particularly attractive due to its potential to produce higher-quality GLC materials with greater surface areas. Though the influence of feedstock composition on GLC material synthesis is not well understood, Table 2 suggests that feedstocks with higher carbon contents are more suitable for producing GLC materials via pyrolysis. Additionally, the characteristics of the final product largely depend on the synthesis method rather than the feedstock material. Table 2 provides elemental and proximate analysis of various bio precursors that have been used to synthesize GLC materials via various methods.

Table 2. Elemental and Proximate Analysis of various bio precursors.

Biomass Sample	Proximate Analysis, wt.%				Ultimate Analysis, wt.%				Ref.
	Moisture Content	Volatile Matter	Fixed Carbon	Ash	C	H_2	N_2	O_2	
Softwood	11.5	67.3	19.5	1.7	44.43	6.16	0.18	49.23	[84]
Hemp	10.7	69.6	18.8	0.9	45.71	5.89	-	48.40	
Rice straw	8.25	72.20	14.44	13.36	45.41	6.28	0.99	47.11	[85]
Pine nutshell	2.12	74.53	22.63	0.94	50.16	5.81	0.28	43.41	[58]
Palm Kernel Shell	14.90	74.68	23.68	1.64	49.90	5.25	0.36	43.54	[83]
Populus wood	-	-	-	-	39.75	6.09	1.52	52.54	[86]
Spent Coffee Beans	-	-	-	-	49.30	3.61	2.24	41.33	[87]
Rice husk	6.81	59.8	13.68	19.71	40.71	4.97	0.49	-	[88]
Sugarcane bagasse	9.51	74.98	13.57	1.94	43.77	6.83	-	47.46	[89]
Orange Peel	-	-	-	3.05	49.59	6.95	0.66	39.7	[90]
Chitosan	-	-	-	-	45.65	7.66	7.6	39.09	[91]

The lignocellulosic contents of several bio precursors previously employed to synthesize GLC materials are shown in Table 3. Each component of lignocellulosic biomass decomposes differently, and the breakdown is affected by temperature, heating rate, and the presence of contaminants [92]. The three components disintegrate at various temperatures, with hemicellulose being the one that would pyrolyze the most easily. Due to the intricate structure and higher resistance to high temperatures than hemicellulose and cellulose, lignin would be the most challenging to pyrolyze [93]. Studies on the influence of cellulose, hemicellulose, and lignin content on the formation of GLC materials via pyrolysis are scarce. According to earlier research, at higher temperatures, cellulose develops significant assemblages of polycyclic aromatic hydrocarbon domains, but lignin and hemicellulose only produce a small amount. That is why early experiments investigated the generation of graphitic carbon from pure cellulose [94]. Additionally, according to a few studies [95,96], lignin is the component of lignocellulosic biomass that is most conducive to the synthesis of laser-induced graphene (LIG). Therefore, future studies may concentrate on the impact of various lignocellulosic components on the quality and characteristics of the GLC materials produced through the pyrolysis process.

Table 3. Lignocellulosic Content of various bio precursors.

Biomass	Cellulose (wt.%)	Hemicellulose (wt.%)	Lignin (wt.%)	Ref.
Hemp	53–91	4–18	1–17	[97]
Rice Husk	32.67	31.68	18.81	[98]
Sugarcane Bagasse	50	25	25	[99]
Empty Fruit Bunches of Palm Oil	37.26	14.62	31.68	[100]
Wheat straw	34.40	20–25	20	[101]
Palm Kernel Shell	27.7	21.6	44	[102]
Bamboo	47.2	23.9	25.3	[103]
Rice Straw	29.2–34.7	12.0–29.3	17.0–19.0	[104]
Switch Grass	30–50	10–40	5–20	[105]
Miscanthus	24	44	17	[106]
Walnut Shell	23.9	22.4	50.3	[107]

8. Correlation between Microwave Pyrolysis Process Conditions and Pyrolyzed Product Characteristics

Microwave pyrolysis depends on the interaction between feedstock and microwave irradiation. The quality and features of the product obtained from microwave pyrolysis of biomass mostly rely on the operational conditions and the properties of biomass feedstocks. In this section, we explore the relationship between the various process parameters of pyrolysis and their effect on the final pyrolyzed product, with the goal of providing insight into how to optimize the production of GLC materials.

The performance of microwave pyrolysis is greatly affected by the microwave power, particle size, and batch size of the feedstock. Figure 9a illustrates that increasing the microwave power increases both the heating rate and the maximum reaction temperature. Additionally, reducing the particle size also increases both the heating rate and maximum reaction temperature, as shown in Figure 9b. This can be attributed to increased bulk density and intra-particle contact area with reduced particle size [85]. Furthermore, reducing the particle size to a specific size (e.g., less than 0.25 mm) has been observed to prevent heat from transferring within the particles, thus slowing the process of pyrolysis [108]. Parthasarathy et al. [109] found that increasing the feedstock particle size increases both the char yield and carbon content of the char. Lastly, the batch size of the raw material has been found to play a crucial role in microwave pyrolysis. Figure 9c shows that using a small amount of starting material (5–15 g) yields higher heating rates and higher residence temperatures at a much lower microwave power [110,111].

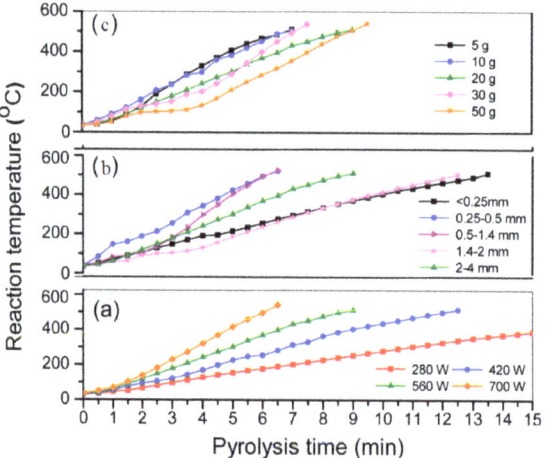

Figure 9. Reaction temperature profiles for various circumstances of microwave pyrolysis: (**a**) microwave power effect, (**b**) effect of particle size, (**c**) effect of initial mass. Reprinted with permission from [108]. Copyright 2015 American Chemical Society.

Low initial moisture levels in biomass are usually preferable for the pyrolysis process. Drying the biomass before subjecting it to pyrolysis improves the energy efficiency of the process [112]. Demirbas observed that increasing moisture content decreases biochar yield and increases the yield of liquid product [113]. However, a few studies have found that increasing the moisture level of the feedstock leads to an increase in char and gas generation [114–116]. Furthermore, the reaction temperature decreases as the moisture content increases, because when the mixture is heated up, heat is expended in the process of drying out the components [117]. Darmstadt et al. [118] observed that feedstock moisture content had a greater impact on softwoods than hardwoods. In addition, Xiaodi Li et al. [119] investigated the effects of preheat temperature on pyrolysis properties and product properties, finding that preheating the feedstock before microwave pyrolysis increased biochar yield, shortened the initial time for rapid temperature rise, and increased the BET surface area. The effect of thermal pretreatment was also reported by Jian et al. [120]. Feedstock moisture content also influences the heating rate. A lower heating rate arises from higher moisture content [121]. Additionally, it was noted that as initial moisture content increased, the specific char surface area increased, though the effect was more pronounced at lower pyrolysis temperatures [122]. Furthermore, it is common practice to dry biomass feedstock prior to pyrolysis, and the feedstock is frequently dried to a moisture level of less than 10% [123].

The reaction temperature of pyrolysis has a significant impact on the char yield and characteristics. According to Mohammad et al. [124], the char yield reduces as the pyrolysis temperature rises. At higher temperatures, the devolatilization process accelerates, which leads to more vapors and gases being produced and a decrease in the char yield [125]. With the help of Raman spectroscopy, Asadullah et al. [126] observed that an increase in temperature causes char to aromatize more quickly. The maximum pyrolysis temperature also affects the surface area, pore structure, and carbon content. According to one study, the BET surface area of a char decreases as the pyrolysis temperature rises. This was found to be quite drastic, with the surface area dropping by a minimum of 200 times when the temperature was raised from 500 to 800 °C. It is believed that the drastic reduction in micropores which occurs between 500 °C and 800 °C is what caused the sharp decline in the surface area [122]. In a different analysis, Fu et al. [127] observed that the surface area of char increases with temperature; however, it decreases slightly if the temperatures surpass 1173 K. Zhao et al. [128] also found that pyrolyzing the rapeseed steam from 200 to 700 °C resulted in an increase in surface area, from 1 to 45 m^2/g, which shows more of a carbonaceous, aromatic structure for the biochar. This behavior can be attributed to the release of volatile gases and the formation of pores at higher temperatures. Furthermore, Lua et al.'s [129] research indicates a general correlation between the BET surface area, micropore surface area, and total pore volume, which increases up to a specific temperature and then begins to decrease gradually. The initial increasing trend could be attributed to the emission of low-molecular-weight gases from the carbon structure. The declining trend might be related to the weakening and liquifying of some of the residual volatiles in the char, which causes an intermediate melt to form in the chars. This intermediate melt obstructs the formation of the char's primitive pore structure by partly sealing some of the pores. Nevertheless, when the pyrolysis temperature was increased further, it caused the pores to grow and develop, thus leading to an increase in the BET surface area, the micropore area, and the total pore volume. The char's intermediate melt undergoing depolymerization and evaporating is the reason behind this phenomenon, which causes the previously sealed pores to open up, as well as the formation of new pores as a result of the disappearance of the heavier volatiles. As the pyrolysis temperature increased further, decreases in BET surface area, micropore surface area, and total pore volume were observed. This was likely due to the compression of pores within the char and the narrowing of the pore openings, which both cause the accessible pore surface area to be reduced. Another potential factor might be the production of secondary melt from high-molecular-weight volatiles, comparable to the previously stated intermediate melt. The fixed carbon content and carbon content of char increase with temperature as a result of deoxygenation and dehydration, indicating greater structural ordering for lowering reaction site concentration [128,130,131]. Furthermore, with increasing pyrolysis temperature, the total volume of pores increases, but the average pore diameter decreases as a result of an increase in the proportion of relatively tiny pores [128].

The char characteristics and yield are significantly Influenced by the heating rate as well. According to Mohammad et al. [124], the char yield reduces as the heating rate rises. Although, the influence of the rate of heating was more noticeable at higher temperatures when it came to the production of char [132]. However, compared to char produced at low heating rates, high-heating-rate char has a smaller surface area [127]. It is believed that this is due to an excessive heating rate that raises the temperature of the char interior and results in partial graphitization and the construction of a graphene structure: neither of which contributes to the development of a large surface area. On the contrary, Zhao et al. [128] found that rapeseed stem surface area increased at first with the increasing heating rate due to a larger extent of thermal decomposition and then slightly decreased. When the heating rate is increased, the carbon content of char decreases slightly while the hydrogen and oxygen content increases. Additionally, at high temperatures, the heating rate impact starts to disappear [133].

Moreover, Parthasarathy et al. [109] reported that the duration of residence in a given environment has an impact on char yield and its carbon content. They observed that increasing the residence period reduces char production while increasing char carbon content. By decreasing the char yield, a longer residence period allows for more time for the reactants in the volatiles to interact with the char and leads to a higher gas yield. Better devolatilization is achieved with a longer residence time, which increases the char's carbon content. Additionally, research has demonstrated that the BET surface area of the char has a direct correlation with the residence time. Initially, the BET surface area increases with the increase in residence time, but this effect levels off after prolonged times. This can be attributed to the sigmoidal-shaped curve of the devolatilization rate [134]. Zhang et al. [135] also observed that the BET surface area of chars increased with the residence time until a certain point, after which it began to decrease. This can be explained by the fact that chars' ability to generate pores might benefit from a fair extension of residence time at high temperatures. However, if the residence period is prolonged too much, the pore structure of the chars may be destroyed, which would then cause deactivation.

Lastly, the use of microwave absorbers to indirectly heat biomass particles during pyrolysis has been shown to increase the reaction temperature at relatively low microwave power. This increase in temperature has a significant effect on the yield and quality of the pyrolysis products [136]. In addition, the use of iron-based catalysts, such as ferric chloride or ferrocene, can help to produce high-quality graphene-like biochar with excellent physicochemical properties. Furthermore, by using ferrocene as a catalyst, it is possible to produce graphene oxide at a much lower temperature [62,68,137].

9. Formation Mechanism of Biochar during Pyrolysis

Recently, pyrolysis has been employed to produce GLC materials from a wide range of biomasses. In order to understand the mechanism of graphene formation during biomass pyrolysis, it is important to understand the mechanism of biochar formation, as the two mechanisms are closely related.

Most biomasses are composed of cellulose, hemicellulose, and lignin, which each degrade at different temperatures and through distinct pathways. For instance, the temperature range for the decomposition of hemicellulose is between 220 and 315 °C, cellulose is between 315 and 400 °C, and lignin is between 160 and 900 °C [138]. Examining the individual pyrolysis mechanisms of these components can provide further insight into the pyrolysis of biomass and the formation of GLC materials.

Figure 10 illustrates the formation mechanism of biochar from cellulose. The pyrolysis of cellulose is initiated by the depolymerization of cellulose into oligosaccharides and the subsequent breaking of glycosidic bonds to form D-glucopyranose. Intramolecular rearrangement of D-glucopyranose leads to the formation of levoglucosan, which can either be converted to levoglucosenone through dehydration or can be exposed to a combination of rearrangements and dehydrations resulting in the formation of hydroxymethylfurfural. Furthermore, levoglucosenone can be chemically altered in various ways, leading to the production of biochar. This includes dehydrating, decarboxylation, aromatizing, and undergoing intramolecular condensation. Additionally, hydroxymethylfurfural can break down to produce bio-oil and syngas that are more volatile, or alternatively become further polymerized, aromatized, and condensed to form biochar [49].

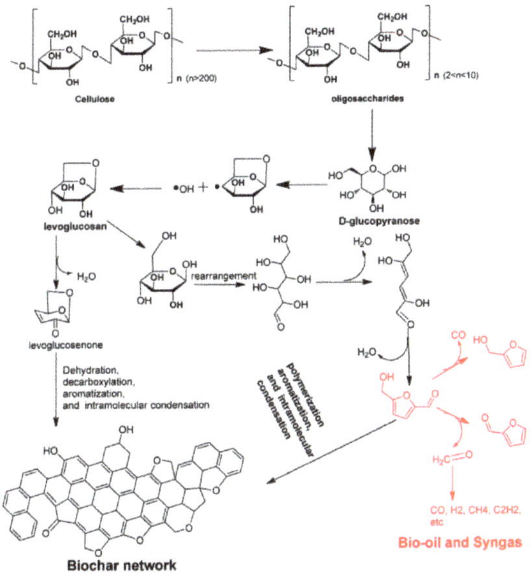

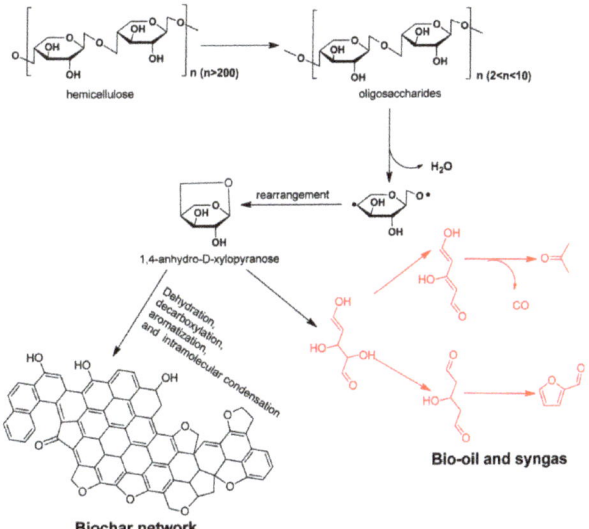

Figure 10. Formation mechanism of biochar by cellulose pyrolysis. Reprinted with permission from [49]. Copyright 2015 American Chemical Society.

The mechanism of hemicellulose pyrolysis is similar to that of cellulose pyrolysis. It is also depolymerized to form oligosaccharides, followed by glycosidic bond breakage and rearrangement of the depolymerized molecules to yield 1,4-anhydro-D-xylopyranose. Furthermore, it may be dehydrated, decarboxylated, aromatized, and intramolecularly condensed to create solid biochar, or it can break down into smaller compounds such as bio-oil, syngas, and low-molecular-weight compounds [49]. Figure 11 depicts the process of producing biochar from hemicellulose.

Figure 11. The formation mechanism of biochar by hemicellulose pyrolysis. Reprinted with permission from [49]. Copyright 2015 American Chemical Society.

The degradation of lignin is relatively intricate compared to cellulose and hemicellulose due to its more intricate structure, which is depicted in Figure 12. The principal mechanism in the pyrolysis of lignin is a reaction involving free radicals, which originate from the cleaving of the β-O-4 bond in the lignin molecules. The radicals are capable of scavenging protons from molecules with weak C-H or O-H bonds, leading to the production of breakdown products such as vanillin and 2-methoxy-4-methylphenol. As the reaction progresses, radicals are transferred to other species, resulting in a chain reaction. Ultimately, the chain reaction is halted when two radicals encounter each other and form a more stable compound. Nevertheless, since the detection of radicals in the pyrolysis process is exceptionally difficult, understanding the precise mechanism of lignin pyrolysis is a major challenge [49].

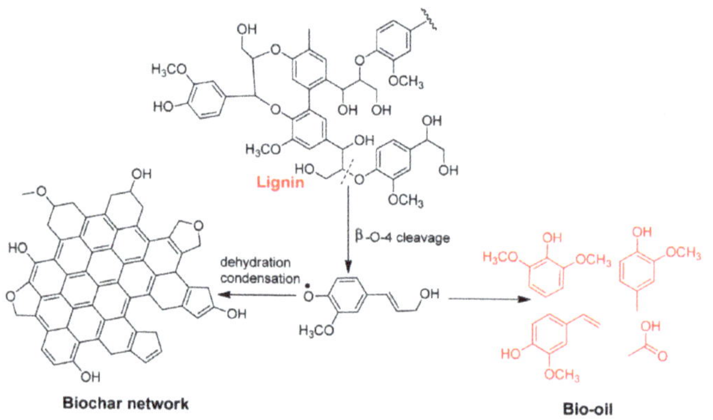

Figure 12. Lignin pyrolysis: biochar formation mechanism. Reprinted with permission from [49]. Copyright 2015 American Chemical Society.

10. Mechanism of GLC Materials Formation during the Biomass Pyrolysis Process

Most biomass is lignocellulosic and contains long chains of carbon, hydrogen, and oxygen compounds. The process of converting lignocellulosic biomass into graphene involves increasing the carbon content and arranging the carbon structures in a graphitic-like form. This process involves two steps: carbonization and graphitization. Carbonization involves the removal of light-molecular-weight compounds through heating, while graphitization is used to arrange the remaining carbon structures into a graphitic-like form. The converted carbon structure may not be similar to pure graphene, but the properties that they possess are somewhat graphene-like [44].

Debbarma et al. [64] synthesized graphene nanosheets from sugarcane bagasse via pyrolysis in the presence of sodium hydroxide. The chemistry behind the formation of graphene nanosheets is presented in Figure 13.

Sugarcane bagasse has a high concentration of cellulose, which is composed of glucose monomers held together by glycosidic bonds. During the pyrolysis of sugarcane bagasse, the breakdown of glucose monomers takes place, and these monomers contain aldehyde and hydroxyl groups. The hydroxyl group on the fifth carbon of the glucose molecule can then bind to the aldehyde group on the first carbon to form a cyclic hemiacetal structure. This structure is similar to pyran and consists of six-membered heterocyclic rings. It is thought that many of the glucose monomers were linked via glycosidic bonds during pyrolysis, and further condensation and aromatization of the cyclic rings occurred to form planar graphitic polyaromatic ring structures. Debbarma et al. [63] also synthesized GO from sugarcane bagasse where heating sugarcane bagasse at different temperatures caused the degradation of glucose monomers, leading to the formation of glycosidic bonds and polyaromatic rings. The presence of air facilitated oxidation, aromatization, and condensation, resulting in the formation of SBGO nanosheets. The mechanism is depicted in Figure 14.

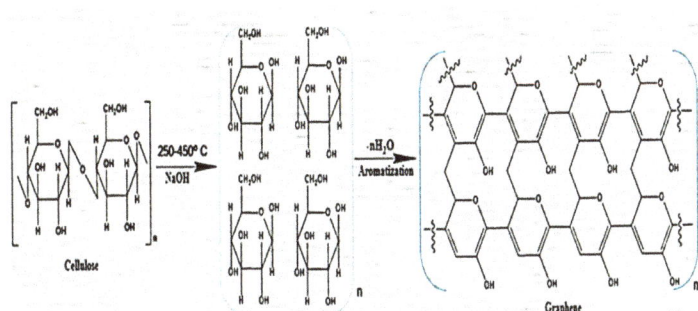

Figure 13. Mechanism of formation of graphene nanosheets at 250–350 °C. Reprinted from [64]. Copyright 2020, Debbarma, J.; Mandal, P.; Saha, M., used under Creative Commons Attribution License (CC BY) (https://creativecommons.org/licenses/by/4.0/ (accessed on 20 January 2023)).

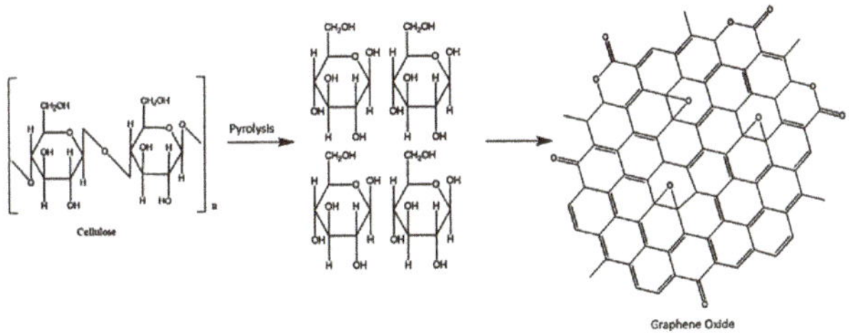

Figure 14. Mechanism of formation of GO from sugarcane bagasse [63]. Reprinted with permission from the publisher (Taylor & Francis Ltd., http://www.tandfonline.com (accessed on 20 January 2023)).

This same research group synthesized nitrogen-doped GO (N-GO), wherein the formation mechanism was the same as stated above, with primary amine groups from the amino acids enhancing the nitrogen content and resulting in the formation of N-GO, as illustrated in Figure 15 [139].

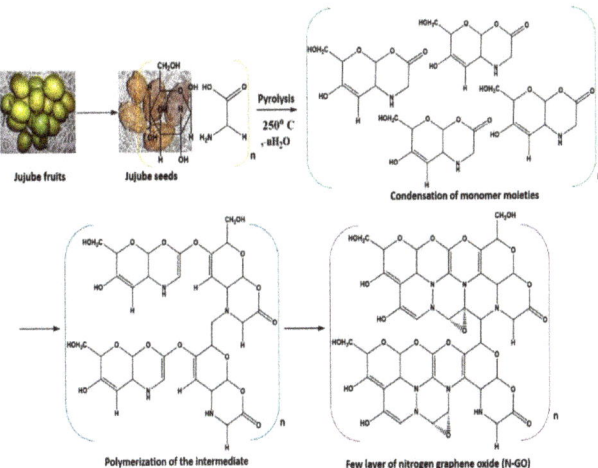

Figure 15. Mechanism of formation of N-GO [139]. Reprinted with permission from the publisher (Taylor & Francis Ltd., http://www.tandfonline.com (accessed on 20 January 2023)).

Roy et al. [76] developed a method for synthesizing graphene from tannic acid, alginic acid, and green tea through a controlled pyrolysis procedure, and proposed a formation mechanism of graphene from alginic acid, illustrated in Figure 16. They proposed that this reaction likely started with the production of radicals, which was then followed by the release of water molecules and CO_2 and the aromatization and intermolecular condensation reactions at a temperature of 1100 °C. They hypothesized that similar processes happened with the polyphenols from green tea and tannic acid based on the presence of carboxyl groups and vulnerable oxygen bonds in their molecular structures.

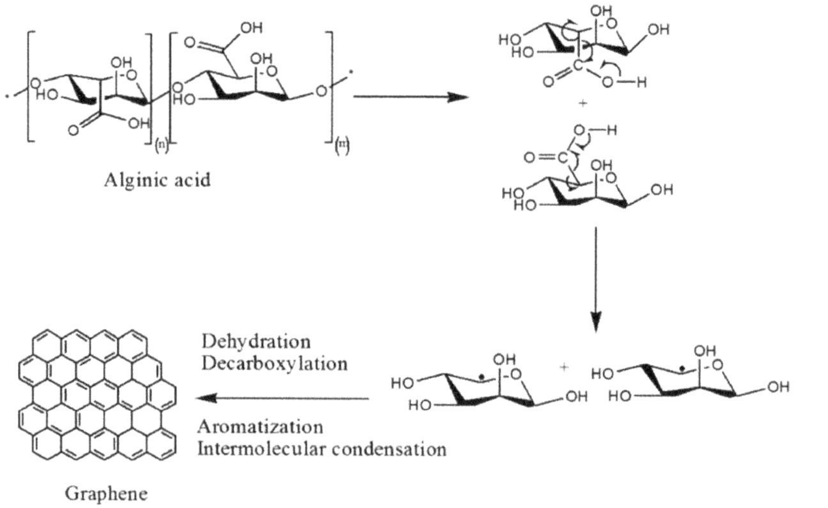

Figure 16. Pyrolysis reaction mechanism for graphene synthesis from alginic acid. Reproduced under the terms of CC BY-NC-ND license [76].

Omoriyekowan et al. [140] explored the process of carbon nanotube (CNT) formation in their most recent investigation, synthesizing CNTs with cellulose taken from PKS. In order to draw out the bio-components from the PKS, two distinct techniques were utilized. Once the extraction process had concluded, cellulose and lignin were then exposed to microwave pyrolysis. The end result of their research suggested that cellulose played a vital part in generating CNTs. Figure 17 illustrates the reaction pathways of the decomposition of cellulose to produce nanotubes. The authors examined bio-oils derived from lignin and cellulose to better understand the role of cellulose. Bio-oils derived from cellulose were high in monosaccharides, while bio-oils derived from lignin were rich in phenols and single-ring hydrocarbons. According to the authors, the breakdown of cellulose resulted in the production of monosaccharides such as D-glucopyranose, which was employed as a carbon source for CNT synthesis. Splitting of the glycosidic bonds in D-glucose generated anhydrides, oligosaccharides, and levoglucosan. Subsequently, these elements underwent degradation, cleavage, and rearrangement, leading to the formation of anhydro sugars and levoglucosan. The splitting of the C-O bonds in levoglucosan was followed by its aromatization, resulting in a formation of graphite layers, as shown in Figure 17.

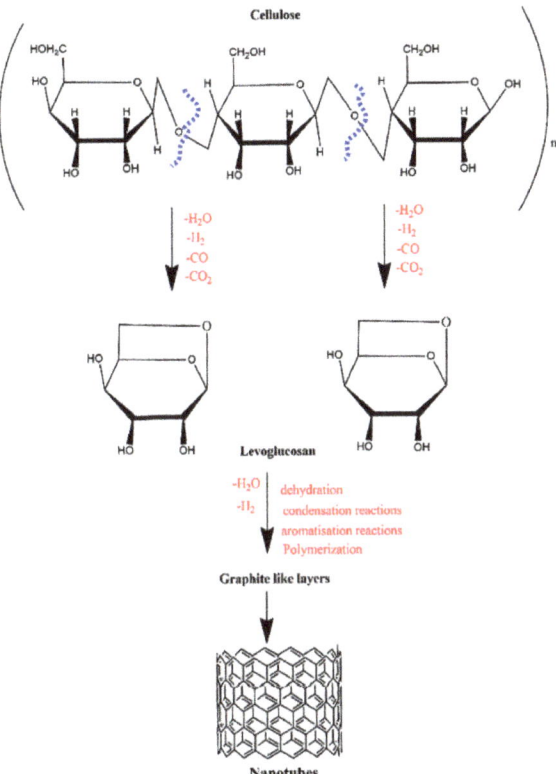

Figure 17. Reaction pathways during the decomposition of cellulose. Reprinted from [141], with permission from Elsevier.

Liu et al. [65] manufactured graphene using commercially accessible kraft lignin (KL) and carefully examined the formation process, structure, and features. They provided a general mechanism of lignin-based graphene that was catalyzed by iron, as demonstrated in Figure 18. The main reactions were pyrolysis and carbonization of KL at temperatures between 250 and 500 °C, wherein polyolefin compounds were converted into amorphous carbon (a-C) with the aid of iron particles via catalytic dehydrogenation. The a-C was in a metastable state and possessed a large amount of energy, so it required less energy to dissolve in iron metal than C atoms. The temperature required for carbon to dissolve into iron is 570 °C. Through the precipitation dissolution mechanism, a-C diffused into the metal particle and then precipitated as graphene on the free surface when the solid solubility limit was reached during cooling. Smaller metal particles and longer annealing times resulted in the migration of activated carbon species to the top surface and the nucleation of graphene. Graphene was observed when the holding time was in the range of 90–105 min, but had notably reduced areas and less graphene identified by Raman when the holding time was longer than 105 min. The probable cause of this was the growth of an sp^2 carbon network along the surface of the iron particles, which accumulated into a graphite shell. Additionally, the contact between the a-C and iron particles at higher temperatures resulted in a catalytic graphitization process. Iron has the ability to catalyze graphitization even at a low temperature due to the decomposition of iron carbide in the insulation stage.

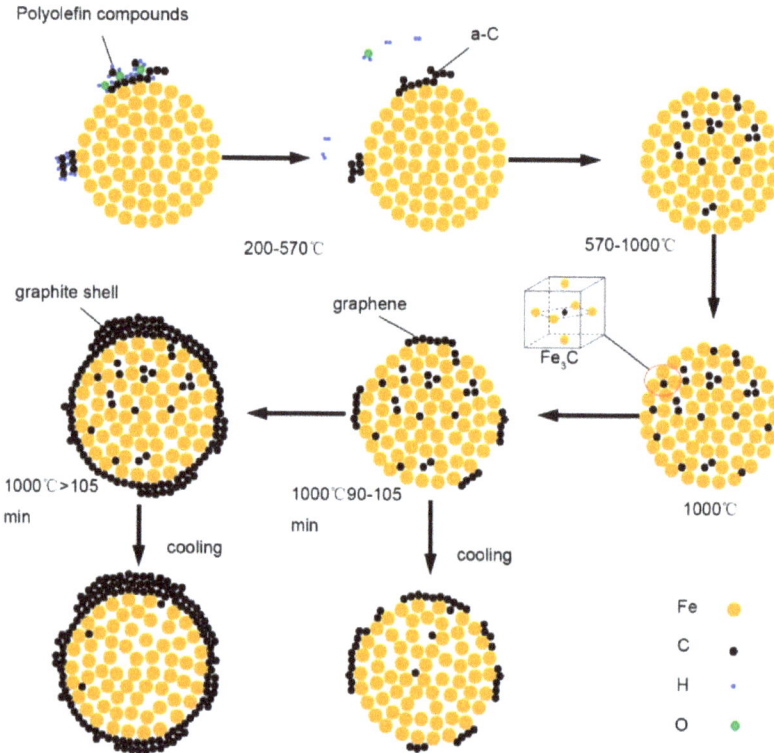

Figure 18. Schematic representation of the graphene synthesis processes from KL using iron particles as catalysts. Reprinted from [65] with permission from author.

The reaction equations were as follows:

$$Fe + C_a \rightarrow Fe_3C$$

$$Fe_3C \rightarrow Fe + C_g,$$

where C_a is amorphous and C_g is graphitic carbon.

The newly produced carbon from the breakdown of iron carbide is active and can be quickly transformed into graphite. On the other hand, too much iron may make the decomposition of Fe_3C more challenging. The formation of graphene is carried out in two phases when iron particles are used as a catalyst. One of these is the precipitation and dissolution of carbon atoms, while the other is the manufacture and disintegration of iron carbide. As a consequence, the retention time has an influence on the formation of graphene.

Reviewing all the mechanisms discussed above, it may be concluded that the pyrolysis process involves the splitting and recombination of molecules. Carbon atoms form single covalent sp^3 bonds with other atoms, but during the graphene formation process, these bonds are broken, allowing the carbon atoms to form sp^2 bonds in the form of benzene rings. This process of nucleation leads to the development of graphene, as illustrated in Figure 19.

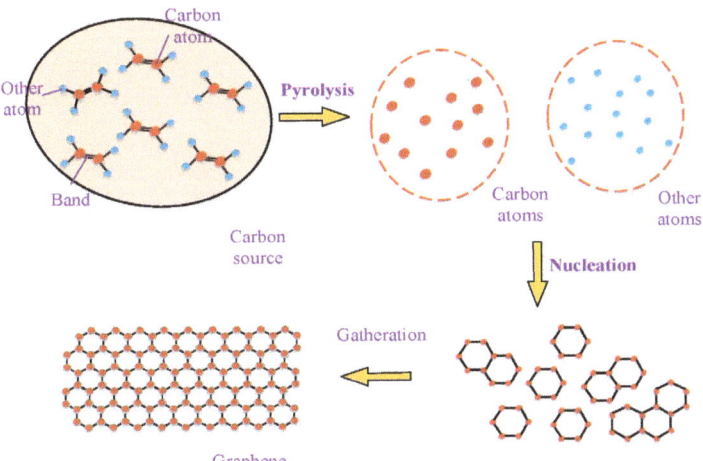

Figure 19. Mechanism of graphene formation via pyrolysis process [142]. Reproduced under the terms of CC BY 4.0 license.

11. Conclusion and Research Outlook

In conclusion, this review paper provides a comprehensive overview of the current state of knowledge of biomass-derived GLC materials and the microwave pyrolysis process for their synthesis. It was revealed that the microwave pyrolysis process is a promising solution for the cost-effective and renewable synthesis of GLC materials from biomass feedstock. Utilizing biomass waste to produce graphene can reduce high-expense production and associated pollution. Despite the fact that several studies have preferred high-temperature pyrolysis methods that use metal precursors along with biomass in order to break down the structure of biomass while also enabling volatile carbon materials to be deposited, future research may have to focus on utilizing lower-temperature thermal treatment to reduce the length of the reaction time. Even though bio-based graphene is not of the highest quality, the green synthesis route can still provide a good amount of multi-layer graphene, GO, and RGO. Further research is necessary to gain a more comprehensive understanding of the formation mechanism of GLC materials from biomass pyrolysis in order to optimize the production process, as well as to improve the efficiency of the microwave pyrolysis process. Additionally, the influence of feedstock particle size on the characteristics of the produced GLC material must be investigated. Furthermore, to gain a better understanding of the formation of GLC materials, individual pyrolysis of cellulose, hemicellulose, and lignin using the same process parameters should be conducted. With further research, biomass-derived GLC materials have the potential to become a viable and renewable alternative to traditional graphene materials for a variety of applications.

Author Contributions: Conceptualization, F.C.A. and G.C.S.; formal analysis, F.C.A.; writing—original draft preparation, F.C.A.; writing—review and editing, G.C.S.; supervision, G.C.S.; project administration, G.C.S.; funding acquisition, G.C.S. All authors have read and agreed to the published version of the manuscript.

Funding: The authors are thankful for the funding provided by the Natural Sciences and Engineering Research Council of Canada (RGPIN-2018-04440) and New Brunswick Innovation Foundation (NBIF-L2M2020-006).

Institutional Review Board Statement: Not applicable.

Informed Consent Statement: Not applicable.

Data Availability Statement: The raw/processed data required to reproduce these findings cannot be shared at this time as the data also form part of an ongoing study.

Conflicts of Interest: The authors declare no conflict of interest.

References

1. Dunn, B.; Kamath, H.; Tarascon, J.-M. Electrical Energy Storage for the Grid: A Battery of Choices. *Science* **2011**, *334*, 928–935. [CrossRef]
2. Demirbaş, A. Global Renewable Energy Resources. *Energy Sources Part A Recover. Util. Environ. Eff.* **2006**, *28*, 779–792. [CrossRef]
3. Arnell, N.W. Climate change and global water resources. *Glob. Environ. Chang.* **1999**, *9*, S31–S49. [CrossRef]
4. Zhou, Y.; He, J.; Chen, R.; Li, X. Recent advances in biomass-derived graphene and carbon nanotubes. *Mater. Today Sustain.* **2022**, *18*, 100138. [CrossRef]
5. Novoselov, K.S.; Geim, A.K.; Morozov, S.V.; Jiang, D.; Zhang, Y.; Dubonos, S.V.; Grigorieva, I.V.; Firsov, A.A. Electric Field Effect in Atomically Thin Carbon Films. *Science* **2004**, *306*, 666–669. [CrossRef]
6. Geim, A.K.; Novoselov, K.S. The rise of graphene. In *Nanoscience and Technology*; Co-Published with Macmillan Publishers Ltd.: London, UK, 2009; Volume 6, pp. 11–19.
7. Novoselov, K.S.; Fal'ko, V.I.; Colombo, L.; Gellert, P.R.; Schwab, M.G.; Kim, K. A roadmap for graphene. *Nature* **2012**, *490*, 192–200. [CrossRef]
8. Balandin, A.A.; Ghosh, S.; Bao, W.; Calizo, I.; Teweldebrhan, D.; Miao, F.; Lau, C.N. Superior Thermal Conductivity of Single-Layer Graphene. *Nano Lett.* **2008**, *8*, 902–907. [CrossRef] [PubMed]
9. Lee, J.-U.; Yoon, D.; Cheong, H. Estimation of Young's Modulus of Graphene by Raman Spectroscopy. *Nano Lett.* **2012**, *12*, 4444–4448. [CrossRef]
10. Zhu, Y.; Murali, S.; Cai, W.; Li, X.; Suk, J.W.; Potts, J.R.; Ruoff, R.S. Graphene and Graphene Oxide: Synthesis, Properties, and Applications. *Adv. Mater.* **2010**, *22*, 3906–3924. [CrossRef]
11. Papageorgiou, D.G.; Kinloch, I.A.; Young, R.J. Mechanical properties of graphene and graphene-based nanocomposites. *Prog. Mater. Sci.* **2017**, *90*, 75–127. [CrossRef]
12. Bizao, R.A.; Machado, L.D.; de Sousa, J.M.; Pugno, N.M.; Galvao, D.S. Scale Effects on the Ballistic Penetration of Graphene Sheets. *Sci. Rep.* **2018**, *8*, 6750. [CrossRef] [PubMed]
13. Sun, L.; Tian, C.; Li, M.; Meng, X.; Wang, L.; Wang, R.; Yin, J.; Fu, H. From coconut shell to porous graphene-like nanosheets for high-power supercapacitors. *J. Mater. Chem. A* **2013**, *1*, 6462–6470. [CrossRef]
14. Jirimali, H.; Singh, J.; Boddula, R.; Lee, J.-K.; Singh, V. Nano-Structured Carbon: Its Synthesis from Renewable Agricultural Sources and Important Applications. *Materials* **2022**, *15*, 3969. [CrossRef] [PubMed]
15. Poorna, A.; Saravanathamizhan, R.; Balasubramanian, N. Graphene and graphene-like structure from biomass for Electrochemical Energy Storage application- A Review. *Electrochem. Sci. Adv.* **2020**, *1*, e2000028. [CrossRef]
16. Laraba, S.R.; Luo, W.; Rezzoug, A.; Zahra, Q.U.A.; Zhang, S.; Wu, B.; Chen, W.; Xiao, L.; Yang, Y.; Wei, J.; et al. Graphene-based composites for biomedical applications. *Green Chem. Lett. Rev.* **2022**, *15*, 724–748. [CrossRef]
17. Daradmare, S.; Raj, S.; Bhattacharyya, A.R.; Parida, S. Factors affecting barrier performance of composite anti-corrosion coatings prepared by using electrochemically exfoliated few-layer graphene as filler. *Compos. Part B Eng.* **2018**, *155*, 1–10. [CrossRef]
18. Jiang, M.-J.; Zhang, Y.; Wu, G.; Chen, S.-C.; Chen, L.; Wang, Y.-Z. NIR light manipulated "paper art" for customizing devices with sophisticated structure from DA-epoxy/graphene composites. *Compos. Part B Eng.* **2019**, *177*, 107369. [CrossRef]
19. Taj, A.; Swamy, R.P.; Naik, K.; Bharath, K.N. Effect of Nano-Filler Aluminum Oxide and Graphene on Flammability Properties of Kenaf Epoxy Composites. *J. Inst. Eng. (India) Ser. D* **2022**, 1–12. [CrossRef]
20. He, G.; Huang, S.; Villalobos, L.F.; Zhao, J.; Mensi, M.; Oveisi, E.; Rezaei, M.; Agrawal, K.V. High-permeance polymer-functionalized single-layer graphene membranes that surpass the postcombustion carbon capture target. *Energy Environ. Sci.* **2019**, *12*, 3305–3312. [CrossRef]
21. Seo, D.H.; Pineda, S.; Woo, Y.C.; Xie, M.; Murdock, A.T.; Ang, E.Y.M.; Jiao, Y.; Park, M.J.; Lim, S.I.; Lawn, M.; et al. Anti-fouling graphene-based membranes for effective water desalination. *Nat. Commun.* **2018**, *9*, 683. [CrossRef]
22. Pandey, S.; Karakoti, M.; Surana, K.; Dhapola, P.S.; SanthiBhushan, B.; Ganguly, S.; Singh, P.K.; Abbas, A.; Srivastava, A.; Sahoo, N.G. Graphene nanosheets derived from plastic waste for the application of DSSCs and supercapacitors. *Sci. Rep.* **2021**, *11*, 3916. [CrossRef]
23. Kittusamy, R.K.; Rajagopal, V.; Felix, P.G. Numerical and experimental investigation on the melting heat transfer of nanographene-enhanced phase change material composites for thermal energy storage applications. *Int. J. Heat Mass Transf.* **2023**, *206*, 123940. [CrossRef]
24. Ibanez-Labiano, I.; Ergoktas, M.S.; Kocabas, C.; Toomey, A.; Alomainy, A.; Ozden-Yenigun, E. Graphene-based soft wearable antennas. *Appl. Mater. Today* **2020**, *20*, 100727. [CrossRef]
25. Hashmi, A.; Nayak, V.; Singh, K.R.; Jain, B.; Baid, M.; Alexis, F.; Singh, A.K. Potentialities of graphene and its allied derivatives to combat against SARS-CoV-2 infection. *Mater. Today Adv.* **2022**, *13*, 100208. [CrossRef]
26. Kurniawan, D.; Mathew, J.; Rahardja, M.R.; Pham, H.; Wong, P.; Rao, N.V.; Ostrikov, K.; Chiang, W.H. Plasma-Enabled Graphene Quantum Dot Hydrogels as Smart Anticancer Drug Nanocarriers. *Small* **2023**, 2206813. [CrossRef]
27. Wang, Y.; Wu, Y.; Zhang, Y.; Li, X.; Min, L.; Cao, Q.; Luo, Y.; Yang, X.; Lu, M.; Zhou, Y.; et al. Graphene oxide coated three-dimensional printed biphasic calcium phosphate scaffold for angiogenic and osteogenic synergy in repairing critical-size bone defect. *J. Mater. Sci. Technol.* **2023**, *145*, 25–39. [CrossRef]

28. Hashim, N.; Abdullah, S.; Yusoh, K. Graphene nanomaterials in the food industries: Quality control in promising food safety to consumers. *Graphene 2D Mater.* **2022**, *7*, 1–29. [CrossRef]
29. Priyadarsini, S.; Mohanty, S.; Mukherjee, S.; Basu, S.; Mishra, M. Graphene and graphene oxide as nanomaterials for medicine and biology application. *J. Nanostructure Chem.* **2018**, *8*, 123–137. [CrossRef]
30. Ollik, K.; Lieder, M. Review of the Application of Graphene-Based Coatings as Anticorrosion Layers. *Coatings* **2020**, *10*, 883. [CrossRef]
31. Cui, G.; Zhang, C.; Wang, A.; Zhou, X.; Xing, X.; Liu, J.; Li, Z.; Chen, Q.; Lu, Q. Research progress on self-healing polymer/graphene anticorrosion coatings. *Prog. Org. Coat.* **2021**, *155*, 106231. [CrossRef]
32. Hoseini-Ghahfarokhi, M.; Mirkiani, S.; Mozaffari, N.; Sadatlu, M.A.A.; Ghasemi, A.; Abbaspour, S.; Akbarian, M.; Farjadain, F.; Karimi, M. Applications of Graphene and Graphene Oxide in Smart Drug/Gene Delivery: Is the World Still Flat? *Int. J. Nanomed.* **2020**, *15*, 9469–9496. [CrossRef]
33. Srimaneepong, V.; Skallevold, H.E.; Khurshid, Z.; Zafar, M.S.; Rokaya, D.; Sapkota, J. Graphene for Antimicrobial and Coating Application. *Int. J. Mol. Sci.* **2022**, *23*, 499. [CrossRef]
34. Mehmood, A.; Mubarak, N.; Khalid, M.; Walvekar, R.; Abdullah, E.; Siddiqui, M.; Baloch, H.A.; Nizamuddin, S.; Mazari, S. Graphene based nanomaterials for strain sensor application—A review. *J. Environ. Chem. Eng.* **2020**, *8*, 103743. [CrossRef]
35. Yan, Y.; Shin, W.I.; Chen, H.; Lee, S.-M.; Manickam, S.; Hanson, S.; Zhao, H.; Lester, E.; Wu, T.; Pang, C.H. A recent trend: Application of graphene in catalysis. *Carbon Lett.* **2020**, *31*, 177–199. [CrossRef]
36. Liu, B.; Zhao, Y.; Chang, Y.; Tai, H.H.; Liang, H.; Chen, T.-C.; Feng, S.; Hou, T.-H.; Lai, C.-S. Implementing Hardware Primitives based on Memristive Spatiotemporal Variability into Cryptography Applications. *Chip* **2023**, 100040. [CrossRef]
37. Liu, Y.; Tian, Y.; Xu, J.; Wang, C.; Wang, Y.; Yuan, D.; Chew, J.W. Electrosorption performance on graphene-based materials: A review. *RSC Adv.* **2023**, *13*, 6518–6529. [CrossRef] [PubMed]
38. Ge, R.; Huo, T.; Gao, Z.; Li, J.; Zhan, X. GO-Based Membranes for Desalination. *Membranes* **2023**, *13*, 220. [CrossRef]
39. de Bortoli, B.; Rodrigues Camargo, M.; de Oliveira Polkowski, R.; de Albuquerque, R. *Graphene: An Overview of Technology in the Electric Vehicles of the Future*; SAE Technical Paper 2022-36-0100; SAE Technical: Warrendale, PA, USA, 2023. [CrossRef]
40. Scalia, T.; Bonventre, L.; Terranova, M.L. From Protosolar Space to Space Exploration: The Role of Graphene in Space Technology and Economy. *Nanomaterials* **2023**, *13*, 680. [CrossRef]
41. Madurani, K.A.; Suprapto, S.; Machrita, N.I.; Bahar, S.L.; Illiya, W.; Kurniawan, F. Progress in Graphene Synthesis and its Application: History, Challenge and the Future Outlook for Research and Industry. *ECS J. Solid State Sci. Technol.* **2020**, *9*, 093013. [CrossRef]
42. Bu, Q.; Morgan, H.; Liang, J.; Lei, H.; Ruan, R. Catalytic Microwave Pyrolysis of Lignocellulosic Biomass for Fuels and Chemicals. *Adv. Bioenergy* **2016**, *1*, 69–123. [CrossRef]
43. Tang, L.; Ji, R.; Li, X.; Teng, K.S.; Lau, S.P. Size-Dependent Structural and Optical Characteristics of Glucose-Derived Graphene Quantum Dots. *Part. Part. Syst. Charact.* **2013**, *30*, 523–531. [CrossRef]
44. Safian, M.T.-U.; Haron, U.S.; Ibrahim, M.N.M. A review on bio-based graphene derived from biomass wastes. *Bioresources* **2020**, *15*, 9756–9785. [CrossRef]
45. Berktas, I.; Hezarkhani, M.; Poudeh, L.H.; Okan, B.S. Recent developments in the synthesis of graphene and graphene-like structures from waste sources by recycling and upcycling technologies: A review. *Graphene Technol.* **2020**, *5*, 59–73. [CrossRef]
46. Zakaria, N.Z.J.; Rozali, S.; Mubarak, N.M.; Ibrahim, S. A review of the recent trend in the synthesis of carbon nanomaterials derived from oil palm by-product materials. *Biomass Convers. Biorefinery* **2022**, 1–32. [CrossRef] [PubMed]
47. Kong, X.; Zhu, Y.; Lei, H.; Wang, C.; Zhao, Y.; Huo, E.; Lin, X.; Zhang, Q.; Qian, M.; Mateo, W.; et al. Synthesis of graphene-like carbon from biomass pyrolysis and its applications. *Chem. Eng. J.* **2020**, *399*, 125808. [CrossRef]
48. Armah, E.K.; Chetty, M.; Adedeji, J.A.; Estrice, D.E.; Mutsvene, B.; Singh, N.; Tshemese, Z. Biochar: Production, Application and the Future. In *Biochar—Productive Technologies, Properties and Application*; IntechOpen: London, UK, 2022. [CrossRef]
49. Liu, W.-J.; Jiang, H.; Yu, H.-Q. Development of Biochar-Based Functional Materials: Toward a Sustainable Platform Carbon Material. *Chem. Rev.* **2015**, *115*, 12251–12285. [CrossRef]
50. Zhang, Y.; Cui, Y.; Liu, S.; Fan, L.; Zhou, N.; Peng, P.; Wang, Y.; Guo, F.; Min, M.; Cheng, Y.; et al. Fast microwave-assisted pyrolysis of wastes for biofuels production—A review. *Bioresour. Technol.* **2019**, *297*, 122480. [CrossRef]
51. Du, Z.; Li, Y.; Wang, X.; Wan, Y.; Chen, Q.; Wang, C.; Lin, X.; Liu, Y.; Chen, P.; Ruan, R. Microwave-assisted pyrolysis of microalgae for biofuel production. *Bioresour. Technol.* **2011**, *102*, 4890–4896. [CrossRef]
52. Parvez, A.M.; Wu, T.; Afzal, M.T.; Mareta, S.; He, T.; Zhai, M. Conventional and microwave-assisted pyrolysis of gumwood: A comparison study using thermodynamic evaluation and hydrogen production. *Fuel Process. Technol.* **2018**, *184*, 1–11. [CrossRef]
53. Fodah, A.E.M.; Ghosal, M.K.; Behera, D. Microwave-assisted pyrolysis of agricultural residues: Current scenario, challenges, and future direction. *Int. J. Environ. Sci. Technol.* **2021**, *19*, 2195–2220. [CrossRef]
54. Chen, P.; Xie, Q.; Addy, M.; Zhou, W.; Liu, Y.; Wang, Y.; Cheng, Y.; Li, K.; Ruan, R. Utilization of municipal solid and liquid wastes for bioenergy and bioproducts production. *Bioresour. Technol.* **2016**, *215*, 163–172. [CrossRef]
55. Wang, N.; Tahmasebi, A.; Yu, J.; Xu, J.; Huang, F.; Mamaeva, A. A Comparative study of microwave-induced pyrolysis of lignocellulosic and algal biomass. *Bioresour. Technol.* **2015**, *190*, 89–96. [CrossRef] [PubMed]
56. Robinson, J.; Binner, E.; Vallejo, D.B.; Perez, N.D.; Al Mughairi, K.; Ryan, J.; Shepherd, B.; Adam, M.; Budarin, V.; Fan, J.; et al. Unravelling the mechanisms of microwave pyrolysis of biomass. *Chem. Eng. J.* **2022**, *430*, 132975. [CrossRef]

57. González, Z.; Acevedo, B.; Predeanu, G.; Axinte, S.M.; Drăgoescu, M.-F.; Slăvescu, V.; Fernandez, J.J.; Granda, M.; Gryglewicz, G.; Melendi-Espina, S. Graphene materials from microwave-derived carbon precursors. *Fuel Process. Technol.* **2021**, *217*, 106803. [CrossRef]
58. Zhang, J.; Tahmasebi, A.; Omoriyekomwan, J.E.; Yu, J. Direct synthesis of hollow carbon nanofibers on bio-char during microwave pyrolysis of pine nut shell. *J. Anal. Appl. Pyrolysis* **2018**, *130*, 142–148. [CrossRef]
59. Gopalakrishnan, A.; Kong, C.Y.; Badhulika, S. Scalable, large-area synthesis of heteroatom-doped few-layer graphene-like microporous carbon nanosheets from biomass for high-capacitance supercapacitors. *New J. Chem.* **2018**, *43*, 1186–1194. [CrossRef]
60. Danafar, H.; Mohammadi, A.; Mousazadeh, N.; Ghaffarlou, M.; Mollasalehi, A.; Sharafi, A.; Barsbay, M.; Nosrati, H. An innovative green approach to the production of bio-sourced and nano-sized graphene oxide (GO)-like carbon flakes. *Curr. Res. Green Sustain. Chem.* **2021**, *4*, 100200. [CrossRef]
61. Somanathan, T.; Prasad, K.; Ostrikov, K.; Saravanan, A.; Krishna, V.M. Graphene Oxide Synthesis from Agro Waste. *Nanomaterials* **2015**, *5*, 826–834. [CrossRef]
62. Hashmi, A.; Singh, A.K.; Jain, B.; Singh, A. Muffle atmosphere promoted fabrication of graphene oxide nanoparticle by agricultural waste. *Fuller. Nanotub. Carbon Nanostructures* **2020**, *28*, 627–636. [CrossRef]
63. Debbarma, J.; Naik, M.J.P.; Saha, M. From agrowaste to graphene nanosheets: Chemistry and synthesis. *Fuller. Nanotub. Carbon Nanostructures* **2019**, *27*, 482–485. [CrossRef]
64. Chemistry and Mechanism of One-Step Formation of Graphene from Agrowaste. *Lett. Appl. NanoBioScience* **2020**, *9*, 1389–1394. [CrossRef]
65. Liu, F.; Chen, Y.; Gao, J. Preparation and Characterization of Biobased Graphene from Kraft Lignin. *Bioresources* **2017**, *12*, 6545–6557. [CrossRef]
66. Wang, Y.; Srinivasakannan, C.; Wang, H.; Xue, G.; Wang, L.; Wang, X.; Duan, X. Preparation of novel biochar containing graphene from waste bamboo with high methylene blue adsorption capacity. *Diam. Relat. Mater.* **2022**, *125*, 109034. [CrossRef]
67. Omoriyekomwan, J.E.; Tahmasebi, A.; Zhang, J.; Yu, J. Synthesis of Super-Long Carbon Nanotubes from Cellulosic Biomass under Microwave Radiation. *Nanomaterials* **2022**, *12*, 737. [CrossRef]
68. Hidalgo, P.; Navia, R.; Hunter, R.; Coronado, G.; Gonzalez, M. Synthesis of carbon nanotubes using biochar as precursor material under microwave irradiation. *J. Environ. Manag.* **2019**, *244*, 83–91. [CrossRef]
69. Xia, J.; Zhang, N.; Chong, S.; Li, D.; Chen, Y.; Sun, C. Three-dimensional porous graphene-like sheets synthesized from biocarbon via low-temperature graphitization for a supercapacitor. *Green Chem.* **2018**, *20*, 694–700. [CrossRef]
70. Widiatmoko, P.; Sukmana, I.F.; Nurdin, I.; Prakoso, T.; Devianto, H. Increasing yield of graphene synthesis from oil palm empty fruit bunch via two-stages pyrolysis. *IOP Conf. Ser. Mater. Sci. Eng.* **2019**, *543*, 012032. [CrossRef]
71. Fu, M.; Huang, J.; Feng, S.; Zhang, T.; Qian, P.-C.; Wong, W.-Y. One-step solid-state pyrolysis of bio-wastes to synthesize multi-hierarchical porous carbon for ultra-long life supercapacitors. *Mater. Chem. Front.* **2021**, *5*, 2320–2327. [CrossRef]
72. Abbas, A.; Tabish, T.A.; Bull, S.J.; Lim, T.M.; Phan, A.N. High yield synthesis of graphene quantum dots from biomass waste as a highly selective probe for Fe3+ sensing. *Sci. Rep.* **2020**, *10*, 21262. [CrossRef] [PubMed]
73. Tatrari, G.; Tewari, C.; Karakoti, M.; Pathak, M.; Jangra, R.; Santhibhushan, B.; Mahendia, S.; Sahoo, N.G. Mass production of metal-doped graphene from the agriculture waste of *Quercus ilex* leaves for supercapacitors: Inclusive DFT study. *RSC Adv.* **2021**, *11*, 10891–10901. [CrossRef]
74. Sebuso, D.P.; Kuvarega, A.T.; Lefatshe, K.; King'Ondu, C.K.; Numan, N.; Maaza, M.; Muiva, C.M. Corn husk multilayered graphene/ZnO nanocomposite materials with enhanced photocatalytic activity for organic dyes and doxycycline degradation. *Mater. Res. Bull.* **2022**, *151*, 111800. [CrossRef]
75. Yeleuov, M.; Daulbayev, C.; Taurbekov, A.; Abdisattar, A.; Ebrahim, R.; Kumekov, S.; Prikhodko, N.; Lesbayev, B.; Batyrzhan, K. Synthesis of graphene-like porous carbon from biomass for electrochemical energy storage applications. *Diam. Relat. Mater.* **2021**, *119*, 108560. [CrossRef]
76. Roy, A.; Kar, S.; Ghosal, R.; Naskar, K.; Bhowmick, A.K. Facile Synthesis and Characterization of Few-Layer Multifunctional Graphene from Sustainable Precursors by Controlled Pyrolysis, Understanding of the Graphitization Pathway, and Its Potential Application in Polymer Nanocomposites. *ACS Omega* **2021**, *6*, 1809–1822. [CrossRef] [PubMed]
77. Primo, A.; Atienzar, P.; Sanchez, E.; Delgado, J.M.; García, H. From biomass wastes to large-area, high-quality, N-doped graphene: Catalyst-free carbonization of chitosan coatings on arbitrary substrates. *Chem. Commun.* **2012**, *48*, 9254–9256. [CrossRef] [PubMed]
78. Huang, B.; Xia, M.; Qiu, J.; Xie, Z.; Qiu, Q. Biomass Derived Graphene-Like Carbons for Electrocatalytic Oxygen Reduction Reaction. *Chemnanomat* **2019**, *5*, 682–689. [CrossRef]
79. Shi, K.; Yan, J.; Lester, E.; Wu, T. Catalyst-Free Synthesis of Multiwalled Carbon Nanotubes via Microwave-Induced Processing of Biomass. *Ind. Eng. Chem. Res.* **2014**, *53*, 15012–15019. [CrossRef]
80. Sha, T.; Liu, J.; Sun, M.; Li, L.; Bai, J.; Hu, Z.; Zhou, M. Green and low-cost synthesis of nitrogen-doped graphene-like mesoporous nanosheets from the biomass waste of okara for the amperometric detection of vitamin C in real samples. *Talanta* **2019**, *200*, 300–306. [CrossRef]
81. Xu, X.; Sun, S.; Luo, J.; Ma, R.; Lin, J.; Fang, L.; Zhang, P.; Chen, Y. Few-layer graphene prepared via microwave irradiation of black sesame for supercapacitor applications. *Chem. Eng. J.* **2021**, *425*, 130664. [CrossRef]
82. Li, T.; Ma, R.; Xu, X.; Sun, S.; Lin, J. Microwave-induced preparation of porous graphene nanosheets derived from biomass for supercapacitors. *Microporous Mesoporous Mater.* **2021**, *324*, 111277. [CrossRef]

83. Omoriyekomwan, J.E.; Tahmasebi, A.; Zhang, J.; Yu, J. Formation of hollow carbon nanofibers on bio-char during microwave pyrolysis of palm kernel shell. *Energy Convers. Manag.* **2017**, *148*, 583–592. [CrossRef]
84. Wallace, C.A.; Afzal, M.T.; Saha, G.C. Effect of feedstock and microwave pyrolysis temperature on physio-chemical and nano-scale mechanical properties of biochar. *Bioresour. Bioprocess.* **2019**, *6*, 33. [CrossRef]
85. Huang, Y.; Kuan, W.; Lo, S.; Lin, C. Total recovery of resources and energy from rice straw using microwave-induced pyrolysis. *Bioresour. Technol.* **2008**, *99*, 8252–8258. [CrossRef] [PubMed]
86. Ekhlasi, L.; Younesi, H.; Rashidi, A.; Bahramifar, N. Populus wood biomass-derived graphene for high CO_2 capture at atmospheric pressure and estimated cost of production. *Process. Saf. Environ. Prot.* **2018**, *113*, 97–108. [CrossRef]
87. Abbas, A.; Eng, X.E.; Ee, N.; Saleem, F.; Wu, D.; Chen, W.; Handayani, M.; Tabish, T.A.; Wai, N.; Lim, T.M. Development of reduced graphene oxide from biowaste as an electrode material for vanadium redox flow battery. *J. Energy Storage* **2021**, *41*, 102848. [CrossRef]
88. Guo, W.; Li, G.; Zheng, Y.; Li, K. Nano-silica extracted from rice husk and its application in acetic acid steam reforming. *RSC Adv.* **2021**, *11*, 34915–34922. [CrossRef]
89. Islam, M.R.; Islam, M.d.N.; Islam, M.N. Fixed bed pyrolysis of sugarcane bagasse for liquid fuel production. In Proceedings of the International Conference on Mechanical Engineering 2003 (ICME2003), Dhaka, Bangladesh, 26–28 December 2003.
90. Zapata, B.; Balmaseda, J.; Fregoso-Israel, E.; Torres-García, E. Thermo-kinetics study of orange peel in air. *J. Therm. Anal. Calorim.* **2009**, *98*, 309–315. [CrossRef]
91. Jumadi, J.; Kamari, A.; Rahim, N.A.; Wong, S.T.S.; Yusoff, S.N.M.; Ishak, S.; Abdulrasool, M.M.; Kumaran, S. Removal of methylene blue and congo red by magnetic chitosan nanocomposite: Characterization and adsorption studies. *J. Phys. Conf. Ser.* **2019**, *1397*, 012027. [CrossRef]
92. Yaman, S. Pyrolysis of biomass to produce fuels and chemical feedstocks. *Energy Convers. Manag.* **2004**, *45*, 651–671. [CrossRef]
93. Shebani, A.; van Reenen, A.; Meincken, M. The effect of wood extractives on the thermal stability of different wood-LLDPE composites. *Thermochim. Acta* **2009**, *481*, 52–56. [CrossRef]
94. Torres, F.G.; Troncoso, O.P.; Rodriguez, L.; De-La-Torre, G.E. Sustainable synthesis, reduction and applications of graphene obtained from renewable resources. *Sustain. Mater. Technol.* **2021**, *29*, e00310. [CrossRef]
95. Yoon, D.; Hwang, J.; Chang, W.; Kim, J. Carbon with Expanded and Well-Developed Graphene Planes Derived Directly from Condensed Lignin as a High-Performance Anode for Sodium-Ion Batteries. *ACS Appl. Mater. Interfaces* **2017**, *10*, 569–581. [CrossRef] [PubMed]
96. Yang, S.; Ling, Y.; Wu, Q.; Zhang, H.; Yan, Z.; Huang, G.; Lin, J.; Wan, C. Lignin-derived porous graphene for wearable and ultrasensitive strain sensors. *J. Mater. Chem. C* **2022**, *10*, 11730–11738. [CrossRef]
97. Liu, M.; Thygesen, A.; Summerscales, J.; Meyer, A.S. Targeted pre-treatment of hemp bast fibres for optimal performance in biocomposite materials: A review. *Ind. Crop. Prod.* **2017**, *108*, 660–683. [CrossRef]
98. Ma'Ruf, A.; Pramudono, B.; Aryanti, N. Lignin isolation process from rice husk by alkaline hydrogen peroxide: Lignin and silica extracted. In *AIP Conference Proceedings*; AIP Publishing LLC.: Melville, NY, USA, 2017; p. 020013.
99. Su, H.; Liu, G.; He, M.; Tan, F. A biorefining process: Sequential, combinational lignocellulose pretreatment procedure for improving biobutanol production from sugarcane bagasse. *Bioresour. Technol.* **2015**, *187*, 149–160. [CrossRef]
100. Sudiyani, Y.; Styarini, D.; Triwahyuni, E.; Sudiyarmanto; Sembiring, K.C.; Aristiawan, Y.; Abimanyu, H.; Han, M.H. Utilization of Biomass Waste Empty Fruit Bunch Fiber of Palm Oil for Bioethanol Production Using Pilot–Scale Unit. *Energy Procedia* **2013**, *32*, 31–38. [CrossRef]
101. Rodriguez-Gomez, D.; Lehmann, L.; Schultz-Jensen, N.; Bjerre, A.B.; Hobley, T.J. Examining the Potential of Plasma-Assisted Pretreated Wheat Straw for Enzyme Production by Trichoderma reesei. *Appl. Biochem. Biotechnol.* **2012**, *166*, 2051–2063. [CrossRef]
102. Zainal, N.H.; Aziz, A.A.; Idris, J.; Mamat, R.; Hassan, M.A.; Bahrin, E.K.; Abd-Aziz, S. Microwave-assisted pre-carbonisation of palm kernel shell produced charcoal with high heating value and low gaseous emission. *J. Clean. Prod.* **2017**, *142*, 2945–2949. [CrossRef]
103. Bai, Y.-Y.; Xiao, L.-P.; Shi, Z.-J.; Sun, R.-C. Structural Variation of Bamboo Lignin before and after Ethanol Organosolv Pretreatment. *Int. J. Mol. Sci.* **2013**, *14*, 21394–21413. [CrossRef]
104. Passoth, V.; Sandgren, M. Biofuel production from straw hydrolysates: Current achievements and perspectives. *Appl. Microbiol. Biotechnol.* **2019**, *103*, 5105–5116. [CrossRef] [PubMed]
105. Muktham, R.; Bhargava, S.K.; Bankupalli, S.; Ball, A.S. A Review on 1st and 2nd Generation Bioethanol Production-Recent Progress. *J. Sustain. Bioenergy Syst.* **2016**, *6*, 72–92. [CrossRef]
106. Jahirul, M.I.; Rasul, M.G.; Chowdhury, A.A.; Ashwath, N. Biofuels Production through Biomass Pyrolysis —A Technological Review. *Energies* **2012**, *5*, 4952–5001. [CrossRef]
107. Albatrni, H.; Qiblawey, H.; Al-Marri, M.J. Walnut shell based adsorbents: A review study on preparation, mechanism, and application. *J. Water Process. Eng.* **2022**, *45*, 102527. [CrossRef]
108. Suriapparao, D.V.; Pradeep, N.; Vinu, R. Bio-Oil Production from Prosopis juliflora via Microwave Pyrolysis. *Energy Fuels* **2015**, *29*, 2571–2581. [CrossRef]
109. Parthasarathy, P.; Sheeba, K.N. Generation of fuel char through biomass slow pyrolysis. *Energy Sources Part A Recover. Util. Environ. Eff.* **2017**, *39*, 599–605. [CrossRef]

110. Huang, Y.-F.; Chiueh, P.-T.; Kuan, W.-H.; Lo, S.-L. Microwave pyrolysis of lignocellulosic biomass: Heating performance and reaction kinetics. *Energy* **2016**, *100*, 137–144. [CrossRef]
111. Liu, H.; E, J.; Deng, Y.; Xie, C.; Zhu, H. Experimental study on pyrolysis characteristics of the tobacco stem based on microwave heating method. *Appl. Therm. Eng.* **2016**, *106*, 473–479. [CrossRef]
112. Kan, T.; Strezov, V.; Evans, T.J. Lignocellulosic biomass pyrolysis: A review of product properties and effects of pyrolysis parameters. *Renew. Sustain. Energy Rev.* **2016**, *57*, 1126–1140. [CrossRef]
113. Demirbas, A. Effect of initial moisture content on the yields of oily products from pyrolysis of biomass. *J. Anal. Appl. Pyrolysis* **2004**, *71*, 803–815. [CrossRef]
114. Liu, H.; Zhang, Q.; Hu, H.; Li, A.; Yao, H. Influence of residual moisture on deep dewatered sludge pyrolysis. *Int. J. Hydrogen Energy* **2014**, *39*, 1253–1261. [CrossRef]
115. Gray, M.R.; Corcoran, W.H.; Gavalas, G.R. Pyrolysis of a wood-derived material. Effects of moisture and ash content. *Ind. Eng. Chem. Process. Des. Dev.* **1985**, *24*, 646–651. [CrossRef]
116. Westerhof, R.J.M.; Kuipers, N.J.M.; Kersten, S.R.A.; van Swaaij, W.P.M. Controlling the Water Content of Biomass Fast Pyrolysis Oil. *Ind. Eng. Chem. Res.* **2007**, *46*, 9238–9247. [CrossRef]
117. Fonseca, F.G.; Funke, A.; Niebel, A.; Dias, A.P.S.; Dahmen, N. Moisture content as a design and operational parameter for fast pyrolysis. *J. Anal. Appl. Pyrolysis* **2019**, *139*, 73–86. [CrossRef]
118. Darmstadt, H.; Pantea, D.; Sümmchen, L.; Roland, U.; Kaliaguine, S.; Roy, C. Surface and bulk chemistry of charcoal obtained by vacuum pyrolysis of bark: Influence of feedstock moisture content. *J. Anal. Appl. Pyrolysis* **2000**, *53*, 1–17. [CrossRef]
119. Li, X.; Peng, B.; Liu, Q.; Zhang, H. Microwave pyrolysis coupled with conventional pre-pyrolysis of the stalk for syngas and biochar. *Bioresour. Technol.* **2022**, *348*, 126745. [CrossRef]
120. Jian, J.; Lu, Z.; Yao, S.; Li, Y.; Liu, Z.; Lang, B.; Chen, Z. Effects of thermal conditions on char yield and char reactivity of woody biomass in stepwise pyrolysis. *J. Anal. Appl. Pyrolysis* **2018**, *138*, 211–217. [CrossRef]
121. Leite, J.C.S.; Suota, M.J.; Ramos, L.P.; Lenzi, M.K.; de Lima Luz, L.F. Depolymerization of Sugarcane Bagasse by Microwave-Assisted Pyrolysis. *Chem. Eng. Trans.* **2022**, *92*, 511–516. [CrossRef]
122. Burhenne, L.; Damiani, M.; Aicher, T. Effect of feedstock water content and pyrolysis temperature on the structure and reactivity of spruce wood char produced in fixed bed pyrolysis. *Fuel* **2013**, *107*, 836–847. [CrossRef]
123. Van de Velden, M.; Baeyens, J.; Brems, A.; Janssens, B.; Dewil, R. Fundamentals, kinetics and endothermicity of the biomass pyrolysis reaction. *Renew. Energy* **2010**, *35*, 232–242. [CrossRef]
124. Safdari, M.-S.; Amini, E.; Weise, D.R.; Fletcher, T.H. Heating rate and temperature effects on pyrolysis products from live wildland fuels. *Fuel* **2019**, *242*, 295–304. [CrossRef]
125. Gerçel, H.F. Bio-oil production from Onopordum acanthium L. by slow pyrolysis. *J. Anal. Appl. Pyrolysis* **2011**, *92*, 233–238. [CrossRef]
126. Asadullah, M.; Zhang, S.; Li, C.-Z. Evaluation of structural features of chars from pyrolysis of biomass of different particle sizes. *Fuel Process. Technol.* **2010**, *91*, 877–881. [CrossRef]
127. Fu, P.; Hu, S.; Xiang, J.; Sun, L.; Su, S.; Wang, J. Evaluation of the porous structure development of chars from pyrolysis of rice straw: Effects of pyrolysis temperature and heating rate. *J. Anal. Appl. Pyrolysis* **2012**, *98*, 177–183. [CrossRef]
128. Zhao, B.; O'Connor, D.; Zhang, J.; Peng, T.; Shen, Z.; Tsang, D.C.W.; Hou, D. Effect of pyrolysis temperature, heating rate, and residence time on rapeseed stem derived biochar. *J. Clean. Prod.* **2018**, *174*, 977–987. [CrossRef]
129. Lua, A.C. A detailed study of pyrolysis conditions on the production of steam-activated carbon derived from oil-palm shell and its application in phenol adsorption. *Biomass Convers. Biorefinery* **2019**, *10*, 523–533. [CrossRef]
130. Kumar, M.; Gupta, R.C. Influence of carbonization conditions on the gasification of acacia and eucalyptus wood chars by carbon dioxide. *Fuel* **1994**, *73*, 1922–1925. [CrossRef]
131. Yu, J.; Sun, L.; Berrueco, C.; Fidalgo, B.; Paterson, N.; Millan, M. Influence of temperature and particle size on structural characteristics of chars from Beechwood pyrolysis. *J. Anal. Appl. Pyrolysis* **2018**, *130*, 127–134. [CrossRef]
132. Zeng, K.; Minh, D.P.; Gauthier, D.; Weiss-Hortala, E.; Nzihou, A.; Flamant, G. The effect of temperature and heating rate on char properties obtained from solar pyrolysis of beech wood. *Bioresour. Technol.* **2015**, *182*, 114–119. [CrossRef] [PubMed]
133. Antal, M.J.; Grønli, M. The Art, Science, and Technology of Charcoal Production. *Ind. Eng. Chem. Res.* **2003**, *42*, 1619–1640. [CrossRef]
134. Lu, G.; Low, J.; Liu, C.; Lua, A. Surface area development of sewage sludge during pyrolysis. *Fuel* **1995**, *74*, 344–348. [CrossRef]
135. Zhang, Y.-L.; Wu, W.-G.; Zhao, S.-H.; Long, Y.-F.; Luo, Y.-H. Experimental study on pyrolysis tar removal over rice straw char and inner pore structure evolution of char. *Fuel Process. Technol.* **2015**, *134*, 333–344. [CrossRef]
136. Ethaib, S.; Omar, R.; Kamal, S.; Biak, D.A.; Zubaidi, S. Microwave-Assisted Pyrolysis of Biomass Waste: A Mini Review. *Processes* **2020**, *8*, 1190. [CrossRef]
137. Fang, Z.; Gao, Y.; Bolan, N.; Shaheen, S.M.; Xu, S.; Wu, X.; Xu, X.; Hu, H.; Lin, J.; Zhang, F.; et al. Conversion of biological solid waste to graphene-containing biochar for water remediation: A critical review. *Chem. Eng. J.* **2020**, *390*, 124611. [CrossRef]
138. Yang, H.; Yan, R.; Chen, H.; Lee, D.H.; Zheng, C. Characteristics of hemicellulose, cellulose and lignin pyrolysis. *Fuel* **2007**, *86*, 1781–1788. [CrossRef]
139. Debbarma, J.; Mandal, P.; Saha, M. N-graphene oxide and N-reduced graphene oxide from jujube seeds: Chemistry and mechanism. *Fuller. Nanotub. Carbon Nanostructures* **2020**, *28*, 702–706. [CrossRef]

140. Omoriyekomwan, J.E.; Tahmasebi, A.; Zhang, J.; Yu, J. Mechanistic study on direct synthesis of carbon nanotubes from cellulose by means of microwave pyrolysis. *Energy Convers. Manag.* **2019**, *192*, 88–99. [CrossRef]
141. Omoriyekomwan, J.E.; Tahmasebi, A.; Dou, J.; Wang, R.; Yu, J. A review on the recent advances in the production of carbon nanotubes and carbon nanofibers via microwave-assisted pyrolysis of biomass. *Fuel Process. Technol.* **2020**, *214*, 106686. [CrossRef]
142. Tan, H.; Wang, D.; Guo, Y. Thermal Growth of Graphene: A Review. *Coatings* **2018**, *8*, 40. [CrossRef]

Disclaimer/Publisher's Note: The statements, opinions and data contained in all publications are solely those of the individual author(s) and contributor(s) and not of MDPI and/or the editor(s). MDPI and/or the editor(s) disclaim responsibility for any injury to people or property resulting from any ideas, methods, instructions or products referred to in the content.

Review

Emerging Bio-Based Polymers from Lab to Market: Current Strategies, Market Dynamics and Research Trends

Tomy Muringayil Joseph [1,*], Aparna Beena Unni [2], K. S. Joshy [3], Debarshi Kar Mahapatra [4,*], Józef Haponiuk [1] and Sabu Thomas [5]

1. Department of Polymer Technology, Faculty of Chemistry, Gdansk University of Technology, G. Narutowicza 11/12, 80-233 Gdansk, Poland
2. Institute of Physics, University of Silesia, 75 Pulku Piechoty 1a, 41-500 Chorzow, Poland
3. School of Energy Materials, Mahatma Gandhi University, Kottayam 686560, Kerala, India
4. Department of Pharmaceutical Chemistry, Dadasaheb Balpande College of Pharmacy, Nagpur 440040, Maharashtra, India
5. International and Inter-University Centre for Nanoscience and Nanotechnology, Mahatma Gandhi University, Kottayam 686560, Kerala, India
* Correspondence: tomy.muringayil@pg.edu.pl (T.M.J.); dkmbsp@gmail.com (D.K.M.)

Citation: Joseph, T.M.; Unni, A.B.; Joshy, K.S.; Kar Mahapatra, D.; Haponiuk, J.; Thomas, S. Emerging Bio-Based Polymers from Lab to Market: Current Strategies, Market Dynamics and Research Trends. *C* 2023, 9, 30. https://doi.org/10.3390/c9010030

Academic Editor: Giuseppe Cirillo

Received: 31 December 2022
Revised: 23 February 2023
Accepted: 28 February 2023
Published: 7 March 2023

Copyright: © 2023 by the authors. Licensee MDPI, Basel, Switzerland. This article is an open access article distributed under the terms and conditions of the Creative Commons Attribution (CC BY) license (https:// creativecommons.org/licenses/by/ 4.0/).

Abstract: Due to the rising worldwide demand for green chemicals, the bio-based polymer market is anticipated to expand substantially in the future. The synthesis of functional polymers has been a burgeoning area of research for decades. The primary driving force behind the development of bio-based polymers has been their compostability and biodegradability, which are critical given the public concern about waste. Significant advancements in the method for refining biomass raw materials towards the creation of bio-based construction materials and products are driving this rise. Bio-based polymers with this chemical structure are more flexible and adaptive, which allows them to attain their intended characteristics and functionalities. In commercial applications and healthcare and biotechnology, where completely manufactured, naturally occurring biomolecules are utilized and such polymers have the greatest impact. At the same time, limitations in polymer architectural control, biostability, and structural dynamics hinder the creation of biocompatible and functionally varied polymers. From this standpoint, the importance of functional biosynthetic polymers in the future years is highlighted, as well as new methods for addressing the aforementioned challenges. The article comprehensively highlighted the current strategies, market dynamics, and research trends of emerging Bio-Based Polymers. In addition, the most recent scientific breakthroughs in bio-based polymers are discussed.

Keywords: polymers; biocompatible; biodegradable; compostable; sustainable; market dynamics; research trends

1. Introduction

Polymeric materials derived from renewable resources have lately garnered more attention as environmentally friendly and sustainable chemistry, demonstrating the variety obtained through synthesis and semi-synthesis [1–3]. The principal source of greenhouse gas emissions is the use of fossil fuels in various industrial operations. Concerns about the increasing carbon footprint of major manufacturing facilities have prompted the industry to switch from biosources to polymers, bringing the possibility of the phasing out of traditional polymers closer to reality. In 1840, Hancock and Rudersdorf developed the first artificial polymers by blending natural rubber with sulfur to create a strong and flexible material [4]. Major advancements in polymer chemistry facilitate the development of purely synthetic or complex polymer materials. The two key advantages of bio-based polymers, according to scientists, have yet to be acknowledged politically. The first advantage is the replacement of fossil carbon in the process of production with renewable carbon from biomass, which is vital for a sustainable and climate-friendly plastics industry. The second benefit is that biodegradable bio-based polymers account for approximately a quarter of all bio-based polymers (depending on the environment). Corn, sugar cane, vegetable oils, soybeans, and cellulose can all be used to produce this polymer. Plant-derived bio-based polymers, also known as the next-generation polymers which have been used to reduce fossil fuel usage and may be a solution for the issue of plastics uncollected and unprocessed in an environment, and such polymers biodegrade without leaving microplastics behind [5,6]. Furthermore, owing to the extensive use of live and controllable polymerization techniques, biocompatible synthetic materials have been reported as one of the most relevant and fruitful disciplines of polymer chemistry in recent decades. Biosynthetic polymers are a kind of material that is utilized in numerous applications, including new biomolecular stabilizers, biosensors, therapeutic agents, biomedical adhesives, drug delivery, biomimetic scaffolds, and vehicle antifouling materials [7]. Bio-based polymer branches are dynamic and versatile, and they are in many stages of development, ranging from research to preliminary market acceptance to application in long-term high-performance polymers such as cellulose and nylon, which indicates market growth.

National policy, technology, raw material costs, competition (fossil fuels and biomass), consumer acceptance, crude oil prices, and accessibility to clear and reliable market data are all factors that influence the rate of expansion of bio-based polymer branches. This article examines a variety of bio-based polymers, focusing on their characteristics, production processes, and commercial uses. The review highlighted the composition and usage of bio-based polymers including polylactic acid, polyhydroxyalkanoate, biopolyethylene, and polysuccinic acid succinate. Furthermore, the process economics of conventional petroleum-based polymers were compared, as well as the process economics of monomer and polymer synthesis. Researchers working on the synthesis of conventional petroleum-derived raw material-based and bio-based polymers may find this paper useful [8]. For example, we described:

(1) Structural dynamics of polymer assemblies,
(2) Architecture control of synthetic components, and
(3) Biostabilization (storage, release, and biotolerance) of therapeutic contents [9].

The authors also point out that the outcome of this review is quite beneficial for researchers. As a result, the research focuses on the benefits of simple formulations for the development of highly functional biosynthetic polymers, as well as current trends, breakthroughs, and potential applications. The merits of simple synthetic polymer formulations include the potential challenges of using bio-based materials in a variety of applications, as well as the solutions and key industry players that may help bring these materials to market. The conclusion offers a predicted position on future growth in the area, which is contingent on the ability to surpass sustainability limitations [10]. The review investigated market conditions, growth rates, market share, market momentum, challenges and opportunities, risk and entry barriers, sales channels, distributors, and future trends. In this context, our creative efforts to go beyond what nature has mastered are showcased.

2. Biopolymer

Sources, variety, and characteristics of biopolymers are still increasing due to the rising interest. Biopolymers are often divided into three categories: bio-based and biodegradable, petroleum-based and biodegradable, and bio-based and non-biodegradable [1]. Traditional petroleum-based polymers, except for poly(ε-caprolactone) (PCL), polybutylene adipate terephthalate (PBAT), and poly(butylene succinate) (PBS), are typically non-biodegradable [11,12].

2.1. Bio-Based Polymers

Bio-based polymers consists of at least one polymer generated from renewable sources and are either biodegradable or non-biodegradable [12–14]. The starting component of a bio-based polymer, for example, could be formed from corn or sugar cane, and the rest could be composed of carbon derived from fossil fuels. Bio-based polymers have a low CO_2 impact and are linked with the sustainability approach. Concerns about the depletion of fossil fuels and global warming are related to the use of petrochemicals. Bio-based polymers are further divided into natural and synthetic polymers. The entire plants and animals are made out of polymers. The most widely used type of polymer is synthetic polymer [15]. Polyethylene, PVC, and synthetic rubber are examples of synthetic polymers. Plants, microbes, and animals all seem to be sources of renewable biomass. Bio-based polymers can be made solely from sustainable biomass resources (such as cellulose) by monomer biosynthesis followed by polymerization (PLA, etc.) or microbial production (PHA, etc.) [16]. Alternatives to petroleum-based polycarbonates include bio-based polycarbonates and isosorbide polycarbonates, and certain bio-based polymers have recently been produced commercially. Bisphenol A (BPA) monomers have been replaced by corn-based isosorbide. Bio-based polybutylene succinate (PBS) resin made from bio-based succinic acid and bio-based 1,4-butanediol can replace biodegradable petrochemical-based PBS (BDO). Green polyethylene is commercialized via bio-based ethylene. In 109 product categories, the United States Department of Agriculture (USDA) has established the minimum content of bio-based components. To be certified as a bio-based material by the BIO preferred program, the reference product must achieve the basic standards. According to the ASTM D6866 standard, testing is necessary [17]. In addition to the BIO preferred label, which itself is based on ASTM D6866 testing, there is a slew of alternative bio-based labels, such as Vincotte's OK bio-based. The composition of bio-based products may be assessed by measuring the ^{14}C concentration, according to ASTM D6866-05 (2005). For disposable tableware manufactured of or coated with plastic, the smallest proportion of bio-based material is 72%. This is attributable to the fact that radiocarbon (^{14}C) from fossil-based carbon depolymerization has a half-life of 5730 years. There is currently no equivalent ISO standard test technique used to determine the bio-based content of materials.

2.2. Synthetic Bio-Based Polymer

Biosynthetic polymers, also known as synthetic bio-based polymers, are polymers based on renewable materials that must be chemically processed into polymers. These are materials that connect synthetic components to biopolymers or parts to generate realistic-looking reproductions [18]. Biopolymers that have been synthetically modified, such as functionalized hyaluronic acid derivatives, make up synthetic bio-based polymers or proteins that have been tagged through cell instruction. In the first instance, site-specific conjugation and subsequent purification are frequently problematic when reactive domains (amine, thiol, hydroxyl, and carboxylic acid) are conservatively present as many copies in biopolymers such as polysaccharides or proteins. Through various grafting processes, other biosynthetic polymers are bimolecularly connected to synthetic polymers, allowing for more exact control over sophisticated topologies, fictionalization, and dynamic function. The main disadvantage of synthetic polymers is that they are by-products of petrochemicals that are used for short-term purposes although they endure for a long time, resulting in a slew of environmental issues such as soil erosion, water pollution, and toxic-gas emissions [19,20].

Many traditional polymers can be produced using renewable resources. Corn starch, for example, is hydrolyzed into fermentative raw ingredients for biological conversion to lactic acid, which is then chemically processed into PLA [21]. Even if the source is renewable, polymers produced in a chemical plant are not considered "natural". The majority of polymers presently in use are renewable bio-based polymers, which are produced from non-sustainable petroleum resources. Synthetic plastic use is estimated to account for 7% of global fossil fuel consumption. People have been compelled to search for green chemicals to create polymer materials from a range of natural resources due to the depletion of petroleum resources and increasing interest in the environment and society. Two main types of natural resources are now being considered.

The first are natural polymers such as cellulose, hemicellulose, lignin, and polysaccharides. These natural polymers are widely used in everyday life and are often recognized among the most plentiful and affordable biodegradable polymers (BDPs). Because of their intricacy and pre-existing macromolecular structures, natural polymers can only be enhanced by physical mixing or minor chemical alterations. The second kind of natural resource is molecular biomass, which comprises fatty acids, lactic acid, vegetable oils, hydroxyalkanoates, and other biomasses. Many synthetic polymers made from petroleum resources may be polymerized with this type of molecular biomass, which can be molecularly transformed into monomer subunits to perform a variety of polymerization operations [22]. Natural molecular biomass is categorized based on the composition of carbon, hydrogen, and oxygen content, which can originate from several sources. Figure 1 illustrates the natural molecular biomass.

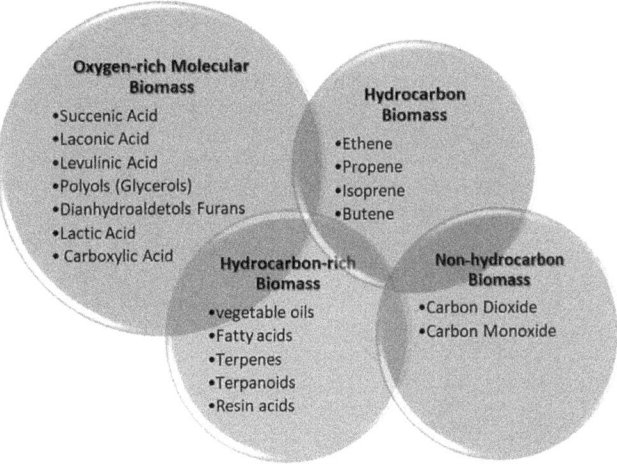

Figure 1. Natural molecular biomass in each category.

(1) Oxygen-rich molecular biomass with a molar ratio C/O of less than 5.0;
(2) Hydrocarbon-rich molecular biomass with a C/O ratio of more than 5.0;
(3) Hydrocarbon molecular biomass;
(4) Non-hydrocarbon molecular biomass.

As a result, natural molecular biomass has a better chemical structure, and the creation of various polymers from biomass with regulated topologies and specific properties has aroused attention recently, emerging as a potential research subject.

2.3. Biodegradable Polymer Plastics

A novel category of biodegradable plastics, biodegradable polymer plastics, are connected to the degradability of polymeric materials. When using these plastics, it is necessary to ensure comparable health and applicable application performance. Following the completion of the procedure, the material can be quickly degraded in a natural setting, providing or destroying environmental pieces. Over time, additional deterioration progressively returns to nature as oxidising products (CO_2 and water). It is critical to research perishable polymer materials in light of environmental pollution brought on by plastic waste, environmental protection, and human requirements. Biodegradable polymers' chemical makeup varies over time and in certain environmental circumstances. Biodegradable plastics may be classified into two groups: biodegradable plastic and photodegradable plastic, depending on what causes the changes in their chemical makeup [23].

2.3.1. Degradation System of Degraded Plastics

Degraded plastic generally refers to a form of plastic that can be broken down into smaller molecules by the action of microorganisms in the soil or by the exposure to sunlight. It complies with the demands of the use of items and is simple to process as a result. Biodegradable characteristics: The process by which sunlight interacts with polymer materials is known as photooxidation degradation because UV light and oxygen in the air interact in a unique way. Consider polyolefin as an illustration of the mechanism of photooxidation degradation. In essence, photooxidation causes the polymer chain to split or crosslink, resulting in the formation of certain functional groups containing oxygen, such as carboxy acids, peroxides, ketones, and alcohols. The primary causes of degradation include catalyst residues in polymers, peroxide production, and carboxyl groups added during processing. The chemical composition of polymers and their molecular weight can vary as a result of being eliminated or metabolically altered by microorganisms (mostly fungus, bacteria, or algae) [24]. There are two primary categories of operation:

(1) Biophysical activity—This physical activity in the polymer generates mechanical damage and results in oligomer fragments due to the stimulation of microbes, organic cell development, and fermentation of polymers, ionisation, or protonation, which are the objective of physical degradation.
(2) Biochemical activity—This activity is brought on by the breakdown of enzymes generated by fungus or bacteria, which causes plastics to degrade or undergo oxidative destruction as well as the breakdown of insoluble polymers into new tiny molecule compounds (CH_4, CO_2, H_2, CO_2, etc.).

There are typically two ideas on the process by which polymer materials degrade through biodegradation. The third concerns the reactive end(s) of polymer chain. Therefore, structural characteristics including composition, main and side chain structure, end group size, and the presence or lack of spatial steric resistance play a significant role in determining the fate of degradation. The primary chain qualities are more significant features among the process of degradation. If the main chain of polymerhas bonds that are simple to hydrate, biodegradation will occur quickly. Then, a flexible spine will degenerate at a very rapid rate, whereas a hard, regular spine would degenerate at a slower rate. By branching and cross-linking, polymer materials are less likely to degrade biologically. For instance, the addition of hydrophobic groups to the polylactic acid (PLA) molecular chain at the end of the chain lowers the rate of erosion during the initial stages of deterioration. This is due to the fact that during the actual disintegration process, erosion rate of PLA mostly depends on the structure of the molecular network, and it decreases when hydrophobic groups are added [25]. Additionally, some researchers have examined the chemical makeup of polymers as well as the relative molecular weight of elements that are crucial to their destruction.

2.3.2. Development of Biodegradable Plastics

The following is the route that biodegradable polymers will follow in the future:

(1) Block copolymerization of biodegradable plastics employing currently available standard polymers, microbial polymers, and natural polymers has been researched and produced in order to create biodegradable plastics by investigating the biodegradation mechanism of the damaged polymers.
(2) Regulation of the pace of breakdown and creation of efficient degradation promoters and stabilisers to boost perishable biodegradable performance of plastics, lower their price, and broaden their market application has been selected as an area of interest.
(3) Development of a common definition of deteriorated plastics, improvement of biodegradation assessment techniques, and a deeper understanding of degradation mechanisms all require further research [26].

3. Characteristics

3.1. Compostability

Compostable materials are described as materials that are biodegradable under aerobic circumstances that convert to biomass, water, carbon dioxide, and inorganic compounds while releasing no noxious chemicals and disintegrating throughout the fermentation phase [27]. Furthermore, biodegradable materials should not create difficulties in the composting process or in the compost itself, according to the CEN standard. BDPs are not compostable, while compostable polymers are deemed biodegradable, resulting in high demand. In this topic, taking into consideration of temporal decay and needs during the composting process that do not create difficulties, as well as compostability. To meet the compostability requirements, polymers must be evaluated according to ISO, ASTM, and EN standards.

EN 13432 is the main European standard for packaging compostability. Industrial compost is the subject of test procedures and certifications [28]. Temperature (58 °C), ventilation, humidity, and the compost cycle must all be carefully controlled in industrial compost (thermophilic period—3 months, maturity period—3 months). Composting at home, on the other hand, typically occurs under distinct circumstances. Temperatures in the compost pile are lower and can change greatly throughout the season. Furthermore, there are no international criteria for home compostability, and the performance of industrial compostable goods (EN 13432) in household composting is unknown. Vincotte certifies "OK Compost Home" compostable household items. The certification passes the EN13432 test standard, although at a lower temperature (20 °C to 30 °C instead of 58 °C) and for a longer duration of time (365 days instead of 180 days). There are other certifications available, such as "compostable".

3.2. Biodegradability

A variety of international (ISO, ASTM, CEN) and national (DIN, etc.) organizations have established biodegradability standards [29]. According to the European Committee for Standardization, biodegradable materials decompose into biomass, CO_2, and/or methane, as well as water (CEN). Other definitions include Directive 94/62/EC of the European Parliament and Council, which states that biodegradable packaging waste must be physically, chemically, thermally, or biologically degraded to convert materials primarily to CO_2, biomass, and water. Different publications on biodegradability standards are available on different issues. These publications describe the complexity of the biodegradation process compared to several polymer detection methods [30].

4. Degradation vs. Biodegradation

Each polymer degrades with time, resulting in a loss of strength, quality, and molecular weight. It is well known that most polymers require 100–1000 years to completely degrade. By-products of BDPs include carbon dioxide, methane, water, and inorganic chemicals, as well as biomass that may be readily scavenged by microbes. They degrade extremely quickly in comparison to non-BDPs [31]. Many pro-oxidants (ferromanganese and cobalt salts) are formulated for use in composting, garbage, and soil treatment polymers. According to scientific recommendations and studies, these additives may regulate the life of plastic films and goods when combined with conventional polymers in the right proportions, enabling plastic waste to be destroyed without having to be collected and disposed of. This oxidized molecular fragment is biodegradable and hydrophilic, according to the study, and the molar mass value is reduced by more than a tenth. It has been reported that polymers containing the above additives are fragmented into smaller parts that are invisible to the naked eye, but it is unclear whether these parts are truly biodegradable; that is, whether they are accessible to microflora (bacteria, fungi, etc.) to convert and assimilate carbon in any substrate, and if so, at what rate [32].

At this point, residue fragmentation and cross-linking with moisture, heat, sunshine, and/or enzymes create persistent residues, which weaken or break the polymer chains.

Biodegradation is the next stage of this process. Only when the fragmented residues are fully absorbed as food or energy by the microorganism at an appropriate pace does biodegradation begin. Because it has yet to be shown, the fragmentation of photodegradable (fragmentable) polymers by biodegradation based on polyethylene with pro-oxidants remains an open concern. It failed because of conflicting biodegradable standards and specifications. The most recent developments are bio-based polymers produced from bacterial fermentation, which are focusing on the creation of monomers made of cellulose, fatty acids, and starch from renewable resources [33].

5. Environmental Biodegradation Steps of the Polymer

The polymer material is employed either during the induction phase, or the polymer chains undergo abiotic degradation (mainly oxidation). Chemicals (impurities, metals, catalyst residues, additives, and so on), energy (light, heat, and sound), the environment, and other factors may all play a role. All of these components may work together throughout fabrication, preservation, in and after service. Exposure to photoirradiation and/or heat energy may accelerate abiotic biodegradation [34]. Figure 2 depicts the environmental biodegradation processes of polymers.

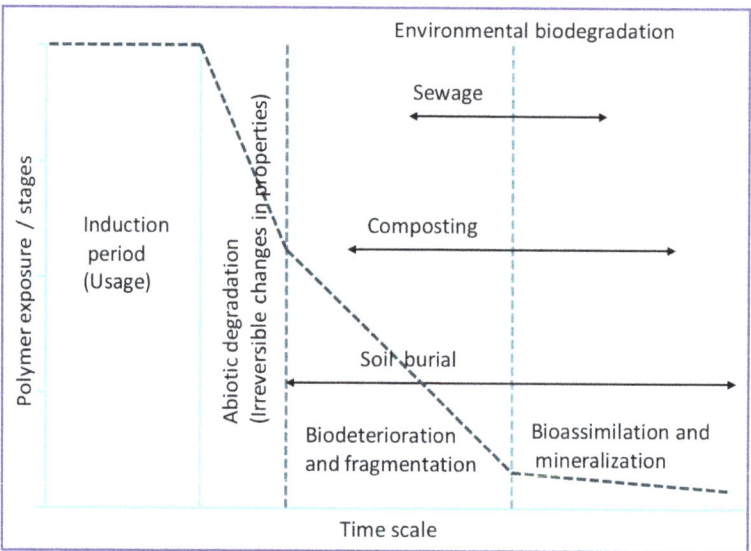

Figure 2. Environmental biodegradation steps of the polymer.

The real biodegradation occurs after that, with the aid of fragmented microorganisms, and the previously segmented polymer chains further scission and fracture. The degrading bacteria absorb and mineralize the segments of the polymer chain, and this process continues [35]. Biodegradation of polymer materials has an impact on the morphological characteristics of the polymer as well as its chemical integrity. Surface regions of amorphous polymers deteriorate faster than crystalline regions, which have less free space and more solid linkages across polymer chains.

5.1. Types

Agro-polymers are biomass products, and BDPs are split into two types based on their fabrication technique [36]. Oil-derived polymers, traditional synthesis from bio-derived monomers, and polymers obtained via microbial processes are the three major types of biopolyesters. Polysaccharides and proteins are two subcategories within the biomass products category [37]. Polysaccharides and other agricultural resources are used to produce agro-polymers. Polysaccharides seem to be the most common macromolecule found naturally (complex carbohydrates). The most well-known polysaccharide is starch, which is derived from grains and tubers. Other polysaccharides include chitosan, chitin, and pectin. Proteins derived from amino acids are the second kind of agricultural polymer. The most essential characteristic of polymers is their capacity to regenerate. Casein, collagen, gelatin, and keratin are examples of animal proteins that are commonly utilized in the food business, whereas wheat protein, soy protein, and maize protein are examples of plant proteins. The main bacterial proteins are lactate dehydrogenase, chymotrypsin, and fumarase [38]. Even though such polymers are primarily utilized in the food business, the packaging industry has yet to discover a method to replace traditional plastics with agro-polymers. Bio-polyesters (BPEs) are the second kind of BDP, and they may be composed of either living or non-living materials (synthetics).

5.2. Examples

There are two types of lactic acid: *D*-lactic acid and *L*-lactic acid are well-known and widely utilized biopolymers in a variety of sectors. They are produced in two ways: biologically and chemically [39]. *Lactobacillus* (bacteria) or fungi are used in the former technique to ferment carbohydrates. Lactic acid may also be formed as a result of chemical chain reactions. When a person's muscles are fatigued, this response may occur in the body. Lactic acid is mostly utilized in the manufacturing of cosmetics in the industrial sector. Polyhydroxyalkanoates (PHAs) are a form of intracellular biopolymer generated by bacteria as intracellular carbon or energy storage granules because of their biodegradability [40]. They are utilized in the packaging and medical sectors, among others. The most typical BDPs are petroleum by-products. Petroleum-based polyesters such as polycaprolactone, aliphatic co-polyesters, and aromatic co-polyesters are soft materials at room temperature.

5.3. Non-Pharmacological Applications of Biopolymers

5.3.1. Bio-Plastics (BPs) for Packaging

Bio-based polymers are slowly reaching mainstream use in today's world, with the packaging being one of the most significant types of it [41]. As yearly sales rise by 20% to 30%, polymers produced from renewable "raw materials" may soon compete with over-the-counter plastics. Packaging is one of the fastest-expanding businesses on the planet, with plastic accounting for almost 40% of all packaging material [42]. Despite this, the growing usage of plastics continues to represent a significant environmental risk. Consumers and package makers are starting to note the significance of balancing ecosystems as a result of the growth of digital media, and they are beginning to embrace sustainable packaging materials, with a preference for ecologically friendly and recyclable materials. As a result of these concerns, demand for BPs that replace conventional plastics, particularly biodegradable BPs, is rising [43]. Films, cover films, bottles, flip-top cartons, shipping bags, trash collection bags, and food service supplies are just a few examples of BP applications. The market for BPs and biopolymers is expected to be worth USD 6.95 billion in 2018 and USD 14.92 billion by 2023, according to estimates. From 2018 through 2023, the annual growth rate is 16.5%. Between 2018 and 2023, this indicates a CAGR of 16.5% [44]. This strong demand is causing the increase in the use of BPs in packaging in the developing nations.

The bio-based polymer industry will be substantially expanded in consumer product applications, and government laws that increasingly concentrate on sustainability and green procurement policies will boost the BP and biopolymer market [45]. Based on their diverse types, BPs and biopolymers are divided into biodegradable and non-biodegradable BPs. Biopolymers and BPs that are not biodegradable are anticipated to be the most popular and fastest-growing. The packaging industry's rising need is causing the increasing demand for non-biodegradable BPs. Bio-PET seems to be the most popular category of non-biodegradable BPs and biopolymers, owing to its cheap cost and simple production method [35]. Automotive and transportation, agriculture and horticulture, packaging, consumer goods, textiles, and other end-use industries are among the BP and biopolymer markets. Packaging is the most significant and fastest-growing application for BPs and biopolymers. Food and beverage sectors, for example, require sustainable solutions. Furthermore, advances in industrial and manufacturing technology are pushing food and beverage businesses to use BPs in their packaging [46]. The use of BP packaging to cover organic foods and high-quality or branded goods is becoming more popular. Natural or organic polymers generated from biomass sources such as carbohydrates, lipids, bacteria, and oils dominate the packaging end-use industry.

Due to its cutting-edge technology and industrial standing, well-established expertise, and creative bio-based product production, Europe now has a market leadership position [47]. However, the fast adoptions of products are hampered by the unclear efficacy and the lack of transparency of market. Although the use, development, and extension of bio-based and compostable products and materials are highly limited within agriculture, progress has been achieved with the expansion of the use of biodegradable (compostable) and bio-based packing materials due to two key considerations: (a) the open debate for biodegradation about testing agricultural biodegradable plastics for in-soil farm use and (b) the open debate on biodegradation about testing agricultural biodegradable plastics [48].

The first stymies work on labelling schemes and appropriate certification, such as when used in tandem with the recently developed labelling system concerning agricultural plastic wastes. For almost the same reason, there is indeed a lot of uncertainty about the ways in which these materials work in real-world soil conditions, which is limiting them from being employed in agricultural applications more generally [49]. When packaging, shipping, distributing, and storing the goods, extreme caution must be exercised. It is required for the transportation and storage of items for a wide range of sectors, including institutional, industrial, and retail. The importance of polymer and the usage of packaging goods provide protection and confinement while also causing a shift in consumer behavior. To wrap various kinds of goods, the packaging business uses a range of materials (polymers, metals, glass, wood, etc.). Plastic constitutes 40% of all packaging. Because of the characteristics of plastics, such as flexibility, impermeability, and high-water resistance, their usage is growing, but new environmental laws have restricted the use of traditional polymers. This is enabled in particular in packaging and handbags, which fuels the need for BPs and biopolymers. Food packaging, pharmaceutical packaging, cosmetic packaging, and shopping bags are just a few of the uses for packaging. Bio-PE is used to pack a variety of cosmetic goods by Procter & Gamble and Johnson & Johnson. Because of its excellent recyclability, PLA is often utilized. For goods such as PLA, PHA, and starch-based polymers, it is difficult to develop eco-friendly packaging manufactured from bio-based polymers for a variety of uses [35,50].

5.3.2. Consumer Goods

Circuit boards, housings, speakers, mobile housings, touch screens, computer housings, keyboard components, laptop mice, and vacuum cleaners are just a few examples of where BPs and biopolymers are used in consumer goods. SUPLA has created a PLA that is specifically designed for use in the electronics sector. Kuender also developed the world's first BP touchscreen computer in conjunction with a Taiwanese OEM/ODM [51]. Excellent high gloss, high impact resistance, and steady and precise processing may be obtained by mixing PLA mixture into the display. Europe is a significant consumer and producer of BPs and biopolymers. It dominates the BP and biopolymer industries and is utilized directly in the packaging end-use sector. These reasons have prompted collaboration in the development of bio-based polymers, which are expected to transform the plastics industry. Market penetration of BPs and biopolymers has been aided by political and economic circumstances [52]. BASF in Germany, Nature Works in the United States, Braskem in Brazil, Toray Industries in Japan, Novamont in Japan, Nova Corbion in Italy, Total Corbion in the Netherlands, Biome BPs in the United Kingdom, Mitsubishi Chemical Corporation in Japan, Plantic Technologies in Australia, Bio-On in the Netherlands, and Mitsubishi Chemical Corporation in Japan are among the leading players in this field [53]. BDPs are a kind of polymer that decomposes rapidly and produces ecologically beneficial by-products, which is a solution to the environmental pollution produced by non-degradable plastics.

5.3.3. Reuse and Recycle

The recyclates are melted and pelletized in mechanical recycling, which is an easy and energy-efficient technique. The majority of plastics nowadays are recovered mechanically. Unfortunately, these materials are frequently of poor quality, possess qualities inferior to virgin materials, and lack permission for use in food. These factors limit the amount of uses for mechanically recycled plastics, which can be found in items such as watering cans, spray bottles, and cleaning supplies. Due to a lack of demand for recycled materials, a recent survey indicated that only 37% of all collected plastic garbage is actually recycled. In France, Germany, the UK, Spain, and Italy, the recycling rates range from 21 to 42%; however, this figure also takes into account exports for recycling in other nations. The possibility of plastic items to be recycled has recently received a lot of attention. Although this development is significant, the key roadblock appears to be defining the uses for recovered plastics. Recyclability is not very useful unless it truly decreases the consumption of materials made from virgin fossil fuels. Several potential techniques for dissolving polymers into monomers, which can then be utilized to create new polymers, are included in chemical recycling which require more energy compared to mechanical recycling. On the contrary, chemical recycling processes are frequently less sensitive to mixed and contaminated plastics, and the resultant polymers exhibit comparable performance to virgin materials. The technological, economic, and ecological elements of chemical recycling technologies, however, need to be examined more closely because they are still in the early stages of becoming widely used [54].

Plastics manufactured from renewable feedstock, such as plants, can be used in part or in full in bio-based products. Furthermore, new materials are created from biological waste, industrial side streams and residues, or even CO_2. They all bind carbon from the atmosphere, which is something they all have in common. The carbon in materials produced from fossil fuels, on the other hand, originates from the geosphere and is released into the atmosphere after the useful lives of these materials. By 2020, just 1% of all plastics would be produced using bio-based feedstock. To lead the shift to bio-based plastics, however, a growing number of businesses have emerged. The recycling and waste management systems also need to take this into consideration. Some bio-based plastics, such as bio-PE and bio-PP, are chemically equivalent to the polymers made from fossil fuels and do not need to be managed differently in the waste stream. On the other hand, PLA is a popular bio-based polymer whose quantities are yet insufficient to establish distinct PLA recycling streams. It is sometimes criticized as "non-recyclable" or even hazardous to current recycling streams because of this; however, the arguments are generally unsupported by evidence. In contrast, research reported that up to three weight percent of PLA was safe to use on injection-moulded PP recycling materials and may therefore be redirected into the PP stream. An economically viable dedicated PLA stream may already be possible at volumes greater than this. Moreover, many PLA-based goods may be naturally recycled in composting facilities; however, at present, this is reliant on the regional guidelines and composting equipment [55].

6. Polymers in Biomedical Applications

As a biocompatible material, BDPs have opened the door to integrating with a variety of medical problems [56]. BDPs are used in the medical sector since they are less hazardous than other materials. Unlike in typical BDP applications, there are several factors to consider when utilizing biodegradable materials as biomaterials (BMs). Non-toxicity, stabilizability, efficacy, and biocompatibility are among these characteristics. Although it is impossible to gain all of the information about most BDPs, scientists, physicians, and engineers have discovered that certain BDPs may be utilized as BMs which closely interact with live cells. There are two common applications of BMs. BMs can be used in waste items, such as blood bags, catheters, and syringes, representing the first application. The second use is in surgical support materials. BMs may be used for tissue replacement prosthetics such as lenses, dental implants, and breast implants, as well as artificial organs such as artificial hearts and kidneys [57]. Table 1 highlights a list of some biopolymers and their emerging applications.

Table 1. Emerging applications of various biopolymers.

BIOPOLYMERS	APPLICATIONS
Cellulose Acetate	Textile, cigarette filter, wound dressing, frames, toothbrushes
Polycaprolactone (PCL)	Metal protection film, sutures, contraceptives, implants, paint
Polyhydroxyvalerate (PHV)	Compost bags, packaging, agriculture, horticulture, film, Rubbermaid
Polylactic acid (PLA)	Bone plates, bone screws, surgical sutures, fibers, films, foam, camera, electronics, molded plastic parts, heat-stable applications
Polyhydroxybutyrate (PHB)	Films, medical, cups, disposable, lids, containers, cutlery, compost bags, Rubbermaid, packaging, agriculture, horticulture
Polyalkylene alkanoate (PBS)	Disposable goods, fibers, pot, molding, injection, fishing gear
Aliphatic-aromatic (co)polyester (PBAT)	Film, sheets, packaging, flower pots, plastic envelope, agriculture, compost bags

BDPs have many surgical applications [56,58]. BDPs with bioabsorbable characteristics are utilized for bonding, closure, separation, scaffolding, and capsulation. BDPs are used in operations for tissue adhesion and sealing. In this instance, liquid polymers are utilized. The liquid will turn to gel and halt bleeding as soon as it is applied to the damaged tissue. The gelled substance will gradually dissolve and integrate into the body as the damaged tissue heals. Bone fixation using biodegradable pins, screws, and wires follows the same procedures. Biodegradable polymers are used in drug delivery systems. This may include both invasive and non-invasive medication delivery techniques. These polymers serve as a transporter in this instance, and after transporting the medication to the appropriate location, they will rapidly integrate into the body without causing damage. The most basic example encountered in everyday life is pill covers and coatings in various colors. The medication is contained inside the cover, which should be eliminated without causing harm to the body. BDPs are used to create these coverings and coatings.

Over the last two decades, significant progress has been made in the development of biocompatible and biodegradable substances for biomedical and industrial applications [58]. Biomedical research aims to develop and characterize artificial materials, or "spare parts", for use in the human body to assess, repair, and enhance physiologic function, as well as to improve survival and quality of life. Inorganic (metals, ceramics, and glasses) and polymeric materials include artificial heart valves (polymeric or carbon-based), synthetic blood vessels, artificial hips (metallic or ceramic), medical adhesives, sutures, dental composites, and polymers for controlled slow drug delivery [59]. When designing innovative biocompatible materials, bioactivity as it relates to interacting with and becoming integrated into the biological environment, as well as other customizable features dependent on the specific in vivo application, are all taken into account.

Biomimetics is a related field that might be characterized as "the abstraction of outstanding design from nature" or, to put it another way, "stealing ideas from nature". The goal is to develop materials inspired by nature and integrated with non-biological technologies or processes for non-biological purposes. This is swiftly emerging as a new topic of research. Biocompatible polymers have been a popular focus of research for regulated medication delivery [6,60,61]. It developed out of a need for more precise and extended pharmaceutical delivery. The goal of controlled release devices is to keep medicine within the prescribed limit with a single dose. Localized drug delivery to a specific physiological compartment lowers systemic drug levels, reduces follow-up care, preserves quickly degraded medicines, and improves patient comfort and/or compliance. The architecture of the system determines release rates, which are virtually independent of environmental variables.

7. Pharmaceutical Applications of Biopolymers

Biopolymers have progressively started to replace conventional materials due to their unique features. Although initially they were primarily employed in the food sector, they swiftly found usage in other adjacent industries. Biopolymers were first employed in the pharmaceutical sector for the same reason they were in the food sector: as thickening and emulsifying agents, host molecules, bulking agents, or fibers. Additionally, their application in cosmetics has grown significantly. By the end of 2021, the global biopolymer market is predicted to have grown by approximately 17% from 2017 to 2021, reaching an estimated value of USD 10 billion. With approximately 41.5% of the worldwide market, Western Europe dominates the largest market sector. Polymers have been effectively employed in biomedicine in both experimental and in vivo settings, including for drug administration, tissue engineering, wound dressing, and the construction of medical devices with electronics, sensors, and batteries. In addition, biopolymers are perfect materials commonly employed in the food and pharmaceutical industries due to their physical, thermal, mechanical, and optical qualities [62].

The internal structure-dependent functional qualities of biopolymers, such as microstructure, permeability, and chargeability, may be obtained by modifying the composition and matrix of the polymer. The ability of particles to aggregate and form bonds inside the biopolymer matrix is influenced by their electrical properties. The biopolymer fractions with a high electrical charge are those that inhibit aggregation. Based on these characteristics, biopolymers are effectively employed to create nanoparticles, nanoemulsions, nanogels, or hydrogels that are used as carrier systems in the biomedical industry. To supplement or replace synthetic materials, polysaccharides are the most often employed class of biopolymers. They can be used alone or in mixes with other biopolymers. In both micro- and macrocapsules, encapsulation entails preventing live cells from being destroyed by trapping them in biopolymer membranes. It is a process wherein one or more materials—representing the active component or core material—are embedded or covered with another material or system, which serves as a mantle, shell, carrier, or encapsulant [63].

The very substantial discrepancy between surface area and volume is a distinguishing characteristic of macrocapsules. As a result, a high quantity of nutrients must be used to provide a proper diffusion gradient for nutrient entry. The crucial nutrients for the cells also fall under this category. Living cells are contained in macrocapsules, which are composed of sizable diffusion chambers that are shaped as flat sheets, hollow fibers, and semi-permeable discs. Both intra- and extra-vascular devices can utilize macrocapsules. Cells are positioned outside the artificial capillaries and linked to the circulation using intravascular devices. They are located close to blood circulation, facilitating the quick movement of beneficial and nourishing materials such as oxygen. Although, the risk of developing thrombosis is the main drawback. To avoid the necessity for direct circulatory access, research is leaning towards use of microcapsules as extra-vascular devices with cells contained within semi-permeable diffusion chambers that are applied transdermally or in the peritoneal cavity. This only requires modest surgery and enables a quick and simple replacement if the graft fails or the transplant needs to be changed for other reasons. Microcapsules efficiently transport helpful compounds and faithfully replicate the release of molecules such as insulin or glucose. The majority of studies concentrate on creating microcapsules with minimal or no inflammatory reactions due to their advantages. Endocrine problems are effectively treated by using this characteristic.

As materials for encapsulation, several biocompatible polymers have been employed. This requires a biopolymer to fulfill the following requirements: (i) stability and absence of interaction with the drug it contains; (ii) non-interference with cellular function or viability; (iii) non-toxicity, low cost, and biodegradability; (iv) molecular weight, solubility properties, glass transition temperature, microstructure, and chemical functionality allowing for proper drug diffusion and release; (v) biosafety; and (vi) biocompatibility [64].

The controlled release techniques may change depending on the mechanism governing the release of active agents from the delivery system. As a result, biopolymer erosion, diffusion, and swelling may occur before diffusion or degradation. To control the release of macromolecules, hydrogels must first be hydrolyzed. Water-insoluble biopolymers must then be solubilized by reactions with groups hanging from the covalently bonded atoms of the polymer, and finally, hydrolytically labile bonds within the covalently bonded atoms of the biopolymer must be broken. The biopolymer present in controlled release device allows an encapsulated medicine or other active agents to pass through the outer membrane of capsules during the diffusion process. Diffusion-controlled systems require a drug delivery system that is stable in the biological setting and retains its size and form despite swelling or deterioration. When additional bioactive chemicals are mixed with biopolymers, for instance, the medicine must be able to permeate through the molecular structure of the polymer or via its pores to reach the biological environment. The biopolymer itself must stay unchanged at this point. Devices with swelling controlled release mechanisms are those that, while initially dry, swell after they enter the body and encounter bodily fluids or water. Changing the environmental conditions of delivery systems can cause the ability of biopolymers to swell. This is one of the most significant and practical properties of biopolymers since it allows for the control of medication or integrated active ingredient release by adjusting temperature or pH. Last but not least, a biodegradation of polymers in the body is a natural procedure that allows the active component to be entirely removed [65].

The use of synthetic polymers as encapsulating agents for various medicinal compounds has long been of interest. Despite having better pharmacokinetics than small-molecule medications, their buildup in the body has led to toxicity concerns. The choice of the appropriate compounds based on the requirement and desired effects are the main challenge with the reorientation of the medical business towards the usage of biopolymers. Not all biopolymers are effective drug encapsulation agents. Since oral or intravenous administration is the two most common routes of administration, the active ingredient must be released to the target location at the appropriate time, safely, and without any negative side effects. The biopolymers based on polysaccharides, such as sodium alginate, chitosan, agar, starch, and cellulose, are the most widely utilized and the subject of this review. They have minimal toxicity, exhibit non-immunogenic behavior, interact synergistically with other biopolymers and polymers, and are compatible with tissues and cells. These polysaccharides are used to create microcapsules, microspheres, or nanocapsules and are stable both in vitro and in vivo. They demonstrated significant levels of biocompatibility and biodegradability when evaluated in vivo, which facilitated treatment, reduced side effects, and enhanced the state of health. Their high solubility makes them ideal for use as disintegrants in tablets that are soluble in water. For instance, the starch and chitosan coating added to tablets enhanced their aesthetic presentation, shielded the medication from deterioration, and masked the compound's bad flavor. Alginate, a vegan alternative to gelatin, or cellulose for hard capsules, was substituted for gelatin when utilized as capsule material [66].

A relevant category of controlled-release systems is the mechanism that governs the release of the chemical in question. The most common approach is diffusion. A reservoir device, in which the bioactive material (drug) is confined and encircled by an inert diffusion barrier, is the first type of diffusion-controlled system. These systems include membranes, capsules, microcapsules, liposomes, and hollow fibers. The active chemical is spread or dissolved in an inert polymer in a monolithic device, on the other hand. The rate-limiting phase in drug release is drug diffusion through the polymer matrix, much as it is in reservoir systems, and release rates are limited by the polymer used and its influence on the diffusing and partitioning ratio of the drug to be released [67,68].

Chemical control may be achieved via bio-erodible or pendant links in chemically controlled systems. The use of bio-erodible (or biodegradable) systems is justified since bio-erodible devices are absorbed by the body and therefore do not require surgical removal. The transition of a water-insoluble material into a water-soluble substance is known as polymer bio-erosion. The drug in a bio-erodible system is disseminated equally throughout the polymer, similar to the way it is in monolithic systems. When the polymer that surrounds it erodes, the drug escapes. Water or enzymes trigger bond scission, releasing the drug which is covalently bonded to the polymer in a pendant chain form [49,69]. In solvent-activated controlled systems, the active component is dissolved or dispersed inside a polymeric matrix and is impossible to permeate through. As the environmental fluid (e.g., water) enters the matrix, the polymer expands and the glass transition temperature drops below the ambient (host) temperature in one kind of solvent-controlled system [70]. As a result, the encapsulant's enlarged polymer becomes stretchy, enabling the medication to pass through. Figure 3 represents various drug delivery devices.

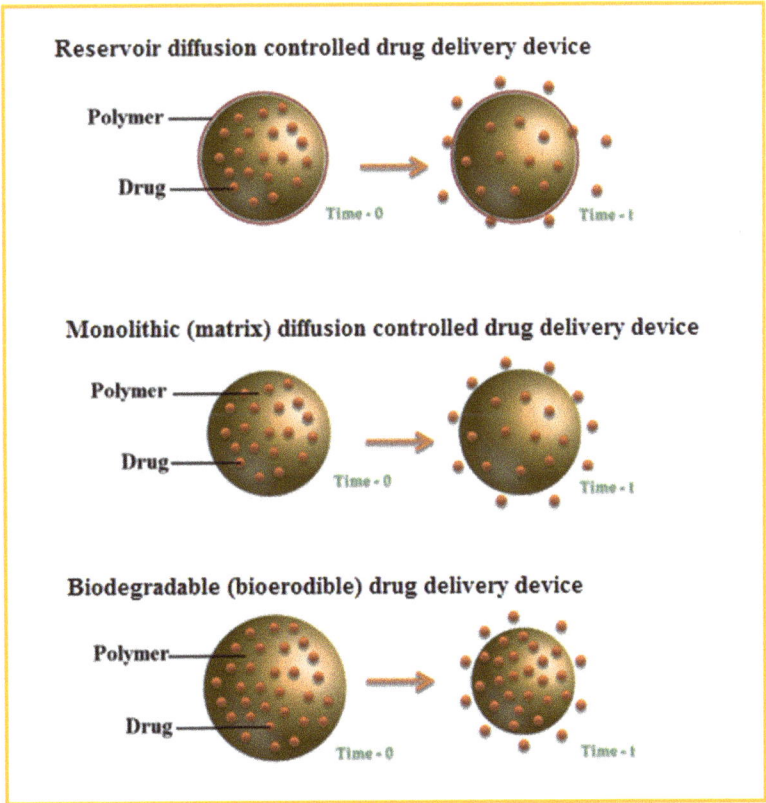

Figure 3. Schematic representation of reservoir diffusion-controlled drug delivery device, monolithic (matrix) diffusion-controlled drug delivery device, and biodegradable (bio-erodible) drug delivery device.

8. Sustainability

There should have been a debate on the limits to biomass usage for energy purposes since the turn of the century to reduce man-made CO_2 emissions. Bioenergy is now subjected to a slew of new regulations. After all these years, how do we see ourselves overcoming the situation?

8.1. Sustainability: From 18thCentury to Present

Bio-based polymers' long-term viability allows them to be used for both general and specialized applications, and deal with any issues or advancements in the polymer sector. However, if humanity abandons accuracy and disregards biomass usage restrictions to reduce CO_2 emissions, its continual use would be impossible. Our world generates approximately 200 billion tonnes of biomass each year. For food production, 9 billion is required; 3 billion tones are used yearly; therefore, there seems to be enough space for some energy generation. However, it is represented worldwide that yearly energy consumption in biomass requires 45 billion tonnes. It is known that burning 4 billion tonnes of biomass, mostly wood, straw, and dung (in particular in basic usage in poor nations) would result in the remaining 40 billion tones being used to provide the same amount of energy as fossil fuels already do (2 tonnes of biomass roughly equals 1 tonne of fossil fuel) [71].

8.2. Controversy on Limits to Biomass Use

It is an ongoing debate about whether biomass should be used for energy production or not. Biomass use has been a preferred plan of climate activists and policymakers since the turn of the century. Almost every sustainable energy concept involving biomass has been accepted. Despite this, there used to be a lack of comprehensive knowledge of the potential of solar, wind, and hydro energy, as well as other sources of energy [72]; regarding genuine options, the general understanding used to be extremely limited. Biomass is a convenient option since it was projected to expand in use at a fast pace, using the size, infrastructure, and long history of food production and commerce [47]. It may have been obvious at the time that for the sake of convenience, humanity would prefer to utilize food components (sugar, maize, and palm oil), but that argument was dismissed. The cultivation of fast-growing crops on empty land is imperative, according to proponents. However, the dispute about whether utilizing farmlands for biofuels enhances the emission of greenhouse gases as a result of land-use change started in 2008 and is still ongoing today. If production of crops such as maize were to be initiated, the outcome would be a decrease in CO_2 emissions from fossil fuels. Because the CO_2 produced during biofuel incineration has already been absorbed by the growing crop, the advantaged and disadvantages sum up to zero (if we disregard diesel oil for tractors and energy used in fertilizer production).

There are other issues with biofuels. The simplest biofuels are prioritized above food production and (once again) tropical forests. Sugar or wheat dietary components are the simplest sources of bioethanol. Vegetable oils are required for biodiesel, such as palm oil (also a food component), for which large areas are burnt down in tropical forests (again) in Indonesia and Malaysia [73]. It is challenging to combat such activities of Western headquarters and cities, since merchants are always looking for the path of least resistance. However, significant policy changes have occurred since 2008, such as the promotion of "second generation" biofuels (produced from straw and other non-food resources) over "first generation" biofuels (produced from food ingredients).

8.3. Outlook and Future Prospects

Sustainable polymers based on renewable resources are now gaining traction, and so, in the coming future, society will seek and require materials that are less environmentally harmful. Initial breakthroughs in the production of sustainable polymers resulted in the creation of commodities which are largely used in packages and as fiber. The discovery of raw materials or structural components which can be easily generated from abundant feedstocks, do not compete for resources with food crops, and do not affect the environment is a major challenge. Innovations in crop growth and harvesting practices such as yield optimization are predicted to increase the economic impact of bio-derived polymers while minimizing their environmental impact. Examples of industrial as well as agricultural waste are corn stover, fruit pulp, forestry debris, and CO_2 emissions which might be employed as monomers. The ubiquitous availability of lignocellulosic biomass offers another substantial opportunity; nevertheless, improved biopolymer separation, degradation, and transformation chemistry, as well as biochemistry, will be required to maximize both monomer yield and affordability. Prioritizing monomer and polymer paths that are interoperable with the present industrial base is an important, but frequently disregarded design factor. Low cost and efficient purity of raw materials and products for highly oxygenated bio-based materials need much greater customization. Retrofitting existing industrial systems to support sustainable polymer manufacturing will remain a significant driver in lowering expenses and accelerating uptake [74].

The end-of-life fate of sustainable polymers used in throwaway applications such as packaging will have a huge impact on the implementation of bio-based materials. Novel recycling, degrading, and disposal options are projected towards becoming increasingly important in the future for preventing new materials from adding to current plastic waste concerns, and policy and law may be able to affect the outcome. Although quantitative assessment and assessment of sustainable polymers with petrochemical-derived equivalents is still in its initial stages, many studies have shown that production impacts are decreased in many cases, particularly in terms of greenhouse gas emissions and fossil resource depletion. Studies should also examine the product's life beyond manufacturing and the consequences of disposal. Only a few polymers have been created that are totally biodegradable; however, aliphatic polyesters such as polylactide are notable examples. Bio-based polymers are currently widely used in packaging, although they are still unable to compete with petrochemical-derived polymers in regard to price. Bio-based polymers should instead compete in high-value, high-performance markets including thermoplastic elastomers, engineering plastics, and composite materials. To ensure success, it is important to customize and improve the properties of these polymers. Heat-resistant polymers, for example, might compete with semi-aromatic polyesters and nylons are available on the market. Elastomers with higher elongation-at-break values could compete with petrochemical-derived polymers. Another area of research is understanding and developing bio-based polymer degradation patterns, such as combining long-term durability with triggered disintegration. The task of broadening the scope and range of sustainable polymers is enormous; to solve these complex problems, researchers from agriculture, biology, biochemistry, catalysis, polymer chemistry, materials science, engineering, environmental impact assessment, economics, and policy fields will need to collaborate. In the future, humanity will require more materials that are effectively produced from natural waste and are recyclable or biodegradable [75].

8.4. Future of Polymers Is a Sustainable Future

Stahl is a firm believer in the use of cutting-edge technology that allows us to produce future-proof, long-lasting, and environmentally friendly polymers for use in coatings, paints, varnishes, sealants, building chemicals, inks, and adhesives. Creation of superior technologies is possible by using water and bio-based technology, as well as APEO-, VOC-, solvent-, and label-free solutions. Legislation and various other efforts have put us on the road to a more prosperous future. These advancements are reshaping our industry's future, paving the way for a new generation of future-proof technology that will contribute to a more sustainable and prosperous sector. Professionals from all sectors are searching for the most recent developments, including the automobile industry, construction chemicals, graphic arts, mechanical engineering, aerospace, and shipbuilding, as well as architecture, particularly in regard to the beautiful use of wood or metal in architectural masterpieces [76].

8.5. Relationship with Sustainable Development Goals of United Nations Organizations (2030 Agenda)

The Sustainable Development Goals of United Nations Organizations (2030 agenda) put forward 17 Sustainable Development Goals and 169 targets which demonstrated the scale and ambition of this new universal Agenda. It commits to lessen the harmful effects of urban activities and polymeric materials that are harmful to the environment and human health, including through environmentally sound management and safe use of polymers, waste reduction, and recycling, and more effective use of water and energy. The main objective states that by 2030, there will be a significant decrease in the number of fatalities and illnesses caused by hazardous non-BDPs, as well as pollution and contamination of the air, water, and soil. The next major target is improving the water quality by decreasing pollution, eliminating dumping, and minimizing the release of dangerous polymeric materials, half the share of untreated wastewater, and greatly boosting recycling and safe reuse internationally. The final objective entails achieving environmentally sound management of polymeric waste and all other chemical waste throughout their life cycles by established international frameworks, and significantly reducing their release to air, water, and soil in order to minimize their detrimental effects on human health and the environment [77].

9. Economic Potential

The market for bio-based polymers is anticipated to reach USD 29.8 billion by 2027, expanding at a CAGR of 18.2% from 2022 to 2027. Sustainable polymers known as "bio-based polymers" are created from renewable resources including biomass, starch, and rubber, among others. Bio-based polymers' starting components originate from plants, animals, enzymes, or microbes. Polylactic acid, also known as polylactide, is considered to be compostable and biodegradable. One of the key factors influencing the market for bio-based polymers during the projected period is the rise in environmental concerns regarding the disposal and recycling of synthetic polymers. The market expansion for bio-based polymers is also being supported by the increasing acceptance of bio-based materials in the end-use sectors, including packaging, pharmaceutical, construction, and consumer products, among others. Chitosan is one of the most popular bio-based polymers because it has amino groups on its polymer backbone, making it a rare natural cationic polymer. Global bioplastics output in 2021 was 2.42 million tonnes, according to figures released by European Bioplastics in partnership with the nova-Institute. Nevertheless, over the projection period of 2022–2027, the market expansion is projected to be hampered by the high cost of bio-based polymers. The market share of polyhydroxyalkanoates, which was over 30% in 2021, is expected to increase significantly over the course of the projected period. PHAs safeguard the environment and aquatic life by degrading in soil, water, and municipal waste treatment facilities. The demand for the polyethylene sector is being driven by the rising need for rigid and flexible packaging across a variety of industries, including food, beverage, and pharmaceutical industries. Numerous packaging materials, including carry bags, plastic films, and bottles, among others, employ polyhydroxyalkanoate. In 2020, 49 million metric tonnes of bottled drinking water will be produced annually, according to DCC China [78].

Approximately 58% of the market in 2021 was accounted for by the packaging segments, which are predicted to expand at a noteworthy CAGR. The packaging sector now has increasing need for bio-based polymers due to consumer lifestyle changes and tightening environmental laws regarding the use of synthetic polymers. The packaging business has boosted demand for polymers because of the great durability and water resistance of plastics. Among additional packaging uses for bio-based polymers are shopping bags, cosmetic and personal care packaging, healthcare packaging, and food packaging. All merchants will now be required to pay a levy for the use of plastic shopping bags starting in July 2020, except for those that are biodegradable and those made with at least 25% renewable plant-based resources. The Coca-Cola Company manufactures over 3 million tonnes of plastic packaging annually, according to the Forbes article. These are a few of the elements influencing the packaging market.

In 2021, the Asia Pacific area accounted for up to 42% of the global market for bio-based polymers, because of the booming packaging, food and beverage, and pharmaceutical sectors. The market expansion in the area is being fueled by the presence of growing countries such as China and India. The area demand is further supported by a large supply of raw materials and a readily available workforce. Leading manufacturers of food, drink, and personal care products, including Nestle, PepsiCo, Coca-Cola, LOreal SA, Procter & Gamble, Beiersdorf AG, Unilever PLC, and others, are present in the region, supporting regional growth during the course of the projection period. The China Chain Store and Franchise Association reported that the country's food and beverage (F&B) market reached over USD 595 billion in 2019. Because bio-based polymers have a high cost of polymerization, their manufacturing costs are typically 20% to 80% higher than those of conventional polymers. In comparison to their petrochemical equivalents, which have been developing quickly for more than 50 years, bio-based products are generally still in the development stage and have not yet reached the same degree of commercialization. Only 1% of the world's total polymers are produced from bio-based materials, as opposed to 99% of the widely accessible conventional polymers. While standard polymers are available for USD 1/kg to US D 2/kg, bio-based polymers range in price from USD 2/kg to US D 5/kg. Because of this, the market expansion during the projected period is being constrained by the high cost of bio-based polymers [79].

10. Current Strategies, Market Dynamics, and Research Trends

10.1. Current Strategies

Material selection is one of the most important tactics for the successful commercialization of bio-based polymers. The right feedstock must be chosen by researchers before bio-based polymers can be produced. The choice of feedstock is critical since it can alter the characteristics of the polymer and the environmental effect. Several feedstocks are available for the manufacturing of bio-based polymers, including sugarcane, maize, potato, and other biomass sources. For instance, maize is used to make polylactic acid (PLA), which has high processing, biodegradability, and mechanical qualities. The choice of feedstock must take into account its accessibility, affordability, and environmental effect. The process of synthesizing polymers is another important tactic. Bio-based polymers must be produced using specialized polymerization processes that are suited to the particular feedstock. To attain the required molecular weight, crystallinity, and other qualities, the reaction conditions must be optimized. In the case of polyhydroxyalkanoates (PHAs), microbial fermentation is used. The resultant polymer is versatile, biodegradable, and has high mechanical characteristics. In order to achieve consistency in quality, cost-effectiveness, and commercial viability, the polymerization process must be optimized.

Another essential tactic for a successful commercialization is the scaling up of the manufacture of bio-based polymers. To guarantee cost-effectiveness and consistency in quality, the production process must be optimized as part of the scaling-up process. The scale-up procedure must take into account a number of variables, such as reactor size, feedstock supply, and downstream processing. In the scale-up process, the deployment of pilot plants is also crucial. Enzymes are used to depolymerize waste PET, which is followed by purification and repolymerization of the resultant monomers to produce bio-based polyethylene terephthalate (PET). This procedure has been successfully scaled up to allow for the commercial production of bio-based PET. Other crucial tactics in the commercialization of bio-based polymers include certification and regulatory compliance. To guarantee their safety and effectiveness, bio-based polymers must adhere to strict regulatory standards. To guarantee that they adhere to the appropriate criteria for biodegradability, toxicity, and other important factors, testing and certification are required. Testing for biodegradability in a variety of settings, including soil, the ocean, and commercial composting facilities, is often required for the certification of bioplastics. Moreover, bio-based polymers need to adhere to a number of rules, such as the REACH regulation of the European Union, which seeks to ensure the safe use of chemicals in the EU [45].

10.2. Market Dynamics

The market dynamics of emerging bio-based polymers are influenced by a variety of factors, including regulations, feedstock availability, cost-effectiveness, and competition from other sustainable materials. One of the primary factors influencing the market dynamics of emerging bio-based polymers is regulatory requirements. Governments across the world are implementing regulations to reduce the use of single-use plastics and promote the use of eco-friendly alternatives. For example, the European Union's Single-Use Plastics Directive aims to reduce the consumption of single-use plastics and promote the use of bio-based and biodegradable materials. Such regulations create a favorable market for bio-based polymers, as companies seek to comply with these requirements.

Another key factor impacting the market dynamics of bio-based polymers is feedstock availability. Bio-based polymers are primarily produced from agricultural crops such as corn and sugarcane. The availability and cost of these feedstocks can significantly impact the cost of bio-based polymers. In recent years, researchers have been exploring alternative feedstocks, such as waste streams from various industries, to reduce the cost of bio-based polymers and increase their availability. Cost-effectiveness is also a significant factor in the market dynamics of bio-based polymers. While bio-based polymers are eco-friendly, they must also be cost-effective to be competitive with traditional plastics. Companies are exploring innovative methods for the production of bio-based polymers to reduce production costs and increase efficiency. For example, researchers are developing microbial fermentation processes for the production of bio-based polymers, which has the potential to reduce production costs significantly.

Competition from other sustainable materials is another factor impacting the market dynamics of bio-based polymers. While bio-based polymers offer a sustainable alternative to traditional plastics, they are not the only option. Companies must compete with other materials, such as recycled plastics and paper-based materials, to provide the most sustainable and cost-effective solutions for their customers. Consumer demand is another significant driver of the market dynamics of emerging bio-based polymers. Consumers are becoming increasingly aware of the environmental impact of traditional plastics and are seeking eco-friendly alternatives. As a result, there is a growing demand for bio-based polymers from consumers who aim to reduce their carbon footprint [80].

10.3. Research Trends

The rapid development of novel feedstocks is one of the key research themes in the creation of bio-based polymers. Many bio-based polymers are now produced from agricultural crops such as maize and sugarcane, but scientists are also examining additional biomass sources such waste streams from different sectors, agricultural and forestry wastes, and algae. These feedstocks may be more affordable, more readily available, and less in conflict with food crops. Researchers are also searching for ways to develop more productive and environmentally friendly production processes in order to maximize the conversion of these feedstocks into bio-based polymers. The alteration of current materials to enhance their qualities is a significant topic of research in the realm of bio-based polymers. For instance, scientists are researching the ways to modify polylactic acid (PLA) in order to increase its flexibility and robustness. Nowadays, PLA is utilized in many different contexts, including packaging, but its brittleness has restricted its use in other contexts. Other bio-based polymers with potential for use in the medical and pharmaceutical sectors, such as polyhydroxyalkanoates (PHAs), are also being studied for methods to enhance their mechanical qualities.

Another significant trend in research is the biodegradability and compostability of bio-based polymers. These characteristics are crucial to the way bio-based polymers affect the environment and their capacity to reduce plastic waste. To increase the biodegradability and compostability of bio-based polymers, researchers are examining the ways to improve their chemical structure and molecular weight. Researchers are also investigating the possibility of creating novel biodegradable materials that may offer a larger variety of applications, including textiles, electronics, and automobile components. A prominent research topic in the area of bio-based polymers is the upcycling and recycling of materials. Recycling and upcycling have the ability to lessen the environmental impact of bio-based polymers just as well as biodegradable and compostable products in terms of plastic waste reduction. In order to recycle bio-based polymers, scientists are considering several strategies, such as using enzymes and microorganisms to break down the materials into their component components, which may then be utilized to create new bio-based polymers. Another method being researched as a way to lessen waste and advance the circular economy is upcycling, which is turning waste materials into goods with higher value.

Improved characteristics, broader potential applications, and cost-effective large-scale production are the main goals of the current research trends in new bio-based polymers. Significant research is conducted in the areas of examining novel feedstocks, modifying current materials, and enhancing biodegradability and compostability. Investigations into recycling and upcycling are especially crucial in the field of bio-based polymers because they have the potential to cut waste and advance the circular economy. These developments in research are essential for the creation and commercialization of bio-based polymers, which have the potential to drastically lessen the environmental effect of plastics and advance the progression of a more sustainable future [81].

11. Conclusions

Future growth of the bio-based polymer sector is anticipated to be significant due to the growing need for green chemicals on a worldwide scale. For decades, scientists have been investigating the production of functional polymers. The primary inspirations for the development of bio-based polymers—which are necessary for a society that cares about reducing waste—are compostability and biodegradability. This increase has been fueled by significant advancements in the production of bio-based construction materials and technology for the processing of biomass raw materials. Because of this, the chemical structure of bio-based polymers is uniquely versatile and flexible, enabling them to achieve the desired properties and functions. Such polymers play a major role in biomedical and biotechnological applications as well as industrial applications, where biomolecules that naturally exist are meticulously produced. At the same time, restrictions on the management of polymer design, biostability, and structural dynamics obstruct the efficient production of biocompatible and functionally diverse polymers despite major advancements. In light of this, the significance of functional biosynthetic polymers in the next years has been highlighted, and innovative methods have been created to address the aforementioned difficulties. The most recent technical developments in bio-based polymers are also discussed. Due to the importance of polymer research in our daily lives, demand is growing fast. The production of polymeric materials that are green, sustainable, energy-efficient, high-quality, and affordable will guarantee that the best solutions are available everywhere, and these polymers will be the foundation of the new millennium. In an effort to provide the greatest goods with sustainable qualities, scientists and researchers keep working to develop this material via studies and research. These people are finding successful careers in the rubber, polymer, petroleum, and oil industries because of their increased polymer engineering skills. Conversely, less attention is directed towards the end-of-life fates of polymers. We suggest shifting more of a focus on chemical recycling in the future; chemical recycling recognizes the value in plastic trash and promotes a closed-loop plastic economy. The synthetic polymer community should regularly seek out the systems viewpoint offered by life cycle assessment to help keep us on the path towards sustainability.

Author Contributions: Conceptualization, T.M.J.; methodology, T.M.J.; validation, T.M.J., K.S.J., A.B.U., D.K.M., J.H. and S.T.; formal analysis, T.M.J., K.S.J., A.B.U. and D.K.M.; investigation, T.M.J.; resources, J.H. and S.T.; data curation, T.M.J. and D.K.M.; writing—original draft preparation, T.M.J.; writing—review and editing, T.M.J. and D.K.M.; supervision, J.H. and S.T.; project administration, J.H. and S.T. All authors have read and agreed to the published version of the manuscript.

Funding: This research received no external funding.

Data Availability Statement: Not applicable.

Acknowledgments: A.B.U. acknowledges NCN Sonata- UMO-2021/43/D/ST5/01114.

Conflicts of Interest: The authors declare no conflict of interest.

References

1. Niaounakis, M. *Biopolymers: Applications and Trends*; William Andrew: New York, NY, USA, 2015; ISBN 978-0-323-35433-2.
2. KI, S.; Nutenki, R.; Joseph, T.M.; Murali, S. Structural, molecular and thermal properties of cardanol based monomers and polymers synthesized via atom transfer radical polymerization (ATRP). *J. Macromol. Sci. Part A* **2022**, *59*, 403–410. [CrossRef]
3. Muringayil Joseph, T.; Murali Nair, S.; Kattimuttathu Ittara, S.; Haponiuk, J.T.; Thomas, S. Copolymerization of Styrene and Pentadecylphenylmethacrylate (PDPMA): Synthesis, Characterization, Thermomechanical and Adhesion Properties. *Polymers* **2020**, *12*, 97. [CrossRef]
4. Hayward, W.W. *The History of Hancock, New Hampshire, 1764–1889*; Vox Populi Press, S. W. Huse & Co.: Philadelphia, PA, USA, 1889.
5. Muringayil Joseph, T.; Mariya, H.J.; Haponiuk, J.T.; Thomas, S.; Esmaeili, A.; Sajadi, S.M. Electromagnetic Interference Shielding Effectiveness of Natural and Chlorobutyl Rubber Blend Nanocomposite. *J. Compos. Sci.* **2022**, *6*, 240. [CrossRef]
6. Nayak, A.K.; Hasnain, M.S.; Pal, D. *Natural Polymers for Pharmaceutical Applications: Volume 1: Plant-Derived Polymers*; CRC Press: Boca Raton, FL, USA, 2019; ISBN 978-1-00-058682-4.
7. Kumar, P.; Mahapatra, D.K.; Kumar, D.; Taleuzzaman, M.; Borikar, S.; Gulecha, V.S.; Zalte, A.G.; Dadure, K.M.; Puranik, M.; Das, M.; et al. *Liposomal Delivery System for the Effective Delivery of Nutraceuticals and Functional Foods*; Apple Academic Press: Palm Bay, FL, USA, 2022; pp. 173–184.
8. Joseph, T.M.; Mahapatra, D.K.; Luke, P.M.; Haponiuk, J.T.; Thomas, S. *Perspectives of Cashew Nut Shell Liquid (CNSL) in a Pharmacotherapeutic Context*; Apple Academic Press: Palm Bay, FL, USA, 2021; pp. 123–133, ISBN 978-1-00-305745-1.
9. Luke, P.M.; Dhanya, K.R.; Joseph, T.M.; Haponiuk, J.T.; Rouxel, D.; Thomas, S. *Developments of Health Care: A Brief History of Medicine*; Apple Academic Press: Palm Bay, FL, USA, 2021; pp. 3–30, ISBN 978-1-00-305745-1.
10. Boichenko, S. *Advance in Petroleum and Gas Industry and Petrochemistry ПРІОРИТЕТ КАТЕГОРІЇ ЯКОСТІ В СУЧАСНИХ УМОВАХ ПАЛИВОЗАБЕЗПЕЧЕННЯ ЕКОНОМІКИ УКРАЇНИ*; Lviv Polytechnic National University: Lviv, Ukraine, 2022.
11. Shokri, Z.; Seidi, F.; Saeb, M.R.; Jin, Y.; Li, C.; Xiao, H. Elucidating the impact of enzymatic modifications on the structure, properties, and applications of cellulose, chitosan, starch and their derivatives: A review. *Mater. Today Chem.* **2022**, *24*, 100780. [CrossRef]
12. Seidi, F.; Arabi Shamsabadi, A.; Ebadi Amooghin, A.; Saeb, M.R.; Xiao, H.; Jin, Y.; Rezakazemi, M. Biopolymer-based membranes from polysaccharides for CO_2 separation: A review. *Environ. Chem. Lett.* **2022**, *20*, 1083–1128. [CrossRef]
13. Zheng, L.; Seidi, F.; Liu, Y.; Wu, W.; Xiao, H. Polymer-based and stimulus-responsive carriers for controlled release of agrochemicals. *Eur. Polym. J.* **2022**, *177*, 111432. [CrossRef]
14. Adampourezare, M.; Hasanzadeh, M.; Dehghan, G.; Hosseinpourefeizi, M.-A.; Seidi, F. An innovative fluorometric bioanalysis strategy towards recognition of DNA methylation using opto-active polymer: A new platform for DNA damage studies by genosensor technology. *J. Mol. Recognit.* **2022**, *35*, e2981. [CrossRef] [PubMed]
15. Joseph, T.M.; Pallikkunnel, M.L.; Mahapatra, D.K.; Kallingal, A.; Thomas, S.; Haponiuk, J.T. Polyurethane–Epoxy Composites: Recent Developments and Future Perspectives. In *Polyurethane Chemistry: Renewable Polyols and Isocyanates*; ACS Symposium Series; American Chemical Society: Washington, DC, USA, 2021; Volume 1380, pp. 257–280. ISBN 978-0-8412-9840-8.
16. Dalton, B.; Bhagabati, P.; De Micco, J.; Padamati, R.B.; O'Connor, K. A Review on Biological Synthesis of the Biodegradable Polymers Polyhydroxyalkanoates and the Development of Multiple Applications. *Catalysts* **2022**, *12*, 319. [CrossRef]
17. Standard Test Methods for Determining the Biobased Content of Solid, Liquid, and Gaseous Samples Using Radiocarbon Analysis. Available online: https://www.astm.org/d6866-22.html (accessed on 30 December 2022).
18. Functionalized Masks: Powerful Materials against COVID-19 and Future Pandemics—Seidi—2021—Small—Wiley Online Library. Available online: https://onlinelibrary.wiley.com/doi/full/10.1002/smll.202102453 (accessed on 30 December 2022).
19. Environmental Problems Caused by Polymers. Available online: https://www.azocleantech.com/article.aspx?ArticleID=1189 (accessed on 30 December 2022).
20. The Harmful Effects of Petrochemicals on the Environment | Sciencing. Available online: https://sciencing.com/harmful-effects-petrochemicals-environment-8771898.html (accessed on 30 December 2022).
21. Balla, E.; Daniilidis, V.; Karlioti, G.; Kalamas, T.; Stefanidou, M.; Bikiaris, N.D.; Vlachopoulos, A.; Koumentakou, I.; Bikiaris, D.N. Poly(lactic Acid): A Versatile Biobased Polymer for the Future with Multifunctional Properties—From Monomer Synthesis, Polymerization Techniques and Molecular Weight Increase to PLA Applications. *Polymers* **2021**, *13*, 1822. [CrossRef]

22. Delidovich, I.; Hausoul, P.J.C.; Deng, L.; Pfützenreuter, R.; Rose, M.; Palkovits, R. Alternative Monomers Based on Lignocellulose and Their Use for Polymer Production. *Chem. Rev.* **2016**, *116*, 1540–1599. [CrossRef]
23. Haider, T.P.; Völker, C.; Kramm, J.; Landfester, K.; Wurm, F.R. Plastics of the future? The impact of biodegradable polymers on the environment and on society. *Ang. Chem. Int. Ed.* **2019**, *58*, 50–62. [CrossRef]
24. Rai, P.; Mehrotra, S.; Priya, S.; Gnansounou, E.; Sharma, S.K. Recent advances in the sustainable design and applications of biodegradable polymers. *Biores. Techol.* **2021**, *325*, 124739.
25. Leja, K.; Lewandowicz, G. Polymer biodegradation and biodegradable polymers-a review. *Pol. J. Environ. Stud.* **2010**, *19*, 255–266.
26. Mohanty, A.K.; Misra, M.A.; Hinrichsen, G.I. Biofibres, biodegradable polymers and biocomposites: An overview. *Macromol. Mater. Eng.* **2000**, *276*, 1–24. [CrossRef]
27. Kyrikou, I.; Briassoulis, D. Biodegradation of Agricultural Plastic Films: A Critical Review. *J. Polym. Environ.* **2007**, *15*, 125–150. [CrossRef]
28. Rujnić-Sokele, M.; Pilipović, A. Challenges and opportunities of biodegradable plastics: A mini review. *Waste Manag. Res.* **2017**, *35*, 132–140. [CrossRef]
29. Bastioli, C. Biodegradable materials—Present situation and future perspectives. *Macromol. Symp.* **1998**, *135*, 193–204. [CrossRef]
30. Harrison, J.P.; Boardman, C.; O'Callaghan, K.; Delort, A.-M.; Song, J. Biodegradability standards for carrier bags and plastic films in aquatic environments: A critical review. *R. Soc. Open Sci.* **2018**, *5*, 171792. [CrossRef]
31. Wojtowicz, A. Biodegradability and Compostability of Biopolymers. In *Thermoplastic Starch*; John Wiley & Sons, Ltd.: Hoboken, NJ, USA, 2009; pp. 55–76. ISBN 978-3-527-62821-6.
32. Briassoulis, D.; Dejean, C. Critical Review of Norms and Standards for Biodegradable Agricultural Plastics Part I. Biodegradation in Soil. *J. Polym. Environ.* **2010**, *18*, 384–400. [CrossRef]
33. Wolter, B.; Hintzen, H.M.T.; Welsing, G.; Tiso, T.; Blank, L.M. Microbes and Plastic—A Sustainable Duo for the Future. In *Good Microbes in Medicine, Food Production, Biotechnology, Bioremediation, and Agriculture*; John Wiley & Sons, Ltd.: Hoboken, NJ, USA, 2022; pp. 294–311, ISBN 978-1-119-76262-1.
34. Eubeler, J.P.; Bernhard, M.; Knepper, T.P. Environmental biodegradation of synthetic polymers II. Biodegradation of different polymer groups. *TrAC Trends Anal. Chem.* **2010**, *29*, 84–100. [CrossRef]
35. Ashter, S.A. *Introduction to Bioplastics Engineering*; William Andrew: New York, NY, USA, 2016; ISBN 978-0-323-39407-9.
36. Mangaraj, S.; Yadav, A.; Bal, L.M.; Dash, S.K.; Mahanti, N.K. Application of Biodegradable Polymers in Food Packaging Industry: A Comprehensive Review. *J. Packag. Technol. Res.* **2019**, *3*, 77–96. [CrossRef]
37. Kahraman Ilıkkan, Ö.; Bağdat, E.Ş.; Yalçın, D. Evaluation of prebiotic, probiotic, and synbiotic potentials of microalgae. *Food Health* **2022**, *8*, 161–171. [CrossRef]
38. Nasrollahzadeh, M.; Sajjadi, M.; Iravani, S.; Varma, R.S. Starch, cellulose, pectin, gum, alginate, chitin and chitosan derived (nano)materials for sustainable water treatment: A review. *Carbohydr. Polym.* **2021**, *251*, 116986. [CrossRef]
39. Lasprilla, A.J.R.; Martinez, G.A.R.; Lunelli, B.H.; Jardini, A.L.; Filho, R.M. Poly-lactic acid synthesis for application in biomedical devices—A review. *Biotechnol. Adv.* **2012**, *30*, 321–328. [CrossRef]
40. Albuquerque, P.B.S.; Malafaia, C.B. Perspectives on the production, structural characteristics and potential applications of bioplastics derived from polyhydroxyalkanoates. *Int. J. Biol. Macromol.* **2018**, *107*, 615–625. [CrossRef]
41. Tolinski, M.; Carlin, C.P. *Plastics and Sustainability Grey Is the New Green: Exploring the Nuances and Complexities of Modern Plastics*; John Wiley & Sons: Hoboken, NJ, USA, 2021; ISBN 978-1-119-59197-9.
42. Hellvig, E.L.F.; Flores-Sahagun, T.H.S. Investments in Braskem Green Polymers: Extraordinary Profits or Decarbonization of the Environment? *Preprints* **2020**, 2020120036. [CrossRef]
43. Chia, W.Y.; Ying Tang, D.Y.; Khoo, K.S.; Kay Lup, A.N.; Chew, K.W. Nature's fight against plastic pollution: Algae for plastic biodegradation and bioplastics production. *Environ. Sci. Ecotechnol.* **2020**, *4*, 100065. [CrossRef]
44. Kabir, E.; Kaur, R.; Lee, J.; Kim, K.-H.; Kwon, E.E. Prospects of biopolymer technology as an alternative option for non-degradable plastics and sustainable management of plastic wastes. *J. Clean. Prod.* **2020**, *258*, 120536. [CrossRef]
45. Moshood, T.D.; Nawanir, G.; Mahmud, F.; Mohamad, F.; Ahmad, M.H.; Abdul Ghani, A. Expanding Policy for Biodegradable Plastic Products and Market Dynamics of Bio-Based Plastics: Challenges and Opportunities. *Sustainability* **2021**, *13*, 6170. [CrossRef]
46. Welz, P.J.; Linganiso, L.Z.; Murray, P.; Kumari, S.; Arthur, G.D.; Ranjan, A.; Collins, C.; Bakare, B.F. Status quo and sector readiness for (bio)plastic food and beverage packaging in the 4IR. *S. Afr. J. Sci.* **2022**, *118*, 1–9. [CrossRef]
47. Scarlat, N.; Dallemand, J.-F.; Monforti-Ferrario, F.; Nita, V. The role of biomass and bioenergy in a future bioeconomy: Policies and facts. *Environ. Dev.* **2015**, *15*, 3–34. [CrossRef]
48. Gerassimidou, S.; Martin, O.V.; Chapman, S.P.; Hahladakis, J.N.; Iacovidou, E. Development of an integrated sustainability matrix to depict challenges and trade-offs of introducing bio-based plastics in the food packaging value chain. *J. Clean. Prod.* **2021**, *286*, 125378. [CrossRef]
49. Mitrus, M.; Wojtowicz, A.; Moscicki, L. Biodegradable Polymers and Their Practical Utility. In *Thermoplastic Starch*; John Wiley & Sons, Ltd.: Hoboken, NJ, USA, 2009; pp. 1–33, ISBN 978-3-527-62821-6.
50. Pal, K.; Paulson, A.T.; Rousseau, D. Biopolymers in controlled-release delivery systems. In *Modern Biopolymer Science*; Academic Press: Cambridge, MA, USA, 2009; pp. 519–557.

51. Chapagain, S. Biodegradable Polymer Synthesis from Renewable Sources. Available online: http://www.theseus.fi/handle/10024/500874 (accessed on 30 December 2022).
52. Mehmood, A.; Raina, N.; Phakeenuya, V.; Wonganu, B.; Cheenkachorn, K. The current status and market trend of polylactic acid as biopolymer: Awareness and needs for sustainable development. *Mater. Today Proc.* **2023**, *72*, 3049–3055. [CrossRef]
53. WORLD BIOPOLYMERS MARKET 2019–2020. Available online: https://ect-center.com/blog/biopolymers-market-2019 (accessed on 30 December 2022).
54. Butenegro, J.A.; Bahrami, M.; Abenojar, J.; Martínez, M.Á. Recent progress in carbon fiber reinforced polymers recycling: A review of recycling methods and reuse of carbon fibers. *Materials* **2021**, *14*, 6401. [CrossRef]
55. Ignatyev, I.A.; Thielemans, W.; Vander Beke, B. Recycling of polymers: A review. *ChemSusChem* **2014**, *7*, 1579–1593.
56. Nanomedicine: Review and Perspectives | Concise Ency. Available online: https://www.taylorfrancis.com/chapters/edit/10.1081/E-EBPPC-70/nanomedicine-review-perspectives-ali-zarrabi-arezoo-khosravi-ali-hashemi (accessed on 30 December 2022).
57. Nicholson, J.W. *The Chemistry of Medical and Dental Materials*; Royal Society of Chemistry: London, UK, 2020; ISBN 978-1-78801-530-1.
58. Kronenthal, R.L. Biodegradable Polymers in Medicine and Surgery. In *Polymers in Medicine and Surgery*; Polymer Science and Technology; Kronenthal, R.L., Oser, Z., Martin, E., Eds.; Springer: Boston, MA, USA, 1975; pp. 119–137. ISBN 978-1-4684-7744-3.
59. Bronzino, J.D.; Peterson, D.R. *Tissue Engineering and Artificial Organs*; CRC Press: Boca Raton, FL, USA, 2016; ISBN 978-1-4200-0387-1.
60. Polymers in Drug Delivery. Available online: https://www.scirp.org/journal/paperinformation.aspx?paperid=62762 (accessed on 30 December 2022).
61. Shokrani, H.; Shokrani, A.; Seidi, F.; Munir, M.T.; Rabiee, N.; Fatahi, Y.; Kucinska-Lipka, J.; Saeb, M.R. Biomedical engineering of polysaccharide-based tissue adhesives: Recent advances and future direction. *Carbohydr. Polym.* **2022**, *295*, 119787. [CrossRef]
62. Udayakumar, G.P.; Muthusamy, S.; Selvaganesh, B.; Sivarajasekar, N.; Rambabu, K.; Banat, F.; Show, P.L. Biopolymers and composites: Properties, characterization and their applications in food, medical and pharmaceutical industries. *J. Environ. Chem. Eng.* **2021**, *9*, 105322.
63. Abdelhak, M.J. A review: Application of biopolymers in the pharmaceutical formulation. *J. Adv. Bio-Pharm. Pharmacovigil.* **2019**, *1*, 15–25.
64. Joseph, T.M.; Kallingal, A.; Suresh, A.M.; Mahapatra, D.K.; Hasanin, M.S.; Haponiuk, J.; Thomas, S. 3D printing of Polylactic acid: Recent Advances and Opportunities. *Int. J. Adv. Manuf. Technol.* **2023**, *125*, 1015–1035. [CrossRef]
65. Joseph, T.M.; Hasanin, M.S.; Unni, A.B.; Mahapatra, D.K.; Haponiuk, J.; Thomas, S. Macromolecules: Contemporary Futurist Thoughts on Progressive Journey. *Eng* **2023**, *4*, 678–702. [CrossRef]
66. Joseph, T.M.; Mahapatra, D.K.; Esmaeili, A.; Piszczyk, Ł.; Hasanin, M.S.; Kattali, M.; Haponiuk, J.; Thomas, S. Nanoparticles: Taking a unique position in medicine. *Nanomaterials* **2023**, *13*, 574.
67. Langer, R. New Methods of Drug Delivery. *Science* **1990**, *249*, 1527–1533. [CrossRef]
68. Kost, J.; Langer, R. Controlled release of bioactive agents. *Trends Biotechnol.* **1984**, *2*, 47–51. [CrossRef]
69. Hollinger, J.O.; Leong, K. Poly(α-hydroxy acids): Carriers for bone morphogenetic proteins. *Biomaterials* **1996**, *17*, 187–194. [CrossRef]
70. Hopfenberg, H.B.; Apicella, A.; Saleeby, D.E. Factors affecting water sorption in and solute release from glassy ethylene—Vinyl alcohol copolymers. *J. Membr. Sci.* **1981**, *8*, 273–282. [CrossRef]
71. Kopetz, H. Build a biomass energy market. *Nature* **2013**, *494*, 29–31. [CrossRef]
72. Poupeau, F.-M. Central-Local Relations in French Energy Policy-Making: Towards a New Pattern of Territorial Governance. *Environ. Policy Gov.* **2014**, *24*, 155–168. [CrossRef]
73. Dauvergne, P. The Global Politics of the Business of "Sustainable" Palm Oil. *Glob. Environ. Polit.* **2018**, *18*, 34–52. [CrossRef]
74. Zhu, Y.; Romain, C.; Williams, C.K. Sustainable polymers from renewable resources. *Nature* **2016**, *540*, 354–362. [CrossRef]
75. Spontak, R.J.; Patel, N.P. Thermoplastic elastomers: Fundamentals and applications. *Curr. Opin. Colloid Interf. Sci.* **2000**, *5*, 333–340.
76. Cortés, M.G.R.; Díaz, B.M.V.; Perilla, J.E.P. Biopolymers as materials for developing products in pharmaceutical applications and biomedical uses. *Ing. E Investig.* **2008**, *28*, 57–71. [CrossRef]
77. Do you know all 17 SDGs? Available online: https://sdgs.un.org/goals (accessed on 20 February 2023).
78. Jemghili, R.; Ait Taleb, A.; Khalifa, M. A bibliometric indicators analysis of additive manufacturing research trends from 2010 to 2020. *Rapid Prototype J.* **2021**, *27*, 1432–1454. [CrossRef]
79. Karnik, N.; Bora, U.; Bhadri, K.; Kadambi, P.; Dhatrak, P. A comprehensive study on current and future trends towards the characteristics and enablers of industry 4.0. *J. Ind. Inf. Integr.* **2022**, *27*, 100294. [CrossRef]
80. Elvers, D.; Song, C.H.; Steinbüchel, A.; Leker, J. Technology trends in biodegradable polymers: Evidence from patent analysis. *Polym. Rev.* **2016**, *56*, 584–606. [CrossRef]
81. Ullah Khan, W.; Ahmed Al-Harthi, M. Research Trends in Carbon Chain Polymers and Their Derivatives: Highlighting the Decade-Long Research. *Chem. Rec.* **2022**, *22*, e202200167. [CrossRef]

Disclaimer/Publisher's Note: The statements, opinions and data contained in all publications are solely those of the individual author(s) and contributor(s) and not of MDPI and/or the editor(s). MDPI and/or the editor(s) disclaim responsibility for any injury to people or property resulting from any ideas, methods, instructions or products referred to in the content.

Article

Customised Microporous Carbon 3D Structures with Good Mechanical Properties and High Nitrogen Content Obtained from Whey Powders

Raúl Llamas-Unzueta [1], Luis A. Ramírez-Montoya [1,2,*], J. Angel Menéndez [1] and Miguel A. Montes-Morán [1]

[1] Instituto de Ciencia y Tecnología del Carbón, INCAR-CSIC, c/Francisco Pintado Fe 26, 33011 Oviedo, Spain; r.llamas@incar.csic.es (R.L.-U.); angelmd@incar.csic.es (J.A.M.); miguel.montes@incar.csic.es (M.A.M.-M.)

[2] Laboratory for Research on Advanced Processes for Water Treatment, Engineering Institute, Universidad Nacional Autónoma de México (UNAM), Campus Juriquilla, Blvd. Juriquilla 3001, Juriquilla 76230, Querétaro, Mexico

* Correspondence: lramirezm@iingen.unam.mx

Abstract: Novel customised carbon monoliths with a high specific surface area were synthesised by carbonisation plus activation of dehydrated whey powders, a biomass byproduct of the dairy industry. The whey powders were casted directly by pouring them into a desired mould. After a pseudo-sintering process promoted by the self-reaction of the whey components (mostly lactose and whey proteins) at moderate temperatures (ca. 250 °C), 3D porous carbons were obtained. The process did not require any binder or external overpressure to prepare the 3D porous carbons. Upon thermal activation with CO_2 or chemical activation with H_3PO_4 and KOH, the shape of the monolithic structure was preserved after the development of a microporous network (S_{BET} up to 2400 m^2/g). Both thermal and chemical activation had little effect on the macroporosity of the monoliths. Activation of these 3D carbons had to be performed with care to avoid heterogeneous skin/core activation and/or overactivation. Highly porous monoliths (S_{BET} of 980 m^2/g; open porosity of 70%) with outstanding compressive strength (10 MPa) could be obtained by thermal activation (CO_2) of whey monoliths at 850 °C for 1.5 h. Additionally, the use of whey as a precursor provided the carbon monolith with a relatively high nitrogen content (ca. 3 wt.%).

Keywords: porous carbon monolith; activation; whey; biomass carbon; mechanical properties

1. Introduction

The manufacture of most carbon materials that are technologically relevant has strongly relied on the fossil fuel industry. The predicted gradual phase-out of this industry is thus expected to have a massive impact on the way technological carbons are made today, with carbon materials corporations being compelled to search for low-carbon alternatives, including precursors. In this scenario, biomass is expected to play a primary role as a precursor in several carbon markets in which its presence has been so far residual or non-existent [1–3].

Biomass is already a significant contributor to the industrial production of porous carbon materials. Activated carbon companies nowadays consume lignocellulosic biomass (wood and coconut shell mainly) to obtain porous carbon powders or granules, which are daily used products in a wide variety of activities including environmental and energy storage applications [4–6]. However, in a number of processes, 3D structured porous carbons (carbon monoliths) offer several advantages over powdered or granular ones [7–9]. Monolithic structures incorporate transport pores to reduce the pressure drop in continuous operations such as industrial filters [10,11]. They are also very relevant materials for heterogeneous catalytic reactors at the industrial level [12,13], and gas storage applications [14–16]. In all these applications, not only the porosity of the carbon material is of paramount importance but also how the 3D carbon structure endures the operating conditions in industrial plants. In other words, 3D structured porous carbons should also have good mechanical properties, which makes the current fossil fuel-derived resins and binders (i.e., pitch) imperative in the manufacture of carbon monoliths [17–19].

There have been significant efforts to cut off ties with this undesirable dependence. Carboxymethyl cellulose (CMC) has been used as a green binder with remarkable results in terms of the mechanical properties of the resulting activated carbon monoliths [20]. However, the use of binders to agglomerate activated carbon powders has traditionally important drawbacks related to pore clogging [21,22]. Strategies based on binder-free monoliths from biomass precursors are scarce. Indeed, a very interesting methodology comprises the formation of slurries from activating agents such as H_3PO_4 or $ZnCl_2$ and lignocellulosic precursors [23–25]. These slurries are prone to be either uniaxially compressed or extruded to obtain monoliths that withstand carbonisation and washing of the activation by-products. On the other hand, sustainable approaches to a greener synthesis of 3D structured porous carbons comprise studies with a number of novel biomass-derived precursors including polysaccharides [26], tannins [27], or novel biomass wastes (i.e., beyond lignocellulosic precursors) [28]. Hydrothermal carbonisation (HTC) and other approaches also looked forward to obtaining 3D structured porous carbons, mainly carbon aerogels [29,30]. However, results in terms of the mechanical integrity of those green carbon monoliths have been, in general, rather limited.

We have recently discovered the possibility of obtaining 3D porous carbons with outstanding mechanical properties by direct carbonisation of whey powders poured on moulds [31]. Whey powders result after the spray-drying of liquid whey, a byproduct of cheese and casein production found in massive quantities in the dairy industry [32]. The relatively high organic matter (mainly lactose) load of liquid whey makes whey management a serious burden. Big dairy companies have invested important resources to valorise whey and an estimated 50% of liquid whey ends nowadays in food and feed industries [33,34]. Still, ca. 100 million tons of liquid whey are wasted each year worldwide [35]. Since, as detailed in [31], the mechanical properties of the 3D whey derived carbons compete with those prepared from conventional resins, whey should be then considered a robust biomass precursor to manufacture green carbon monoliths that can be used, for example, as bone scaffolds [36].

The porosity of the 3D whey-derived carbons have been also characterised in previous studies. The carbonisation of whey powders renders materials that are mainly macroporous, with limited meso- and micropore development, as determined by N_2 adsorption at cryogenic temperatures [31]. The aim of this work is to study the activation of whey carbon monoliths to improve their micro- and mesoporosity, which is compulsory for their use as adsorbents or catalysts/catalyst supports. Although KOH activation of whey powders has been reported before [37], as it has been that of non-fat powdered milk, a precursor similar to whey [38], the final product of these studies are activated carbon powders, which is not the purpose of this work.

2. Materials and Methods

2.1. Synthesis and Activation of the Whey-Derived Carbon Monoliths

The preparation of the custom-made microporous carbon 3D structures from whey comprises two steps: (i) pre-conformation of the 3D structure; (ii) carbonisation and/or activation.

For the pre-conformation of the 3D structures, whey powder (W) (CAPSA-Food, Granda, Spain) was poured into a 12 mm inner diameter glass tube. Details of W including its chemical composition have been reported elsewhere [31]. The pre-conformation step was carried out at 150 °C for 1 h in a lab stove. After this step, the W powder transformed into a solid rod of 12 mm diameter. The rod was then cut in cylinders of 30 mm height using a circular saw.

The cylinders were further carbonised at different temperatures in a tubular furnace (Carbolite, Hope Valley, UK) under N_2 (100 mL/min) with a heating rate and dwell time of 10 °C/min and 1.5 h, respectively. The carbonised samples were labelled as CW_X, with X being the carbonisation temperature. Some of the carbonised whey cylinders were demineralised (samples CW_X_D) by acid (HCl) plus water washing in an ultrasonic bath.

Thermally activated samples (TAW) were first heat treated at up to 850 °C for 1.5 h under 100 mL/min N_2. After carbonisation, the temperature of the oven was set to the activation temperature (800–850 °C), the N_2 flux was switched to 50 mL/min CO_2 and kept at that activation temperature for 1–3 h. Finally, the monolith was cooled down in N_2 (100 mL/min). The TAW_X_Y samples' labelling code includes the activation temperature (X) followed by the dwelling time (Y).

Chemically activated samples (CASs) were prepared using both KOH and H_3PO_4 as activating agents (CH). KOH powder, H_3PO_4 (85% in water) and fuming HCl were purchased from Sigma-Aldrich (Schnelldorf, Germany). Milli-Q water was used for washing and impregnation procedures. The direct chemical activation of the pre-conformed monoliths was not feasible and a thermal stabilisation step was required prior activation. In the case of H_3PO_4 activation, the pre-conformed cylinders were heated up to 250 °C in the tubular furnace under N_2 (100 mL/min) for 1.5 h, whereas for KOH activation, the stabilisation temperature was set at 450 °C. The thermally stabilised monoliths (S) were then impregnated with different proportions of either KOH or H_3PO_4. This impregnation step was carried out by sonicating the monolith in 100 mL of a water solution of the CH at 80 °C for 2 h and then dried at 100 °C. It is important to mention that the integrity of the monolith remained unaltered after this stage. Afterwards, the impregnated monolith was placed in an alumina crucible and thermally treated at different temperatures for 1.5 h with a heating ramp of 10 °C/min under N_2 atmosphere (100 mL/min). After cooling down in N_2, the CAS samples obtained using KOH were washed with 1 M HCl under US during three cycles of 15 min each followed by vacuum filtration for each batch. Then, samples were washed with distilled water in an ultrasonic bath and filtrated. This procedure was performed for three consecutive 15 min washing cycles. The samples were finally dried at 110 °C in a convection oven overnight. For phosphoric activated samples, five washing cycles were performed using only sonicated water. The resulting monoliths were labelled as CAS_X_CH_Z, X being the heat treatment temperature, CH the activating agent (K for KOH and P for H_3PO_4) and Z the S:CH impregnation ratio.

2.2. Characterisation Methods

The thermogravimetric (TG) analyses of the W powders were carried out in a TGA Q5000 device from TA Instruments (New Castle, DE, USA) under N_2 and CO_2 atmospheres (20 mL/min) using a heating rate of 10 °C/min from room temperature to 1000 °C with sample masses of ca. 20 mg. For ash and moisture contents of the carbonised and activated monoliths (also ca. 20 mg), TGs were obtained using 20 mL/min of N_2 up to 900 °C and then switching to air (20 mL/min) for 30 min at that temperature. The moisture contents were calculated from the loss of weight at 110 °C. The ash content was estimated by the difference of the total loss of weight of the sample after combustion at 900 °C on air. The elemental analysis of the carbon monoliths was carried out in LECO (Geleen, The Netherlands) apparatuses (LECO CHNS-932, and LECO VTF-900 for oxygen content determination). The point of zero charge of the monoliths (pH_{PZC}) was determined by mass titration [39].

Nitrogen adsorption–desorption isotherms at −196 °C were performed in a Micromeritics Tristar II volumetric adsorption system, after outgassing the samples at 120 °C overnight. The specific surface area (S_{BET}) and micropore volume (V_{micro}) were calculated using the Brunauer–Emmett–Teller (BET) and Dubinin–Radushkevich (DR) models, respectively, while the pore size distributions (PSD) were calculated using the Density Functional Theory (DFT) model. The helium densities (ρ_{He}) were determined in an AccuPyc 1330 pycnometer (Micromeritics, Norcross, GA, USA). For the Hg intrusion technique, an AutoPore IV porosimeter (Micromeritics, GA, USA) was used. The Hg densities (ρ_{Hg}) and corresponding PSDs were determined from the Hg intrusion data up to a maximum operating pressure of 227 MPa. For both He and Hg densities determination, samples were outgassed at 120 °C overnight.

Compressive strengths (S_u) were evaluated in an Instron Model 8562 (Instron, Norwood, MA, USA) device with a cell load of 10 kN and a test velocity of 0.5 mm/min. The final dimensions of the tested specimens were 12.25 ± 0.2 mm diameter and 22.5 ± 0.2 mm height. The compressive strength value is the maximum stress (load/area) that the specimen stood during the test. The elastic compressive moduli (E) were obtained from the slope of the linear elastic region of the stress vs. strain curve. For a given sample, the S_u and E values reported are the average of three compressive tests.

3. Results and Discussion

3.1. Customised 3D Structures Made of Carbonised Whey Powders

As shown in Figure 1, 3D monolithic porous carbons with different shapes and sizes were obtained by the simple carbonisation of pre-conformed W powders, depending on the selected mould (cubes, cylinders, spheres, hexagons, etc.). Even more sophisticated geometries, such as nuts or hollow cylinders, are possible. Also, the mechanical stability of the whey-derived carbon monoliths bears machining processes to obtain customised materials for a variety of applications. As pointed out in previous works [31], whey particles stick to each other by the self-reaction between the precursor constituents (lactose and whey proteins) during the heat treatment. Whey particles do not melt with temperature and both the pre-conformation and carbonisation/activation of the 3D structures is carried out in absence of any external template, binders or pressure. It should be finally mentioned that quantities of inorganic matter (6–13% ash contents, depending on the W batch and/or brand) remain in the carbon monoliths, mainly composed of Ca, K, Na, and P [31,40].

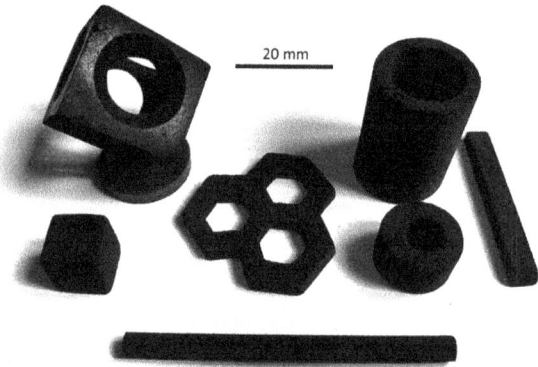

Figure 1. Examples of customised whey-derived carbon 3D structures made by simple moulding, or by moulding + machining.

3.2. Thermal Activation of the Whey-Derived 3D Carbons

As mentioned in the Introduction, whey carbon monoliths are essentially macroporous materials (maximum of the Hg intrusion PSD centred at 20–40 µm, see Section 3.4) and show no significant micro or mesoporosity, as measured by N_2 adsorption at −196 °C (Figure 2a). However, the adsorption of CO_2 at 0 °C on the same materials (Figure 2b) is significant and reveals the ultra-microporosity (pore sizes below 0.7 nm) of these materials. The ultra-microporosity is incipient in the low-temperature carbonised monoliths CW_450 and increases substantially as the carbonisation temperature rises up to 750 and 850 °C. Finally, higher carbonisation temperatures (CW_1000) result in a decrease in the volume of CO_2 adsorbed, thus suggesting the collapse of the narrow micropores. The relevant CO_2 adsorption on these materials, especially CW_750 and CW_850, opens the possibility of transforming the whey-derived carbon 3D structures into activated carbon monoliths by thermal activation.

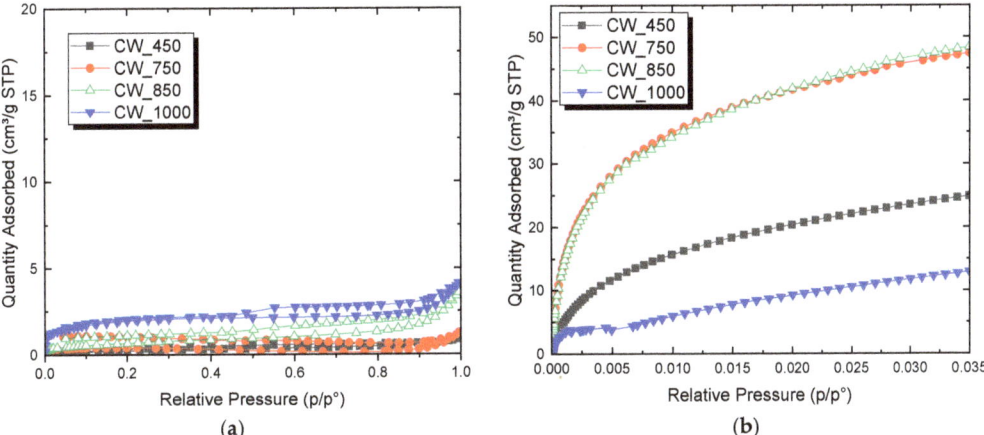

Figure 2. (**a**) N_2 adsorption isotherms (-196 °C); and (**b**) CO_2 adsorption isotherms (0 °C) on whey monoliths carbonised at different temperatures.

This was further confirmed by the TG analysis of the precursor W. Figure 3 shows the TG and DTG profiles of W heated in both N_2 and CO_2 atmospheres. Discussing first the TG profile of whey carbonisation, the TG curve is characterised by five stages. The first stage with the DTG peak at ca. 80 °C is attributed to the evaporation of the moisture remaining in the W sample. The second stage (130–160 °C) corresponds to the loss of crystalline water present in the α-lactose monohydrate (spray dried whey particles are a matrix of amorphous lactose in which lactose monohydrate crystals, proteins, fats and minerals are embedded). The reactions occurring in this temperature range are critical for the formation of 3D structures from W that withstand carbonisation. The release of crystalline water in this second stage is expected to boost the formation of melanoidins through the Maillard reactions between the lactose and the proteins in W. The successive thermal events (from 160 °C onwards) are thus complex and comprise the thermal evolution of unreacted lactose and whey proteins, on one side [31], and melanoidins, on the other [41,42].

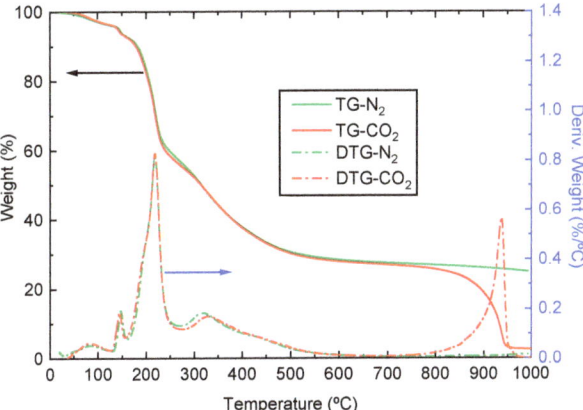

Figure 3. TG and DTG of the whey powders under N_2 and CO_2 atmospheres.

The TG of W in CO_2 is, as expected, virtually identical to that in N_2 during most of the temperature scan. Gasification starts at 720 °C, and the quick mass drop, also reflected in the sharpness of the corresponding CO_2 DTG band with maximum at 935 °C, indicates the relatively high reactivity of the W char when compared, for example, to conventional biomass chars such as coconut char [43]. This high reactivity of the whey char is a consequence of two contributions: (i) the high ultra-micropore volume of the CWs obtained at carbonisation temperatures above 700 °C (Figure 2b); and (ii) the existence of mineral matter catalysing the gasification. Analysis of the TG/DTG CO_2 profile of W suggests that the best temperature range for thermal activation of W chars is 800–850 °C, temperatures at which the gasification would be kept under (relative) control.

The activation of the W chars (CW_850) was thus carried out at 800 °C and 850 °C. Before showing and discussing the results obtained, a fundamental comment regarding the thermal activation of these monoliths should be pointed out. The thermal activation of the stocky 3D structures of carbonised whey in the tubular oven was quite tricky. Considering that, even when working with powders, the standard procedures of CO_2 activation recommend the deposit of thin layers over the crucible in order to facilitate the contact between the gas and the carbon material, the activation of a monolith represents a challenge. The CO_2 needs to diffuse into the whole cylinder in order to obtain a homogeneous activation, i.e., to minimise the differences of activation between the skin and the core of the monolith [44]. N_2 isotherms (-196 °C) of the monolith skin and core were thus measured by removing the outermost (approx. 1 mm) crust of the cylinders after activation. Although we were confident that the open macroporosity of the whey-derived chars would ease the diffusion of the CO_2 inside the cylinders, the initial experiments failed to provide homogeneous activation of the pieces. Regardless of the changes in the composition (relative proportion of CO_2 and N_2) and flux of the inlet gas, those initial experiments brought about either non-activated monoliths or only skin-activated monoliths. It was necessary to plug a substantial amount of quartz wool at the tube inlet (approx. 10 cm length), that helped to spread the laminar flow of CO_2, to succeed with differences between the skin and core being less than 10%. The results shown in Figure 4 and Table 1 correspond to the measurements of the whole (skin + core) monolith.

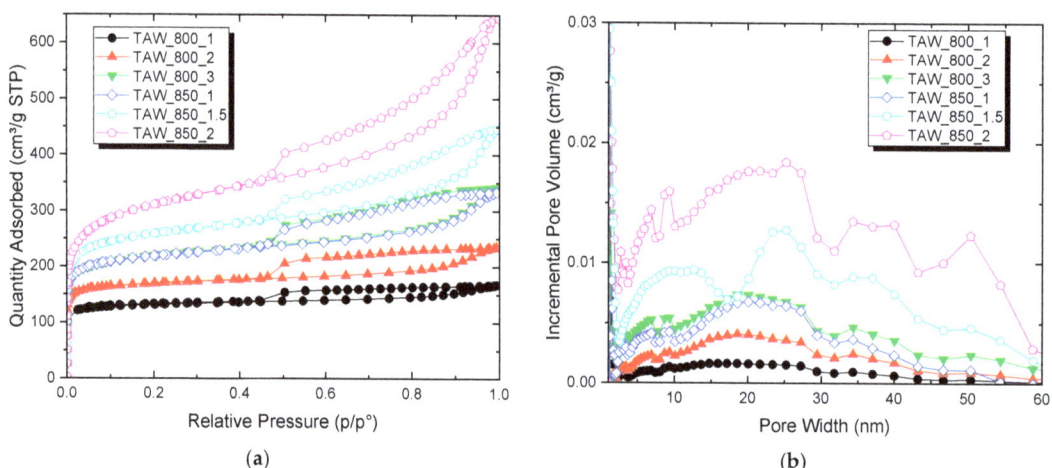

Figure 4. (**a**) N_2 adsorption isotherms (-196 °C); and (**b**) DFT pore size distributions of the thermally (CO_2) activated monoliths derived from whey.

Table 1. Selected textural parameters and yields of the thermally activated carbon monoliths.

Sample	S_{BET} (m²/g)	V_{total} [a] (cm³/g)	V_{micro} [b] (cm³/g)	V_{meso} [c] (cm³/g)	ρ_{He} (g/cm³)	Yield [d] (%)	Burn-Off [e] (%)
CW_850	<10	0.001	-	-	2.00	22.4	0
TAW_800_1	500	0.287	0.205	0.082	2.07	17.2	23.2
TAW_800_2	663	0.363	0.261	0.102	2.00	15.0	33.0
TAW_800_3	834	0.529	0.332	0.197	1.98	12.3	45.1
TAW_850_1	839	0.516	0.331	0.185	2.01	12.3	45.1
TAW_850_1.5	981	0.689	0.377	0.312	2.01	10.1	54.9
TAW_850_2	1140	0.995	0.442	0.553	1.99	7.1	68.3

[a] Calculated at $p/p^0 = 0.99$; [b] calculated with the DR method; [c] $V_{meso} = V_{total} - V_{micro}$; [d] expressed as $(mass_{monolith}/mass_{whey}) \times 100$; [e] expressed as $100 - [(yield_{CW\_850}/yield_{monolith}) \times 100]$.

Thermal activation is effective at both temperatures, rendering materials that combine micro and mesoporosity (isotherm Type IV, Figure 4a). The monoliths activated at 800 °C show a limited development of the microporosity, with S_{BET} maximum values of 840 m²/g after 3 h of dwell time. This very same value is obtained after only 1 h of CO_2 activation at 850 °C, whereas 2 h activation at this higher temperature brings about materials with S_{BET} values well above 1100 m²/g. Due to the high reactivity of the whey-derived carbon, this relatively high surface area is attained at the expense of a high burn-off (68.3%, Table 1) of the whey char, which strongly limits the mechanical integrity of the resulting 3D structure (Figure 5). Profusion of cracks in the whey monoliths makes the activation conditions (2 h at 850 °C) unfeasible for their practical use, thus establishing the temperature and dwell time boundaries for thermal activation of these 3D structures with CO_2. On the other hand, looking at the textural parameters of the activated monoliths at 850 °C (Table 1), there is a considerable gap between the specific surface area of the TAW_850_1 and TAW_850_2. We decided then to perform an additional activation experiment with a 1.5 h dwell time (sample TAW_850_1.5); the results are also included in Figure 4 and Table 1. The TAW_850_1.5 monoliths were free of external crevices.

Figure 5. An example of a monolith with crevices due to overactivation. This picture corresponds to TAW_850_2, but it is also representative of the chemically activated monoliths.

The gasification of the whey monoliths under the tested conditions encompasses not only the formation of micropores but also an enhancement of the mesoporosity. The pore size distribution in the mesopore range of these materials is very similar for all materials except TAW_850_1.5 (Figure 4b). They are wide distributions in which several maxima can be spotted at ca. 20, 35, and 50 nm. In the case of the TAW_850_1.5 mesopore size distribution, the maximum at 20 nm seems to further split into two bands at 10–12 nm and 25 nm. The evolution of the micropore and mesopore volumes (Table 1) as the activation conditions become harsher confirms that a 850 °C and 2 h dwell is a point of inflection with $V_{meso} > V_{micro}$, thus suggesting that gasification of the char is widening the micropore network.

Figure 6 shows the relation between the burn-off and the specific surface area and micropore volume of the thermally activated samples. A good linear correlation exists in both cases.

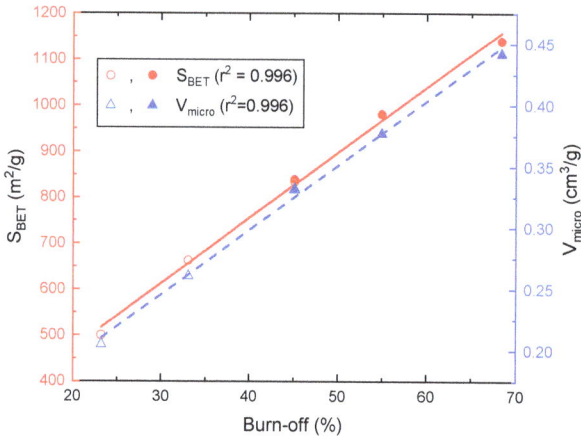

Figure 6. Specific surface area (S_{BET}) and micropore volume (V_{micro}) as a function of the thermal activation burn-off of the whey-derived activated carbon monoliths. Void and solid symbols correspond to the TAW_800 and TAW_850 monoliths, respectively.

3.3. Chemical Activation of the Whey-Derived 3D Structures

KOH and H_3PO_4 were selected as activating agents (CH) due to their well-known capacity to generate a microporous network into the carbon matrix. As detailed in the Section 2, the chemical activation was carried out over thermally stabilised monoliths. The direct activation of the whey brought about melted structures that did not keep the shape of the mould containing them, obtaining either powders (KOH activation) or foamy structures (H_3PO_4 activation). On the other hand, the issue that was pointed out when discussing the results of the thermal activation, regarding the homogeneity of the activated materials, was much less critical in the case of the chemical activation of the thermally stabilised monoliths. The macroporosity already present in these stabilised monoliths was well interconnected and the CH impregnation step was effective in reaching the inner parts of the stabilised monoliths. In addition, in the case of KOH it is also expected that the melting of the salt at temperatures above 360 °C will help the infiltration of CH into the monoliths and thus lead to a more homogeneous activation.

3.3.1. KOH Activation

Figure 7 shows the N_2 adsorption isotherms (−196 °C) of the KOH-activated samples and Table 2 summarises the principal textural parameters calculated from those isotherms. All N_2 isotherms are Type I with well-defined plateaus corresponding to highly microporous materials with slit-shaped pores in the micropore range, with the exception of CAS_750_K_1:1, which is a mixed Type I and IV isotherm with a hysteresis loop denoting the presence of mesopores (V_{meso} = 0.137 cm^3/g, Table 2).

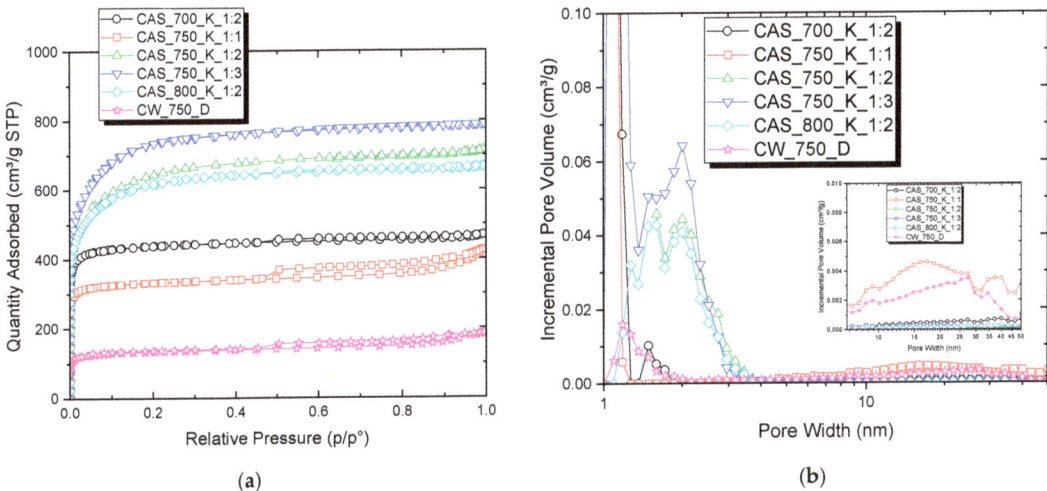

Figure 7. (a) N$_2$ adsorption isotherms (−196 °C); and (b) DFT pore size distributions of the chemically (KOH) activated monoliths derived from whey. Inset of (b) is a zoom of a selected area (5–50 nm).

Table 2. Selected textural parameters and yields of the chemically (KOH) activated carbon monoliths.

Sample	S_{BET} (m^2/g)	V_{total} [a] (cm^3/g)	V_{micro} [b] (cm^3/g)	V_{meso} [c] (cm^3/g)	ρ_{He} (g/cm^3)	Yield [d] (%)
CW_750	<10	-	-	-	1.91	27.1
CW_750_D	528	0.287	0.205	0.082	1.93	26.1
CAS_700_K_1:2	1777	0.722	0.690	0.032	2.08	14.2
CAS_750_K_1:1	1346	0.655	0.518	0.137	2.02	21.3
CAS_750_K_1:2	2440	1.100	1.072	0.028	2.12	13.9
CAS_750_K_1:3	2815	1.224	1.216	0.008	2.36	10.9
CAS_800_K_1:2	2356	1.029	1.017	0.012	2.44	13.0

[a] Calculated at $p/p^0 = 0.99$; [b] calculated with the DR method; [c] $V_{meso} = V_{total} - V_{micro}$; [d] expressed as (mass$_{monolith}$/mass$_{whey}$) × 100.

The effect of the monolith/CH ratio on the KOH activation was as expected. For a given temperature (750 °C), increasing the KOH content increases the total pore volume of the monoliths from 0.655 to 1.224 cm^3/g. The microporosity is enhanced as one moves from CAS_750_K_1:1 to CAS_750_K_1:3, whereas the mesopore volume diminishes quickly (Table 2). A close look to the PSD of these materials (Figure 7b) shows how the initially wide mesopores (>5 nm) present in CAS_750_K_1:1 disappear as the relative CH ratio increases (i.e., samples CAS_X_K_1:2 and CAS_X_K_1:3, with X being 750 or 800), with narrow mesopores with an average diameter of around 2.8 nm evolving in those samples. The origin of the wide mesopores in sample CAS_750_K_1:1 should be related to the precursor porosity [45]. Since the carbonised whey CW_750 shows no significant N$_2$ adsorption at cryogenic temperatures, the origin of the wide mesoporosity of CAS_750_K_1:1 should be related to the washing carried out after the KOH activation to remove byproducts (mainly carbonates). To check this possibility, a CW_750 monolith was thoroughly washed with HCl and water following the same methodology used for the KOH-activated monoliths. The N$_2$ adsorption isotherm and textural parameters of the CW_750_D monolith are also included in Figure 7 and Table 2. It is clear that (i) a simple demineralisation process increases significantly the porosity (as determined by the N$_2$ adsorption) of the whey carbon monoliths, with mineral species in the CWs blocking the entrance to pores; and (ii) mesopores are a significant part of CW_750_D porosity, with a PSD resembling that of CAS_750_K_1:1 (see inset in Figure 7b).

As for the effect of the activation temperature [45], for a given monolith/CH ratio (1:2), results fulfilled the expectances only partially. Increasing the heat treatment temperature from 700 °C to 750 °C leads to an important increment in the porosity of the monoliths (Figure 7 and Table 2). However, a further rise to 800 °C has little effect in either the specific surface area or pore volumes.

The effectiveness of the KOH activation when compared to thermal activation is clearly demonstrated with materials reaching S_{BET} values up to 2815 m^2/g and a V_{total} = 1.224 cm^3/g for the maximum chemical agent ratio tested (1:3). Moreover, the yields of the activated monoliths per specific surface area are much higher in the case of the KOH-activated samples (Tables 1 and 2). In spite of this latter observation, the harshness of the alkali activation caused at least the presence of cracks that where evident after ocular inspection of the monoliths (Figure 5). In some cases, CAS_750_K_1:3 and CAS_800_K_1:2, the breakdown of the monoliths in several pieces occurred, thus limiting strongly the feasibility of this type of activation to obtain customised 3D microporous carbons.

3.3.2. H$_3$PO$_4$ Activation

As in the case of the KOH activation, we explored the effect of the CH relative proportion and the activation temperature variables on the outcome. Figure 8 shows the N$_2$ adsorption isotherms (−196 °C) of the H$_3$PO$_4$ activated samples, with Table 3 collecting the principal textural parameters calculated. All phosphoric activated monoliths isotherms are a combination of Type I and Type IV for microporous materials with a certain amount of mesopores and hysteresis loop type H4 for slit-shaped pores. The isotherms have open "knees" in the micropore relative pressure range, which indicates a lower contribution of microporosity to the total pore volume. This contrasts with the KOH-activated samples showing isotherms with very close "knees". Thus, comparing the samples with similar V_{total} values (i.e., CAS_750_K_1:1 and CAS_800_P_1:2, Tables 2 and 3, respectively), the KOH-activated samples exhibit significantly higher S_{BET} and V_{micro} than the H$_3$PO$_4$-activated samples. As a consequence, the contribution of mesoporosity to the V_{total} is the highest for this type of activation. The N$_2$ PSDs of all the phosphoric-activated samples (Figure 8b) point out that a relatively sharp, unimodal distribution in the mesoporosity range is common to all monoliths, with maxima located at ca. 3 nm (i.e., narrow mesopores).

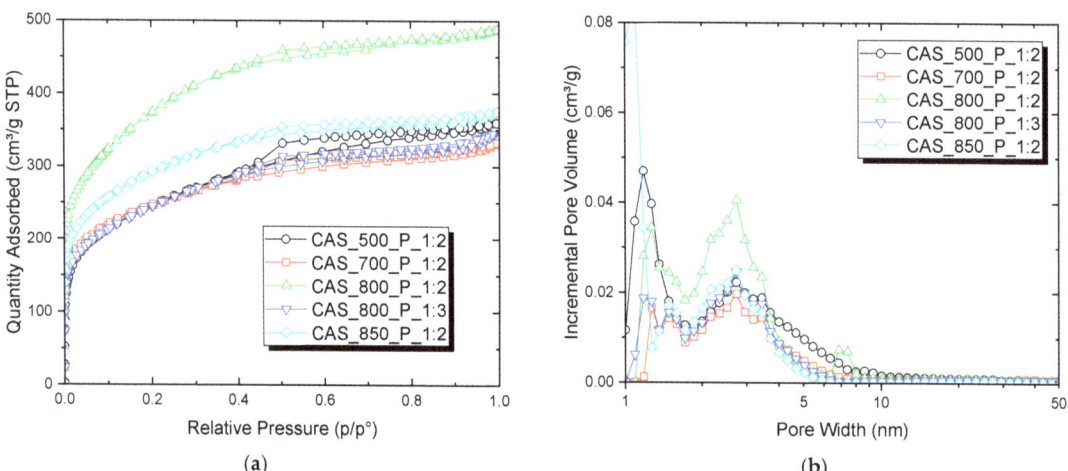

Figure 8. (a) N$_2$ adsorption isotherms (−196 °C); and (b) DFT pore size distributions of the chemically (H$_3$PO$_4$) activated monoliths derived from whey.

Table 3. Selected textural parameters and yields of the chemically (H_3PO_4) activated carbon monoliths.

Sample	S_{BET} (m^2/g)	V_{total} [a] (cm^3/g)	V_{micro} [b] (cm^3/g)	V_{meso} [c] (cm^3/g)	ρ_{He} (g/cm^3)	Yield [d] (%)
CAS_500_P_1:2	893	0.554	0.329	0.225	1.64	33.7
CAS_700_P_1:2	891	0.511	0.303	0.208	1.83	31.8
CAS_800_P_1:2	1354	0.751	0.438	0.313	2.00	23.7
CAS_800_P_1:3	886	0.539	0.297	0.242	1.93	26.4
CAS_850_P_1:2	1062	0.578	0.372	0.206	1.93	22.1

[a] Calculated at p/p^0 = 0.99; [b] calculated with the DR method; [c] V_{meso} = V_{total} − V_{micro}; [d] expressed as (mass$_{monolith}$/mass$_{whey}$) × 100.

The possibility of using different thermally stabilised whey:H_3PO_4 ratios was very limited and restricted in practice to a 1:2 ratio. The thermally stabilised whey: H_3PO_4 ratios of 1:1 produced little to no pore development, whereas the use of higher ratios (1:3) led to the overactivation of the monoliths. This latter observation is exemplified with the samples CAS_800_P_1:2 and CAS_800_P_1:3 in Figure 8 and Table 3. Increasing the relative proportion of the CH brings about a monolith with lower textural parameters. It has been reported that, when using high phosphoric acid impregnation ratios at relatively high (800 °C) activation temperatures, the progressive formation of polyphosphates and C-O-P bonds could block the pore entrances and lead to some loss of porosity, since these phosphate-like species are not easily removed in the washing process [46]. This would also justify the higher yield of CAS_800_P_1:3 when compared to that of CAS_800_P_1:2 (Table 3).

Regarding the effect of the activation temperature, although the activation with H_3PO_4 works better at lower temperatures for lignocellulosic precursors [47,48], in this case, results show that higher temperatures (800 °C) are necessary to obtain an optimal porous development (Figure 8), with a maximum S_{BET} for sample CAS_800_P_1:2 of 1354 m^2/g (Table 3). This could be attributed to the different mechanisms involved in the activation with phosphoric acid. In the activation at low temperatures (500 °C) of the thermally stabilised (250 °C, in this case) whey monoliths, the porosity is mainly produced by the elimination of volatile matter (Figure 3) catalysed by the dehydration of the phosphoric acid, which promotes bond cleavage reactions and crosslinking formation via processes such as cyclisation and condensation [47,48]. Activation at higher temperatures (>500 °C) provokes, in some cases, the collapse of the porous network. However, when the activation temperature rises to 700 °C, the structure of the whey carbon is strong enough to maintain the microporosity developed at earlier stages (hence the similar pore development of samples CAS_500_P_1:2 and CAS_700_P_1:2), with a slight densification taking place as shown by the significant increase in the real density (ρ_{He}) of the CAS_700_P_1:2 sample.

As just mentioned, the most important pore development occurs at 800 °C, sample CAS_800_P_1:2. The activation mechanism of phosphoric acid at such relatively high temperatures is different to that already discussed. At this temperature, H_3PO_4 dehydrates and transforms into P_4O_{10} [40]. This latter moiety behaves in the absence of water as an oxidant reacting with the carbon matrix as follows [48]:

$$P_4O_{10} + 2C \rightarrow P_4O_6 + 2CO_2 \quad (1)$$

The formation of volatile phosphorous-containing compounds as a result of the present phosphate reduction and the introduction of phosphorus species into the carbon structure through C-O-P bonds led to the formation of new pores and the widening of the existent pores. At the same time, the reaction between the evolved CO_2 and the carbon skeleton results in an additional increase in porosity. Depending on the precursors, some authors report the presence of PH_4 and elemental P_4 as products of the high-temperature phosphoric activation of carbons [48–50]. These phosphorous species are easily removed in the washing process thus leaving pores available. Finally, further rising the activation temperature up to 850 °C (sample CAS_850_P_1:2) backtracks in terms of the pore development, thus suggesting that some of the micro- and narrow mesoporosity start to collapse.

Although the porous development of the H_3PO_4-activated samples is poorer compared with the KOH-activated samples, the yields are considerably higher, higher even than the thermally activated ones. This would make this activation very interesting, although we could not obtain samples free of cracks (Figure 5) under the activation conditions tested.

3.4. Further Characterisation of Selected Samples

Five monolithic samples combining high specific surface area and (in principle) good mechanical integrity, namely CW_850, CW_750_D, TAW_850_1.5, CAS_750_K_1:2, and CAS_800_P_1:2 were further characterised using different techniques.

To complete the characterisation of the porosity of the monoliths, the selected samples were analysed by Hg intrusion. Figure 9 shows the PSD obtained by Hg intrusion with the corresponding main textural parameters collected in Table 4. All monoliths show an important macroporous development with the maxima of the Hg PSD centred at ca. 27 µm, except in the case of TAW_850_1.5, for which the Hg PSD is slightly shifted to smaller pore sizes values (Table 4). The percentage of porosity of the activated samples is >70%, with sample CAS_750_K_1:2 showing an outstanding 82% porosity value. The widening of the PSD of CAS_750_K_1:2 in the lower pore sizes region is also evident. This might suggest that the intrusion of Hg is damaging the sample. The same observation also stands when analysing the PSD of the monoliths in the mesoporous region. The relatively high V_{meso} (as determined with Hg intrusion) of CAS_750_K_1:2 was totally unexpected in the light of its N_2 adsorption isotherm (Figure 7). Actually, the only monoliths that show significant V_{meso} values in the Hg intrusion characterisation are the activated ones (Table 4), thus pointing out that the monoliths might not be bearing such massive pressures.

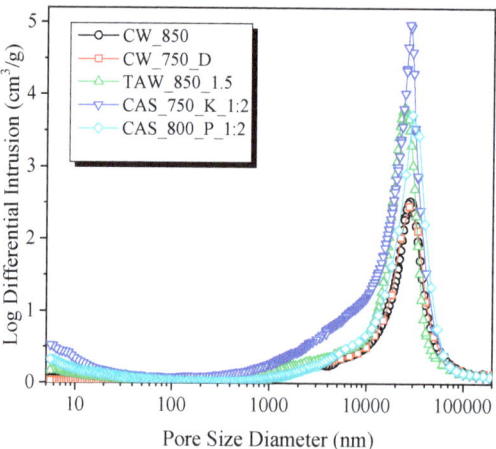

Figure 9. Pore size distributions (PSD) of selected monoliths as measured by Hg intrusion.

The elemental analyses, moisture and ash contents of the selected monoliths are collected in Table 5. Both CW_850 and TAW_850_1.5 samples have the highest ash contents. The demineralisation of the carbonised samples and the cleaning of the by-products after the chemically activated monoliths remove most of the ashes leaves only <4% leftover. Those two samples also contain the highest moisture values and oxygen contents, and the lowest carbon contents, all of which being related to their high ash content. It is also remarkable the high nitrogen content that remains in the carbon structure of the activated monoliths (≥3%), regardless of the activation procedure. Specifically, chemical activation had little effect on the final nitrogen content of the monoliths. This is especially relevant for a broad spectrum of electrochemical applications with special interest in natural and induced nitrogen carbon materials [51–53]. This is very interesting since all this nitrogen content is naturally present in the whey powders, i.e., there is no need for an extra nitrogen source for N-doping. Furthermore, the N content of these materials surpasses even those reported for N-doped mesoporous carbons derived from tannins, as well as the microporous carbons derived from various natural biowaste fibres that have been modified with different nitrogen source species, such as urea or melamine [54,55]. Finally, monolithic materials with very different surface chemistry in terms of acidity/basicity, as determined by the pH_{PZC}, were obtained.

Table 4. Textural parameters of selected monoliths obtained from Hg intrusion data.

Sample	V_{Hg} [a] (cm^3/g)	V_{macro} [b] (cm^3/g)	V_{meso} [c] (cm^3/g)	d_{max} [d] (μm)	ρ_{Hg} (g/cm^3)	s [e] (wt.%)
CW_850	1.435	1.435	-	27.3	0.85	57
CW_750_D	1.493	1.469	-	27.1	0.85	59
TA_850_1.5	1.947	1.854	0.094	23.7	0.62	70
CAS_750_K_1:2	2.378	2.134	0.245	27.0	0.38	82
CAS_800_P_1:2	2.038	1.884	0.145	28.8	0.57	72

[a] Volume of intruded Hg; [b] volume of Hg intruded in pores with sizes > 50 nm; [c] $V_{meso} = V_{Hg} - V_{macro}$; [d] pore size of the PSD maximum; [e] open porosity, $s = [1 - (\rho_{Hg}/\rho_{He})] \times 100$.

Table 5. Elemental analysis, moisture and ash contents, and pH_{PZC} of selected monoliths.

Sample	C [a] (wt.%)	H [a] (wt.%)	N [a] (wt.%)	S [a] (wt.%)	O [a] (wt.%)	Moisture [b] (wt.%)	Ash [b] (wt.%)	pH_{PZC}
CW_850	68.0	1.2	2.3	0.2	13.4	12.2	6.1	11.1
CW_750_D	88.5	0.8	3.0	0.2	6.6	6.2	2.7	5.2
TA_850_1.5	64.5	0.8	3.0	0.2	14.0	13.7	18.1	11.3
CAS_750_K_1:2	88.3	0.5	3.2	0.2	6.7	9.6	3.0	4.8
CAS_800_P_1:2	82.3	1.2	3.1	0.0	11.4	7.8	3.8	2.4

[a] Elemental analysis, dry basis; [b] proximate analysis, wet basis.

3.5. Mechanical Properties of Selected Monoliths

Mechanical properties of the selected samples were evaluated by compressive tests. Figure 10a shows typical strain/stress curves of the tested materials. Two of the monoliths, the chemically activated CAS_750_K_1:2 and CAS_800_P_1:2, could not bear any load during the compressive test, i.e., the structures shattered as soon as the load started to build up. The crevices on the surface of the chemically activated monoliths acted as fracture initiators that propagated quickly and destroyed the structures. Conversely, the compressive test curves of the CW_850 and CW_850_D monoliths increased monotonically within the elastic linear zone of the curve until the experiment reached the point of failure. This behaviour is characteristic of ceramic materials presenting brittle failure. Finally, a clear difference is observed in the stress/strain curve of the thermally activated monolith TA_850_1.2. In this case, the curve progresses with some breaks or jumps, which could indicate small fractures or imperfections in the monolith. After these discontinuities in the curve, the stress recovers almost immediately and the monoliths is capable of bearing loads up to higher strain values, when compared to CW_850 and CW_850_D. The imperfections/fractures change the monolith failure mode to non-brittle, and if they exacerbate, as in the case of the chemically activated monoliths and the thermally activated monoliths under more severe conditions, thus leading to surface crevices (Figure 5), the monolith cannot bear a substantial load.

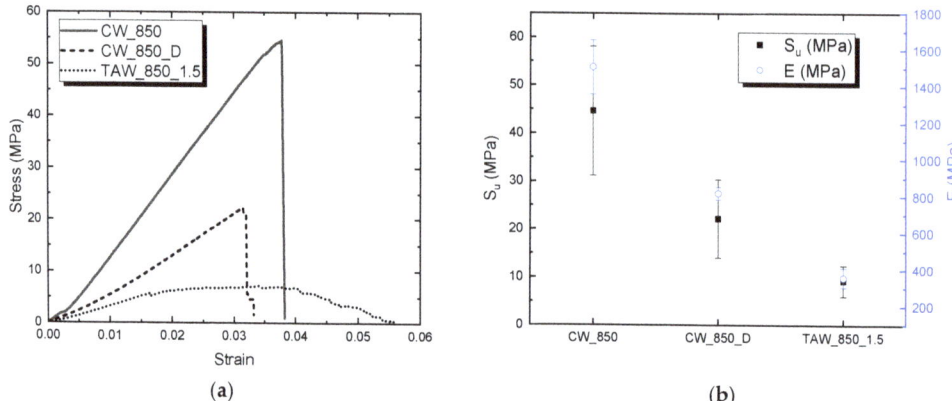

Figure 10. (**a**) Examples of stress/strain curves; and (**b**) strength (S_u) and modulus (E) of selected carbon monoliths.

Figure 10b shows the average values (and standard deviations) of the compressive strength (S_u) and modulus (E) of CW_850, CW_750_D, and TAW_850_1.5. The monolith elastic modulus goes down with porosity, from 1.5 GPa to 350 MPa for CW_850_D and TAW_850_1.5, respectively. This latter value is, however, much higher than the elastic modulus reported for hierarchical carbon monoliths made from glucose (1 MPa) [56]. In the case of CW_850, the S_u value is over 40 MPa, which is superior to other porous carbons and even ceramic materials with similar porosities. As the porosity of the monoliths increases, the compressive strength decreases (Figure 11). In this way, CW_850_D presents values around 20 MPa, while for TAW_850_1.5 the value of S_u drops to ca. 10 MPa. However, all these values are still very remarkable and considerably higher than those of other activated carbons found in the literature (Figure 11, red symbols) [19,20,57–62], specifically those made from coal (red triangles) [19], or those that are resin-based [57]. The red star of Figure 11 corresponds to a ceramic (cordierite)/carbon monolith [62].

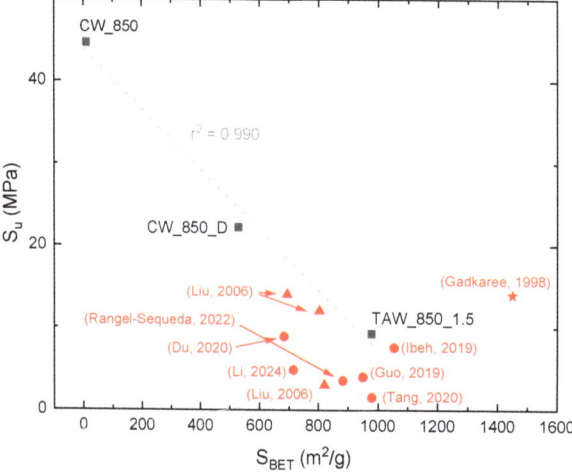

Figure 11. Correlation between S_u and S_{BET} of the selected carbon monoliths made from whey (black squares). The red symbols are results of other activated carbon monoliths found in the literature: (Liu, 2006) [19]; (Rangel-Sequeda, 2022) [20]; (Du, 2020) [57]; (Li, 2024) [58]; (Ibeh, 2019) [59]; (Guo, 2019) [60]; (Tang, 2020) [61]; (Gadkaree, 1998) [62].

4. Conclusions

A simple and novel method is presented for synthesising highly porous 3D carbon structures. These carbon monoliths can be easily moulded to specific shapes and machined for specific applications. They are easily produced by pseudo-sintering dehydrated whey powders at high temperatures, which promote self-reaction between the precursor components (specifically lactose and whey proteins) with no need for external binders, overpressure, or templates. A mechanically strong porous carbon 3D structure with a hierarchical macro/meso/microporosity can be obtained simply by demineralising the whey carbon monoliths. The monolithic shape is preserved upon thermal activation with CO_2 and chemical activation with KOH or H_3PO_4, resulting in high porosities and S_{BET} up to 2400 m^2/g, although chemical activation affects critically the mechanical properties of the 3D structures. The final carbonised, demineralised, and thermally activated monoliths have exceptional mechanical properties and an interconnected porous structure. In addition, they have a notable natural nitrogen content of up to 3 wt.%.

The possibility of using whey as a precursor of activated carbon monoliths constitutes a modest alternative for whey valorisation, in addition to lessening their dependence on fossil-fuel derived precursors. These two outcomes are significant in a circular economy model of materials/goods production and consumption.

Author Contributions: Conceptualization, L.A.R.-M., J.A.M. and M.A.M.-M.; methodology, all authors; investigation, R.L.-U. and L.A.R.-M.; writing—original draft preparation, L.A.R.-M.; writing—review and editing, all authors; funding acquisition, J.A.M. and M.A.M.-M. All authors have read and agreed to the published version of the manuscript.

Funding: This research was funded by the Spanish Ministerio de Ciencia e Innovación (MCIN/AEI/10.13039/501100011033) (Project PID2020-115334GB-I00) and Principado de Asturias (FICYT)—European Union (FEDER) (Project PCTI-Asturias IDI/2021/000015). L.A.R.-M. thanks CONACYT, Mexico, for a postdoctoral grant (CVU No. 330625, 2022).

Data Availability Statement: The data presented in this study are available on request from the corresponding author.

Acknowledgments: J. Angel Menéndez and Miguel Montes are members of the CSIC Interdisciplinary Thematic Platform (PTI+) for Sustainable Plastics towards a Circular Economy (PTI-SusPlast+).

Conflicts of Interest: The authors declare no conflict of interest. The funders had no role in the design of the study; in the collection, analyses, or interpretation of the data; in the writing of the manuscript; or in the decision to publish the results.

References

1. Lan, G.; Yang, J.; Ye, R.-P.; Boyjoo, Y.; Liang, J.; Liu, X.; Li, Y.; Liu, J.; Qian, K. Sustainable carbon materials toward emerging applications. *Small Methods* **2021**, *5*, 2001250. [CrossRef] [PubMed]
2. Yap, Y.W.; Mahmed, N.; Norizan, M.N.; Abd Rahim, S.Z.; Ahmad Salimi, M.N.; Abdul Razak, K.; Mohamad, I.S.; Abdullah, M.M.A.-B.; Mohamad Yunus, M.Y. Recent advances in synthesis of graphite from agricultural bio-waste material: A review. *Materials* **2023**, *16*, 3601. [CrossRef] [PubMed]
3. Wang, Y.; Zhang, M.; Shen, X.; Wang, H.; Wang, H.; Xia, K.; Yin, Z.; Zhang, Y. Biomass-derived carbon materials: Controllable preparation and versatile applications. *Small* **2021**, *17*, 2008079. [CrossRef] [PubMed]
4. Chakraborty, R.; Vilya, K.; Pradhan, M.; Nayak, A.K. Recent advancement of biomass-derived porous carbon based materials for energy and environmental remediation applications. *J. Mater. Chem. A* **2022**, *10*, 6965–7005. [CrossRef]
5. Chen, Q.; Tan, X.; Liu, Y.; Liu, S.; Li, M.; Gu, Y.; Zhang, P.; Ye, S.; Yang, Z.; Yang, Y. Biomass-derived porous graphitic carbon materials for energy and environmental applications. *J. Mater. Chem. A* **2020**, *8*, 5773–5811. [CrossRef]
6. Sun, L.; Gong, Y.; Li, D.; Pan, C. Biomass-derived porous carbon materials: Synthesis, designing, and applications for supercapacitors. *Green Chem.* **2022**, *24*, 3864–3894. [CrossRef]
7. Crittenden, B.; Patton, A.; Jouin, C.; Perera, S.; Tennison, S.; Botas Echevarria, J.A. Carbon monotihs: A comparison with granular materials. *Adsorption* **2005**, *11*, 537–541. [CrossRef]
8. Vilaplana-Ortego, E.; Alcañiz-Monge, J.; Cazorla-Amorós, D.; Linares-Solano, A. Activated carbon fibre monoliths. *Fuel Process. Technol.* **2002**, *77–78*, 445–451. [CrossRef]
9. Moreno-Castilla, C.; Pérez-Cárdenas, A.F. Carbon-based honeycomb monoliths for environmental gas-phase applications. *Materials* **2010**, *3*, 1203–1227. [CrossRef]

10. Yu, F.D.; Luo, L.A.; Grevillot, G. Adsorption isotherms of VOCs onto an activated carbon monolith: Experimental measurement and correlation with different models. *J. Chem. Eng. Data* **2002**, *47*, 467–473. [CrossRef]
11. Yates, M.; Martín-Luengo, M.A.; Vega Argomaniz, L.; Nogales Velasco, S. Design of activated carbon–clay composites for effluent decontamination. *Microporous Mesoporous Mater.* **2012**, *154*, 87–92. [CrossRef]
12. McIntosh, S.L.; Herkes, F.E.; Keller, J.H. Activated carbon monolith catalysts (ACMC): A new and novel catalyst system. *Top. Catal.* **2010**, *53*, 1091–1095. [CrossRef]
13. Gatica, J.M.; Gómez, D.M.; Vidal, H. Monolithic honeycomb design applied to carbon materials for catalytic methane decomposition. *Appl. Catal. A* **2013**, *458*, 21–27. [CrossRef]
14. Marco-Lozar, J.P.; Kunowsky, M.; Suárez-García, F.; Carruthers, J.D.; Linares-Solano, A. Activated carbon monoliths for gas storage at room temperature. *Energy Environ. Sci.* **2012**, *5*, 9833–9842. [CrossRef]
15. Reljic, S.; Cuadrado-Collados, C.; Farrando-Perez, J.; Jardim, E.O.; Martinez-Escandell, M.; Silvestre-Albero, J. Carbon-based monoliths with improved thermal and mechanical properties for methane storage. *Fuel* **2022**, *324*, 124753. [CrossRef]
16. Romero-Anaya, A.J.; Kunowsky, M.; Rufete-Beneite, M.; Lillo-Ródenas, M.Á.; Linares-Solano, Á. Novel monoliths prepared from sucrose avoiding binder and thermal treatment. *Microporous Mesoporous Mater.* **2019**, *284*, 78–81. [CrossRef]
17. Liu, L.; Liu, Z.; Huang, Z.; Liu, Z.; Liu, P. Preparation of activated carbon honeycomb monolith directly from coal. *Carbon* **2006**, *44*, 1598–1601. [CrossRef]
18. Jordá-Beneyto, M.; Lozano-Castelló, D.; Suárez-García, F.; Cazorla-Amorós, D.; Linares-Solano, A. Advanced activated carbon monoliths and activated carbons for hydrogen storage. *Microporous Mesoporous Mater.* **2008**, *112*, 235–242. [CrossRef]
19. Liu, L.; Liu, Z.; Yang, J.; Huang, Z.; Liu, Z. Effect of preparation conditions on the properties of a coal-derived activated carbon honeycomb monolith. *Carbon* **2007**, *45*, 2836–2842. [CrossRef]
20. Rangel-Sequeda, J.F.; Loredo-Cancino, M.; Águeda Maté, V.I.; De Haro-Del Rio, D.A.; Dávila-Guzmán, N.E. 3D printing of powdered activated carbon monoliths: Effect of structuring on physicochemical and mechanical properties and its influence on the adsorption performance. *Mat. Today Commun.* **2022**, *33*, 104758. [CrossRef]
21. Lozano-Castelló, D.; Cazorla-Amorós, D.; Linares-Solano, A.; Quinn, D.F. Activated carbon monoliths for methane storage: Influence of binder. *Carbon* **2002**, *40*, 2817–2825. [CrossRef]
22. Ramos-Fernández, J.M.; Martínez-Escandell, M.; Rodríguez-Reinoso, F. Production of binderless activated carbon monoliths by KOH activation of carbon mesophase materials. *Carbon* **2008**, *46*, 365–389. [CrossRef]
23. Nakagawa, Y.; Molina-Sabio, M.; Rodríguez-Reinoso, F. Modification of the porous structure along the preparation of activated carbon monoliths with H_3PO_4 and $ZnCl_2$. *Microporous Mesoporous Mater.* **2007**, *103*, 29–34. [CrossRef]
24. Vargas, D.P.; Giraldo, L.; Moreno-Piraján, J.C. CO_2 adsorption on activated carbon honeycomb-monoliths: A comparison of Langmuir and Tóth models. *Int. J. Mol. Sci.* **2012**, *13*, 8388–8397. [CrossRef] [PubMed]
25. Ibeh, P.O.; García-Mateos, F.J.; Ruiz-Rosas, R.; Rosas, J.M.; Rodríguez-Mirasol, J.; Cordero, T. Acid mesoporous carbon monoliths from lignocellulosic biomass waste for methanol dehydration. *Materials* **2019**, *12*, 2394. [CrossRef]
26. Budarin, V.L.; Shuttleworth, P.S.; White, R.J.; Clark, J.H. From polysaccharides to starbons®. *RSC Green Chem.* **2015**, *32*, 53–81. [CrossRef]
27. Castro-Gutiérrez, J.; Sanchez-Sanchez, A.; Ghanbaja, J.; Díez, N.; Sevilla, M.; Celzard, A.; Fierro, V. Synthesis of perfectly ordered mesoporous carbons by water-assisted mechanochemical self-assembly of tannin. *Green Chem.* **2018**, *20*, 5123–5132. [CrossRef]
28. Dodson, J.R.; Budarin, V.L.; Hunt, J.; Shuttleworth, P.S.; Clark, J.H. Shaped mesoporous materials from fresh macroalgae. *J. Mater. Chem. A* **2013**, *1*, 5203–5307. [CrossRef]
29. Titirici, M.-M.; Antonietti, M. Chemistry and materials options of sustainable carbon materials made by hydrothermal carbonization. *Chem. Soc. Rev.* **2010**, *39*, 103–116. [CrossRef]
30. Antonietti, M.; Fechler, N.; Fellinger, T.-P. Carbon aerogels and monoliths: Control of porosity and nanoarchitecture via sol-gel routes. *Chem. Mater.* **2007**, *19*, 4205–4212. [CrossRef]
31. Llamas-Unzueta, R.; Menéndez, J.A.; Ramírez-Montoya, L.A.; Viña, J.; Argüelles, A.; Montes-Morán, M.A. 3-D structured porous carbons with virtually any shape from whey powders. *Carbon* **2021**, *175*, 403–412. [CrossRef]
32. Prazeres, A.; Carvalho, M.d.F.; Rivas, J. Cheese whey management: A review. *J. Environ. Manag.* **2012**, *110*, 48–68. [CrossRef]
33. González-Weller, D.; Paz-Montelongo, S.; Bethencourt-Barbuzano, E.; Niebla-Canelo, D.; Alejandro-Vega, S.; Gutiérrez, A.J.; Hardisson, A.; Carrascosa, C.; Rubio, C. Proteins and minerals in whey protein supplements. *Foods* **2023**, *12*, 2238. [CrossRef]
34. Zadow, J. *Whey and Lactose Processing*; Elsevier Science Publishers Ltd.: Essex, UK, 1992.
35. Barba, F.J. An integrated approach for the valorization of cheese whey. *Foods* **2021**, *10*, 564. [CrossRef] [PubMed]
36. Llamas-Unzueta, R.; Suárez, M.; Fernández, A.; Díaz, R.; Montes-Morán, M.A.; Menéndez, J.A. Whey-derived porous carbon scaffolds for bone tissue engineering. *Biomedicines* **2021**, *9*, 1091. [CrossRef] [PubMed]
37. Tan, S.; Chen, X.; Zhai, S.; Ebrahimi, A.; Langrish, T.; Chen, Y. Spray drying assisted synthesis of porous carbons from whey powders for capacitive energy storage. *Energy* **2018**, *147*, 308–316. [CrossRef]
38. Pokrzywinski, J.; Keum, J.K.; Ruther, R.E.; Self, E.C.; Chi, M.; Meyer, H.; Littrell, K.C.; Aulakh, D.; Marble, S.; Ding, J.; et al. Unrivaled combination of surface area and pore volume in micelle-templated carbon for supercapacitor energy storage. *J. Mater. Chem. A* **2017**, *5*, 13511–13525. [CrossRef]
39. Menéndez, J.A.; Illán-Gómez, M.; León, C.Y.; Radovic, L. On the difference between the isoelectric point and the point of zero charge of carbons. *Carbon* **1995**, *33*, 1655–1657. [CrossRef]

40. Llamas-Unzueta, R.; Menéndez, J.A.; Suárez, M.; Fernández, A.; Montes-Morán, M.A. From whey robocasting to 3D porous carbons. *Add. Manufact.* **2022**, *59*, 103083. [CrossRef]
41. Zhang, Q.; Chen, M.; Emilia Coldea, T.; Yang, H.; Zhao, H. Structure, chemical stability and antioxidant activity of melanoidins extracted from dark beer by acetone precipitation and macroporous resin adsorption. *Food Res. Int.* **2023**, *164*, 112045. [CrossRef] [PubMed]
42. De la Rosa Arranz, J.M.; González-Vila, F.J.; López-Capel, E.; Manning, D.A.C.; Knicker, H.; González-Pérez, J.A. Structural properties of non-combustion-derived refractory organic matter which interfere with BC quantification. *J. Anal. Appl. Pyrol.* **2009**, *85*, 399–407. [CrossRef]
43. Llamas-Unzueta, R.; Montes-Morán, M.A.; Ramírez-Montoya, L.A.; Concheso, A.; Menéndez, J.A. Whey as a sustainable binder for the production of extruded activated carbon. *J. Environ. Chem. Eng.* **2022**, *10*, 107590. [CrossRef]
44. Lopez, M.; Labady, M.; Laine, J. Preparation of activated carbon from wood monolith. *Carbon* **1996**, *34*, 825–827. [CrossRef]
45. Linares-Solano, A.; Lozano-Castelló, D.; Lillo-Ródenas, M.A.; Cazorla-Amorós, D. Carbon activation by alkaline hydroxides preparation and reactions, porosity and performance. *Chem. Phys. Carbon* **2008**, *30*, 2-64–62. [CrossRef]
46. Liou, T.-H.; Wu, S.-J. Characteristics of microporous/mesoporous carbons prepared from rice husk under base- and acid-treated conditions. *J. Hazard. Mater.* **2009**, *171*, 693–703. [CrossRef]
47. Jagtoyen, M.; Derbyshire, F. Activated carbons from yellow poplar and white oak by H_3PO_4 activation. *Carbon* **1998**, *36*, 1085–1097. [CrossRef]
48. Li, Y.; Zhang, X.; Yang, R.; Li, G.; Hu, C. The role of H_3PO_4 in the preparation of activated carbon from NaOH-treated rice husk residue. *RSC Adv.* **2015**, *5*, 32626–32636. [CrossRef]
49. Myglovets, M.; Poddubnaya, O.I.; Sevastyanova, O.; Lindström, M.E.; Gawdzik, B.; Sobiesiak, M.; Tsyba, M.M.; Sapsay, V.I.; Klymchuk, D.O.; Puziy, A.M. Preparation of carbon adsorbents from lignosulfonate by phosphoric acid activation for the adsorption of metal ions. *Carbon* **2014**, *80*, 771–783. [CrossRef]
50. Puziy, A.M.; Poddubnaya, O.I.; Socha, R.P.; Gurgul, J.; Wisniewski, M. XPS and NMR studies of phosphoric acid activated carbons. *Carbon* **2008**, *46*, 2113–2123. [CrossRef]
51. Bejjanki, D.; Banothu, P.; Kumar, V.B.; Kumar, P.S. Biomass-derived N-doped activated carbon from eucalyptus leaves as an efficient supercapacitor electrode material. *C* **2023**, *9*, 24. [CrossRef]
52. Liao, Y.; Shang, Z.; Ju, G.; Wang, D.; Yang, Q.; Wang, Y.; Yuan, S. Biomass derived N-doped porous carbon made from reed straw for an enhanced supercapacitor. *Molecules* **2023**, *28*, 4633. [CrossRef]
53. Inagaki, M.; Toyoda, M.; Soneda, Y.; Morishita, T. Nitrogen-doped carbon materials. *Carbon* **2018**, *132*, 104–140. [CrossRef]
54. Liu, F.; Gao, Y.; Zhang, C.; Huang, H.; Yan, C.; Chu, X.; Xu, Z.; Wang, Z.; Zhang, H.; Xiao, X.; et al. Highly microporous carbon with nitrogen-doping derived from natural biowaste for high-performance flexible solid-state supercapacitor. *J. Coll. Interf. Sci.* **2019**, *548*, 322–332. [CrossRef]
55. Zhao, J.; Shan, W.; Zhang, P.; Dai, S. Solvent-free and mechanochemical synthesis of N-doped mesoporous carbon from tannin and related gas sorption property. *Chem. Eng. J.* **2020**, *381*, 122579. [CrossRef]
56. Estevez, L.; Dua, R.; Bhandari, N.; Ramanujapuram, A.; Wang, P.; Giannelis, E.P. A facile approach for the synthesis of monolithic hierarchical porous carbons-high performance materials for amine based CO_2 capture and supercapacitor electrode. *Energy Environ. Sci.* **2013**, *6*, 1785–1790. [CrossRef]
57. Du, J.; Li, W.C.; Ren, Z.X.; Guo, L.P.; Lu, A.H. Synthesis of mechanically robust porous carbon monoliths for CO_2 adsorption and separation. *J. Energy Chem.* **2020**, *42*, 56–61. [CrossRef]
58. Li, D.; Tian, Y.; Qiao, Y.; Wen, L. Conversion of powdered active carbon into monoliths without reducing specific surface area using H_3PO_4-impregnated waste sawdust. *Mater. Lett.* **2024**, *125*, 175–178. [CrossRef]
59. Ibeh, P.O.; García-Mateos, F.J.; Rosas, J.M.; Rodríguez-Mirasol, J.; Cordero, T. Activated carbon monoliths from lignocellulosic biomass waste for electrochemical applications. *J. Taiwan Inst. Chem. Eng.* **2019**, *97*, 480–488. [CrossRef]
60. Guo, L.P.; Li, W.C.; Qiu, B.; Ren, Z.X.; Dua, J.; Lu, A.H. Interfacial assembled preparation of porous carbon composites for selective CO_2 capture at elevated temperatures. *J. Mater. Chem. A* **2019**, *7*, 5402–5408. [CrossRef]
61. Tang, S.H.; Zaini, M.A.A. Development of activated carbon pellets using a facile low-cost binder for effective malachite green dye removal. *J. Clean. Prod.* **2020**, *253*, 119970. [CrossRef]
62. Gadkaree, K.P. Carbon honeycomb structures for adsorption applications. *Carbon* **1998**, *36*, 981–989. [CrossRef]

Disclaimer/Publisher's Note: The statements, opinions and data contained in all publications are solely those of the individual author(s) and contributor(s) and not of MDPI and/or the editor(s). MDPI and/or the editor(s) disclaim responsibility for any injury to people or property resulting from any ideas, methods, instructions or products referred to in the content.

Article

Production, Characterization, and Activation of Biochars from a Mixture of Waste Insulation Electric Cables (WIEC) and Waste Lignocellulosic Biomass (WLB)

Roberta Mota-Panizio [1,2,*], Ana Assis [1], Luís Carmo-Calado [1], Catarina Nobre [1], Andrei Longo [2], José Silveira [3], Maria Margarida Goncalves [1,2] and Paulo Brito [1]

1. VALORIZA—Research Center for Endogenous Resource Valorization, Polytechnic Institute of Portalegre, Campus Politécnico 10, 7300-555 Portalegre, Portugal
2. MEtRICs, Chemistry Department, NOVA School of Science and Technology (FCT NOVA), Universidade NOVA de Lisboa, Campus Caparica, 2829-516 Caparica, Portugal
3. Associated Laboratory of Guaratinguetá, Institute of Bioenergy Research, IPBEN-UNESP, São Paulo State University, São Paulo 12516-410, Brazil
* Correspondence: rpanizio@ipportalegre.pt

Citation: Mota-Panizio, R.; Assis, A.; Carmo-Calado, L.; Nobre, C.; Longo, A.; Silveira, J.; Goncalves, M.M.; Brito, P. Production, Characterization, and Activation of Biochars from a Mixture of Waste Insulation Electric Cables (WIEC) and Waste Lignocellulosic Biomass (WLB). *C* **2023**, *9*, 49. https://doi.org/10.3390/c9020049

Academic Editors: Dimitrios Kalderis, Indra Neel Pulidindi, Pankaj Sharma and Aharon Gedanken

Received: 15 February 2023
Revised: 18 April 2023
Accepted: 6 May 2023
Published: 9 May 2023

Copyright: © 2023 by the authors. Licensee MDPI, Basel, Switzerland. This article is an open access article distributed under the terms and conditions of the Creative Commons Attribution (CC BY) license (https://creativecommons.org/licenses/by/4.0/).

Abstract: Waste insulation electrical cables (WIEC) currently do not have an added value, due to their physical–chemical characteristics. Carbonization is known to enhance feedstock properties, particularly fuel and material properties; as such, this article aimed to study the production and activation of biochars using WIEC and lignocellulosic biomass wastes as feedstock. Biochars were produced in a ceramic kiln with an average capacity of 15 kg at different temperatures, namely 300, 350 and 400 °C. After production, the biochars were further submitted to a washing process with water heated to 95 °C ± 5 °C and to an activation process with 2 N KOH. All biochars (after production, washing and activation) were characterized regarding an elemental analysis, thermogravimetric analysis, heating value, chlorine removal, ash content, apparent density and surface area. The main results showed that the increase in carbonization temperature from 300 to 400 °C caused the produced biochars to present a lower amount of oxygen and volatile matter, increased heating value, greater chlorine removal and increased ash content. Furthermore, the activation process increased the surface area of biochars as the production temperature increased. Overall, the carbonization of WIEC mixed with lignocellulosic wastes showed potential in enhancing these waste physical and chemical properties, with prospects to yield added-value products that activates biochar.

Keywords: WIEC; carbonization; biochar; pretreatment; activation

1. Introduction

Nowadays, the problems related to humanity's carbon footprint have become an important theme, and carbon neutrality has been pointed out as the only viable pathway to solve our current environmental issues. As such, it is necessary to research new materials that can be considered similar to fossil fuels but which are characterized by a reduced or zero contribution to the overall balance of carbon dioxide [1]. The search for alternative fuels with low emissions and renewable characteristics is crucial to mitigate environmental impacts and meet the increased global energy demand [2]. In this sense, wastes (e.g., biomass wastes, municipal solid wastes, mixed wastes and construction and demolition wastes) are presented as valuable and sustainable energy sources to generate clean energy instead of fossil fuels, presenting a fundamental role in sustainable development strategies [3–5].

There are significant amounts of plastic and mixed wastes that cannot be recycled and are not adequate for direct combustion or gasification due their heterogeneity and high chlorine contents. These wastes usually end up being landfilled instead of being recognized for their potential as energy sources. Wastes from electrical equipment (e-waste or WEEE, which includes waste insulation electric cables—WIEC) are the fastest-growing waste type and are becoming a major environmental problem, mainly for developing and emerging economies [6–9]. This increase is associated with the consumption of electrical equipment (EEE), which increases annually around 2.5 Mt, resulting from economic and technological development [10]. The amount of WEEE generated is increasing year by year, thus making it difficult for recycling to keep up with this growth [11]. In the last 5 years, the global amount of WEEE has increased by more than 9.0 Mt, producing an amount of 53.9 Mt in 2019, with Europe being the third-largest producer of these wastes with annual rates of 2% and estimating that less than 40% of these wastes are recycled [6]. In Portugal, in 2017, 43.5% of WEEEs were recycled. This increasing trend is clearly related to a higher consumption of electrical equipment associated with a short lifetime and limited repair options for the equipment [12,13].

Plastics assume different proportions in WEEEs, varying according to size and characteristics. The plastic component can serve as housing, casing, insulation, internal shelves or lining. The amount of plastic by weight represents a range between 3.5% and 45%, presenting a great challenge to the recycling of WEEE derived from the mixture of plastics that is used for each component and the plastic's mechanical properties [14–16]. The characteristics of different types of plastics must be taken into account for WEEE management to provide a solid benchmark of environmental performance in recycling WEEE plastics [17].

Considering that wastes—in particular, biomass and mixed wastes—have very particular properties regarding their chemical characteristics, an understanding and comprehension of the various processes that integrate the energy recovery from them is essential. To enhance the process of converting wastes into products with value, a pretreatment is usually necessary. To further use these wastes in thermochemical conversion processes, adsorption processes, as an additive in construction materials or for agricultural applications, carbonization can be applied to deal with the inherent recalcitrance present in the components of these materials, being an alternative to homogenizing the material, increasing their energy density or even their porosity. Carbonization is a thermochemical process carried out at atmospheric pressure, temperatures between 300 and 500 °C in a nonoxidizing atmosphere, with slow heating rates [18–21]. This process has emerged as one of the approaches to treat biomass and wastes before submitting the feedstock to gasification processes, for example, to avoid the excessive production of tars and harmful compounds such as chlorine and yield better-quality syngas [22,23]. The main product of this thermal process is known as biochar. Currently, there is a great interest in biochar production due to the possibility of energy storage and possible use as anodes, filters and biofuels. The use of different reaction conditions and different feedstock allow the production of biochar with customized physical and chemical properties [24–26]. When the feedstock is a biomass, the lignin content has a positive effect on the biochar yield and its physical–chemical properties [27,28] The lignin content of a biomass also influences the particle size, porosity and aromatic carbon content of the biochar products [29].

The production of biochar brings many benefits to the environment, because it can be produced from a broad range of renewable sources and has low production costs and various applications [30]. Biochars can also be produced from biomass or mixed wastes, but conversion conditions must be optimized in order to maximize the energy efficiency and production yield, as these wastes are usually heterogeneous and their composition is variable, depending on the source and collection period [31].

This work aimed to produce biochar with waste insulation electric cables (WIEC) and waste lignocellulosic biomass (WLB) using the carbonization process. The main goal was to evaluate the characteristics of the produced biochars and assess their use in further thermochemical conversion processes (e.g., as feedstock for gasification) and their use as activated carbon precursors. Overall, applying carbonization can be one solution to achieve better properties for these wastes to find them added-value applications to decrease the large amounts that are currently being landfilled.

2. Materials and Methods

2.1. Feedstock

Lignocellulosic biomass wastes (WLB) were supplied by a biomass waste management company (CMC Biomassa) located in Portugal. This company collects lignocellulosic waste from deforestation, furniture and other wood wastes, which are later recycled into fractions that can be recovered as raw material and transformed into pellets, briquettes, flooring, new pallets, etc.

Waste insulation electric cables (WIEC) were supplied by a company dedicated to the management, recycling and recovery of waste, namely the recovery of noble metals (AFCarreto). This recovery from the pickling of electrical cables leads to the generation of a large amount of polymeric wastes, which are part of the coating of electrical cables corresponding to a mixture of polyvinyl chloride (PVC) and polyethylene (PE) and noble metals in small quantities.

Detailed characterizations of both WIEC and WLB can be found in [32]. For this work, a mixture with 50% WIEC and 50% WLB was prepared for the carbonization tests, as shown in Figure 1.

Figure 1. Mixture of WIEC and WLB used for the carbonization tests on a pilot scale.

2.2. Carbonization Experiments

The carbonization experiments were based on the carbonization study carried out by [32], whose main objective was to optimize the carbonization process to obtain biochar with higher chemical and energy properties for energy recovery. Using the results of the best mixture, temperature and residence time, carbonization was carried out on a pilot scale. Briefly, an electric oven (KS 72L ceramic oven) with a capacity between 12 and 20 kg per test (depending on the properties of the raw material) was used. The oven had a temperature and heating rate control panel and capacity of 72 L in the carbonization zone. For biochar production, 15 kg of the mixture (50% WIEC + 50%WLB) was placed in capped clay vessels (to ensure a nonoxidizing atmosphere) that were heated at a heating rate of 10 °C/min until reaching the final carbonization temperatures (300, 350 and 400 °C) and remained in the oven for 2 h. A general schematic for the used oven can be seen in Figure 2.

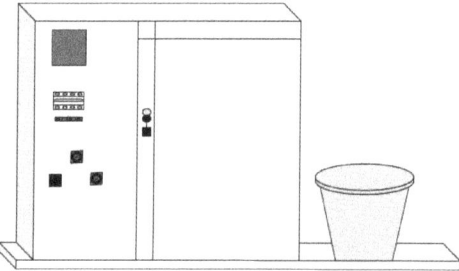

Figure 2. Schematic representation of the carbonization system on a pilot scale.

2.3. Biochar Washing Process

The produced biochars were crushed and sieved, and the fraction lower than 425 μm was used. This fraction was washed in heated water for the removal of water-soluble compounds based on the works of [33,34]. The biochars were placed in glass containers under heating and stirring plates with a ratio of 100 g/200 mL of deionized water and heated to the temperature of 95 ± 5 °C, remaining after temperature stabilization for 30 min. The biochars were then allowed to cool to room temperature and filtered. Afterwards, the biochars were dried in an oven (Holelab Greenhouse) at 105 °C until reaching a constant mass, which took about 24 h [32,35].

2.4. Biochar Activation Process

The activation process involved mixing 2 g of biochar (already washed and dried as described above) with 500 mL of a KOH 2 N solution under agitation for 1 h. After this process, the biochar samples were left at rest for 30 min and filtered for recovery. After the filtration process, the biochars were washed in deionized water, and the resulting solution was neutralized with the addition of HCl 1 N. The biochar samples were then filtered again and dried in an oven at 105 °C for 12 h prior to the characterization analysis.

2.5. Biochar Characterization

Table 1 presents the conditions for obtaining the Table 1 different biochar samples according to the processes described in Sections 2.2–2.4.

Table 1. Processing conditions for each biochar sample.

Sample	Temperature	Features
B300	300	
B350	350	Biochars were produced at different temperatures.
B400	400	
B300-L	300	
B350-L	350	Biochars were washed in hot water, filtered and dried.
B400-L	400	
B300-A	300	
B350-A	350	Biochars were washed in hot water, filtered and dried and were submitted to an activation process with KOH 2 N.
B400-A	400	

To evaluate the efficiency of the carbonization process, the biochar yield (Equation (1)), energy yield (Equation (2)) and energy density (Equation (3)) were determined.

$$\text{Biochar Yield (wt. \%)} = \frac{M_1}{M_0} \times 100\% \quad (1)$$

where M_1 is the final mass in g, and M_0 is the initial mass in g.

$$\text{Energy Yield (wt. \%)} = \frac{M_1 \times HHV_1}{M_0 \times HHV_0} \times 100\% \quad (2)$$

where M_1 is a final mass in g, M_0 is the initial mass in g, HHV_1 is the final higher heating value in MJ/kg and HHV_0 corresponds to the initial higher heating value in MJ/kg.

$$\text{Energy Density (\%)} = \frac{\text{Energy Yield}}{\text{Char Yield}} \times 100\% \quad (3)$$

2.5.1. Elemental Analysis

The elements of interest included carbon (C), hydrogen (H), nitrogen (N), sulfur (S) and oxygen (O). The amounts of C, H, N, S and O were determined using a Thermo Fisher Scientific Flash 2000 CHNS-O analyzer. Oxygen was determined by the difference in a dry base.

2.5.2. Thermogravimetric Analysis

A thermogravimetric analysis (TGA) was used to determine the moisture content, volatile matter and fixed carbon combined with ash. The tests were performed in triplicate with sample weights between 3 and 4 mg. A PerkinElmer thermogravimetric analyzer, STA 6000, using a heating rate of 20 °C/min was used. The content of each proximate analysis parameter was taken from the thermogravimetric profiles (sample mass variation versus temperature), considering the inflection points of the mass derivative on the function of time.

2.5.3. High Heating Value and Low Heating Value

For the high heating value (HHV) of the biochar samples, IKA C2000 calorimetry equipment was used, which performed the complete combustion of the samples in an adiabatic environment. For this measurement, a sample with 0.5 ± 0.1 g of each sample was placed in the calorimeter, and its total combustion was carried out. The measurements for each biochar sample were performed in triplicate, and the presented values represent the average values. The lower heating value (LHV) was determined using Equation (4) as follows:

$$LHV = HHV - 2.26 \times \frac{9H}{100} \qquad (4)$$

where LHV is a lower heating value in MJ/kg, HHV is a higher heating value in MJ/kg and H is hydrogen.

2.5.4. Chlorine Content and Mineral Composition

The chlorine content of the produced biochars was determined through X-ray fluorescence (Niton XL 3T Gold++).

The biochars' complete mineral composition (Al, B, Ba, Ca, Cr, Fe, K, Mg, Na, Ni and Zn) was determined through ICP-AES (Inductively Couple Plasma—Atomic Emission Spectrometer, Horiba Jobin-Yvon, Ultima), after ashing and acid digestion of the biochar samples. All the measurements were carried out in triplicate, and the results shown are the average values.

2.5.5. Ash Content

For the determination of the ash content, biochar samples were placed in porcelain crucibles in a muffle furnace at 500 °C $\pm$ 5 °C until the total burning of the organic matter. The ash content was calculated using Equation (5):

$$m_{ashes} = \frac{m_f - m_{crucible}}{m_{sample}} \qquad (5)$$

where m_{ashes} is the mass of the ashes in g, m_f final mass in g, $m_{crucible}$ is the mass of the calcinated empty crucible in g and m_{sample} is the mass of the initial sample in g.

2.5.6. Apparent Density

The determination of the apparent density is the relationship between the mass of a sample and its occupied volume. This determination was performed with a beaker and expressed in g/cm^3, according to Equation (6).

$$D = \frac{M}{V} \qquad (6)$$

where M corresponds to the mass in g and V to the volume in cm^3.

2.5.7. Fourier-Transform Infrared

Fourier-transform infrared spectra (FTIR) were obtained as an average of 128 scans at a resolution of 4 cm^{-1} using an ATR-FTIR spectrometer (Thermo Scientific Nicolet iS10), with a range of 4000 to 400 cm^{-1}. The biochar samples were powdered and placed in the ATR diamond crystal and compacted using a vertical screw to the plane to perform the analysis.

2.5.8. Nitrogen Adsorption at 77 K

Nitrogen adsorption at 77 K was used to measure the specific area of the samples. For biochar samples, Micromeritics ASAP 202 Plus equipped with sensors and a vacuum system was used. The vacuum is controlled by a high vacuum pump with a 1 mmHg transducer. This equilibrium has two independent vacuum systems, which allows the preparation of two samples and the analysis of a third simultaneously.

3. Results and Discussion

3.1. Biochar Yield, Enery Yield and Energetic Densification

The production of biochar can be affected by several factors: the characteristics of the materials used, the method, formation of byproducts, application, economic aspects and environmental impact. The chemical composition, particle size, pH value, moisture level, calorific value and volatility are important characteristics of materials that must be considered as they directly affect the biochar results [36]. The results obtained for the biochar samples (without washing or activation) regarding the biochar yield, energy yield and energy density are shown in Figure 3.

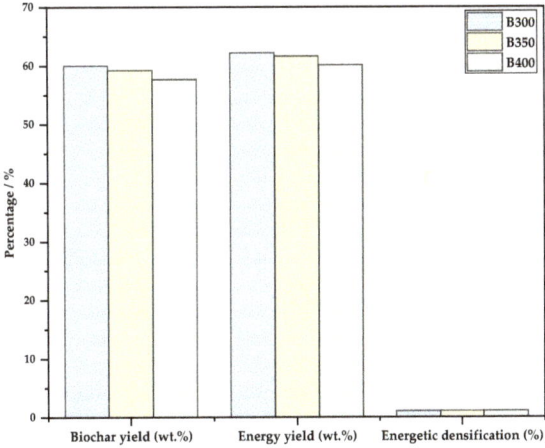

Figure 3. Evaluation parameters of the biochar production process.

Temperature is one of the main factors that affect the biochar structure and physicochemical properties, as it affects the decomposition, formation and transformation of the biomass and wastes [37]. With higher temperatures, there is a favoring of free radical reactions such as decarboxylation, decarbonylation, dehydration, aromatization and intermolecular rearrangement, among others [27]. From the obtained results, it was possible to observe that, for the three biochar production conditions, the biochar yield was between 58 and 63%. Sample B300 consistently presented higher values for the parameters depicted in Figure 3, indicating that higher temperatures promote sample decomposition via the above-described reactions, entailing lower mass and energy yields. On the other hand, energy densification was similar for the three studies, and considering the type of wastes used as feedstock, higher temperature were most likely needed to reach a higher densification factor.

3.2. Biochar Characterization

3.2.1. Elemental Analysis and Heating Value

After the biochar production process at temperatures of 300, 350 and 400 °C, part of the biochar was washed, and another part was washed and activated. Table 2 presents the results for the elemental analysis, HHV and LHV of the produced biochars.

Table 2. Elemental analysis, HHV and LHV of the original mixture and the different biochar samples.

Parameters	WIEC/WLB	B300	B350	B400	B300-L	B350-L	B400-L	B300-A	B350-A	B400-A
C (wt.%, db)	52.3	40.77	42.64	43.95	42.38	43.06	43.14	41.59	35.22	47.44
H (wt.%, db)	2.5	4.02	3.74	2.83	2.81	3.33	3.76	4.34	3.56	5.12
N (wt.%, db)	0.2	5.08	4.24	4.01	12.9	11.38	10.41	0.9	0.85	0.51
S (wt.%, db)	<d.l.	<d.l.	<d.l.	<d.l.	<d.l.	<d.l.	<d.l.	<d.l.	<d.l.	<d.l.
O (wt.%, db)	45.0	50.13	49.38	49.21	41.91	42.23	42.69	53.17	60.37	46.93
HHV (MJ/kg, db)	21.23	18.15	18.27	18.45	19.6	19.67	19.71	-	-	-
LHV (MJ/kg, db)	19.88	15.98	16.25	16.92	18.08	17.87	17.68	-	-	-

The main elements that are present in the WIEC/WLB mixture are C (52.3%) and O (42.7%). Concerning the produced biochars without treatment (B300, B350 and B400) and washed biochars (B300-L, B350-L and B400-L), it is possible to observe that the amount of C is not very variable, 42.85% ± 4%, but the amount of O was reduced by an average of 30% for the washed biochars. On the other hand, the amount of ash for the mixture was 3%, and after carbonization, the ash content increased to an average of 37% (results shown below. With the increase in the amount of ash in biochar, there is a reduction in the calorific value of the material by an average of 2 MJ/kg.

Activated carbons are very similar to other biochar samples, differing only in the amount of oxygen, which is 10% higher than other biochars. The activation process is carried out to produce a biochar that can be used for other methods, such as being used as an adsorbent for gas or liquid effluents.

The carbonization process promotes dehydration, thus promoting the elimination of H, resulting in an accumulation of C in the biochar. The content and proportions of these elements are the most important factors influencing the stability or carbon sequestration capacity of biochar. The lability of biochar is directly related to the oxygen content and indirectly to the C content present in the biochar [38]. The O/C and H/C ratios are indicative of biochar structures and provide an intrinsic measure of biochar stability. These relationships are negatively correlated with the percentage of aromatic C in the biochar [39]. Figure 4 shows the van Krevelen diagram for the obtained biochars.

The oxygen content plays an important role in the chemical behavior of the biochar surface; this factor is associated with a close relationship with the number and composition of substituted functional groups, and these functional groups constitute an important driver for the degradation potential. The H/C molar ratio can be used to evaluate the thermochemical alterations that produce fused aromatic ring structures in the material. A lower amount of H/C means that higher fused aromatic ring structures provide greater stability. The H/C ratio is considered an index of aromaticity and resistance of char to microbial and chemical degradation. As biochar is mainly composed of some aromatic compounds, the amount of C present is an important factor in determining the stability of biochar. According to Spokas [40], Budai et al. [41] and the European Biochar Certificate (EBC), the O/C and H/C molar ratios are indicators of biochar stability. Moreover, the upper limit for the O/C ratio is 0.4, and for the H/C ratio, the established limit is 0.7 [31].

Based on Figure 4, it is possible to observe that all the biochar samples that were washed have a stability below 0.2, and sample B400-A biochar is also considered stable for this parameter. Regarding the H/C ratio, it is possible to observe that the biochars are relatively similar, with little variation [42]. Using the two parameters to define the most stable biochar, it is possible to verify that B300-L and B400-A are the most stable.

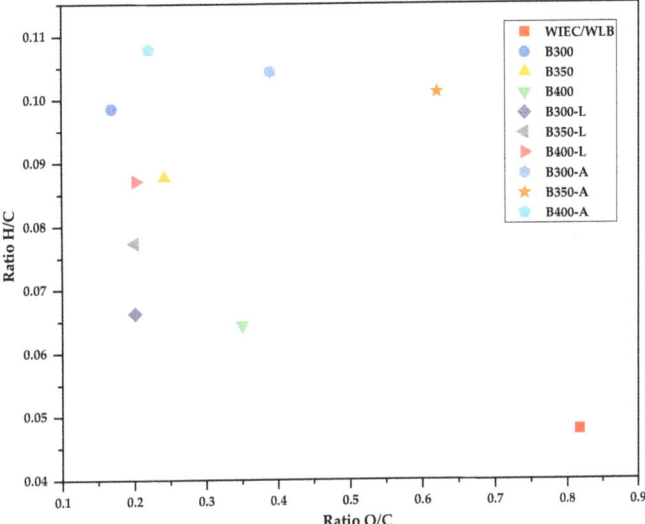

Figure 4. The van Krevelen diagram of the produced biochars.

3.2.2. Thermogravimetric Analysis

Figure 5 shows the results for the TGA analysis of the produced biochars compared with the original feedstock.

The moisture content is an important parameter and is negatively correlated with the heating value of fuel and with the potential for biological degradation during storage. Materials with a relatively low moisture content become more prone to self-ignition at room temperature [43]. In the obtained TGA profiles (Figure 5), the WIEC/WLB mixture (original feedstock) has a low moisture content, less than 0.5%, which means that it is suitable for thermochemical conversion processes, as they contain less than 10% moisture [44]. However, the produced biochars showed higher moisture contents, between 8 and 15% (B300, B350 and B400). These results can be associated with the absorption of atmospheric moisture and the tars that are on the surface of the biochar [45]. As for the biochars that were washed and activated, despite having been dried in an oven before carrying out the tests, the values for the moisture contents can also be associated with atmospheric humidity. The moisture content variations of the washed biochars were between 2 and 7% and, for the activated biochars, less than 5%, and in all the studied biochars, the loss of mass in the first stage was associated with dehydration and the decomposition of the hydrated compounds and light volatiles, which occurs up to 200 °C [46].

The main degradation occurred after dehydration, causing a sharp decrease at 200 °C and ending close to 600 °C, reaching the peak of T_{max} at 475 °C. This peak corresponds to a loss in volatile matter, being 45–50% of the weight for the biochars. This decrease in volatiles may be related to the decomposition of the biomass fraction (hemicellulose, cellulose and lignin), which occurs in a temperature range up to 400 °C, but also to the decomposition of some polymeric monomeric units with a higher degree of unsaturation [46].

The second curve can be divided into the depolymerization of biodegradable materials up to approximately 400 °C and degradation of less reactive and high molecular weight components [46,47]. The degradation of polystyrene (PS), polypropylene (PP) and polyethylene terephthalate (PET) occurs between 350 and 500 °C; this phenomenon can be observed in biochars produced at 400 °C, where the amount of organic matter present in the biochar is lower because they have already been previously degraded [48].

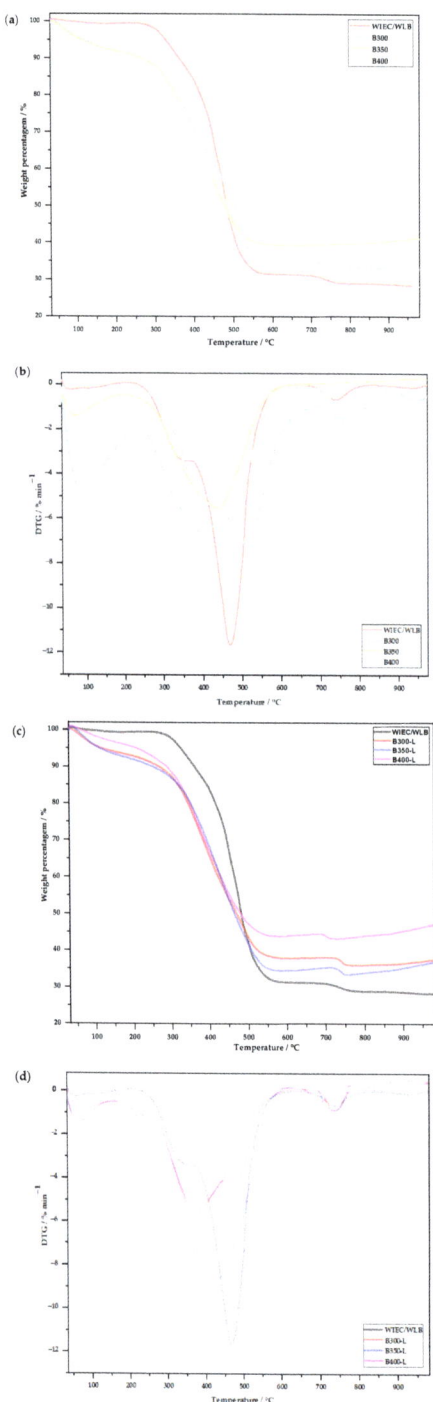

Figure 5. *Cont.*

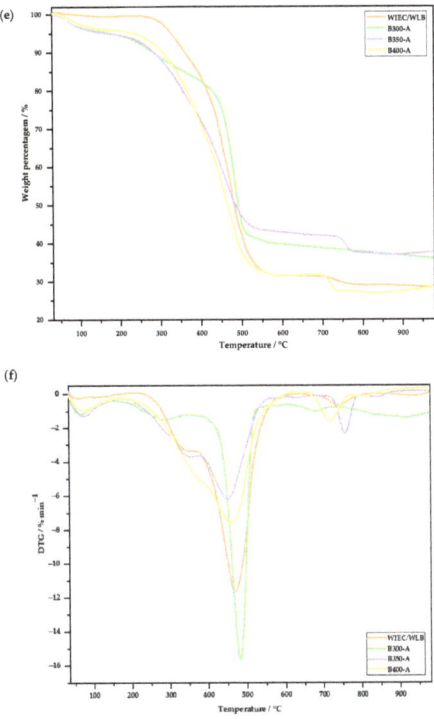

Figure 5. Thermogravimetric profiles. (**a**) TGA—raw biochars, (**b**) DTG—raw biochars, (**c**) TGA—washed biochars, (**d**) DTG—washed biochars, (**e**) TGA—activated biochars and (**f**) DTG—activated biochars.

The third peak, which started after 600 °C, corresponds to the decomposition of inorganic substances such as inorganic carbonate and lignin aromatic rings [49,50], varying up to 850 °C for B300, B350 and B400 biochars, and for biochars that have been washed and/or activated up to 800 °C.

3.2.3. Mass Yield, Ash Content, Chlorine Removal Potential and Apparent Density of the Biochars

Figure 6 shows the results for the mass yield, ash content, chlorine removal potential and apparent density for all the produced biochars (without treatment, washed and activated).

It is possible to observe that, when compared with the nontreated biochar samples, the washed and activated samples show greater mass yields, particularly samples B300-A, B350-A and B400-A.

For the ash content results, there was an increase in the ash content when the carbonization temperature was increased from 300 to 400 °C. When washing the biochar, there was a reduction of the ash content, and with the activation of biochars B300-A and B350-A, the ash content increased. For samples B400-L and B400-A, the differences between the ash contents after the different processes were not noticeable. The ash contents obtained in biochars produced at 300 and 350 °C were very similar and differed significantly from B400, with very close mass yields (70–75%).

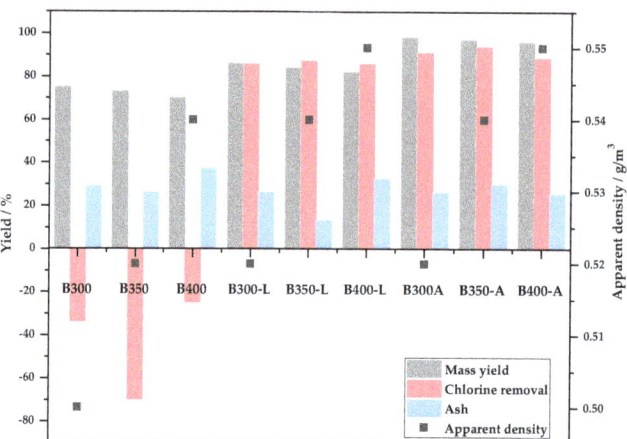

Figure 6. Mass yield, ash content, chlorine removal potential and bulk density for different biochars.

The chlorine removal potential for biochars B300, B350 and B400 were negative, which means that there was a concentration effect, caused by the mass loss, associated with the chlorine volatilization and, further, deposited onto the biochar surface. When the biochars were washed with hot water, the chlorine removal potential showed very significant values greater than 85%. This removal potential is an important indicator that demonstrates the further need to treat biochars that are produced from chlorine-containing wastes, since the chlorine content is extremely important in further thermal conversion processes. Excess chlorine is known not only for producing harmful emissions (such as HCl or PCDD/Fs) but also to cause equipment damage due to corrosion phenomena [51]. Activated biochars have shown to be very stable in terms of the mass yield, chlorine and ash removal. These biochars differ mainly regarding the apparent density, as the higher the production temperature, the greater the density of the biochar. Biochar samples B400-L and B400-A presented chlorine removal potentials of 86 and 89%, respectively, mass yields above 82% and ash contents of 33% for B400-L and 26% for B400-A, thus presenting the best carbonization and treatment conditions.

Overall, the apparent density of the produced biochars increased as the production temperature increased, varying between 0.50 and 0.54 g/m^2. However, as it was a measure with experimental observations, this difference was not very significant. For the biochars that were washed (B300-L, B350-L and B400-L) and for the activated ones (B300-A, B350-A and B400-A), the values for the apparent density when compared to the nontreated biochars at the same production temperature were the same.

3.2.4. Mineral Composition

Figure 7 shows the mineral compositions for the different biochars.

It is possible to observe that the concentrations of aluminum in the ashes of biochars B300, B300-L, B300-A, B400, B400-L and B400-A are decreasing. These results can be related to the fact that, with washing, the aluminum concentration was reduced and, with the activation process, the reduction compared to the initial concentration was, on average, 50%. Regarding the calcium in all washed biochars, it was possible to observe that, when the production temperature increased, the concentration of this element was lower, and when the biochars were activated, the calcium concentration had no significant variation.

The concentrations of magnesium and silica for the different biochars that did not undergo treatment (B300, B350 and B400) and for the biochars that were subjected to a washing process (B300-L, B350-L and B400-L) also showed no relevant variations. For the activated biochars (B300-A, B350-A and B400-A), the concentrations of magnesium, copper, potassium and sodium were similar and higher when compared to the other studied biochars.

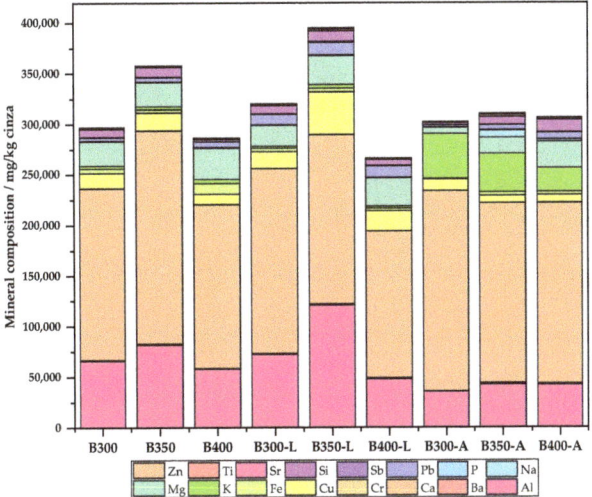

Figure 7. Mineral compositions of the produced biochars expressed in mg/kg of ash.

3.2.5. Fourier-Transform Infrared Spectroscopy

In a general observation of the FTIR spectra presented in Figure 8, it is possible to observe that there was a reduction in the intensity of the peaks between biochar samples B300, B350 and B400 for the biochars that were washed and activated.

The FTIR analysis provided us with an absorption spectrum of chemical clusters and not substances. Despite being an analysis that hardly identifies a substance, in the PP, PS and PVC spectra, it is possible to observe a higher concentration of chemical groups up to 1800 cm^{-1}. When polymeric materials undergo a thermochemical process, it is possible to observe that these chemical groups are no longer evident.

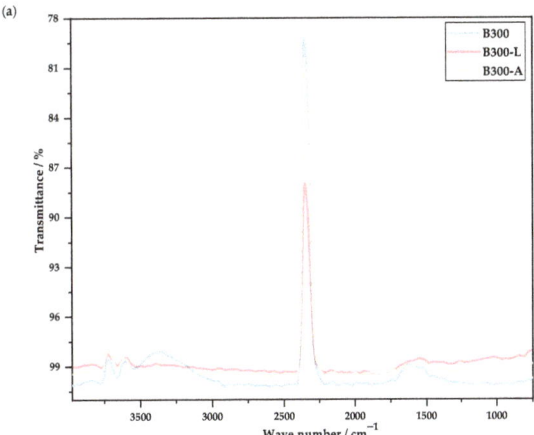

Figure 8. *Cont.*

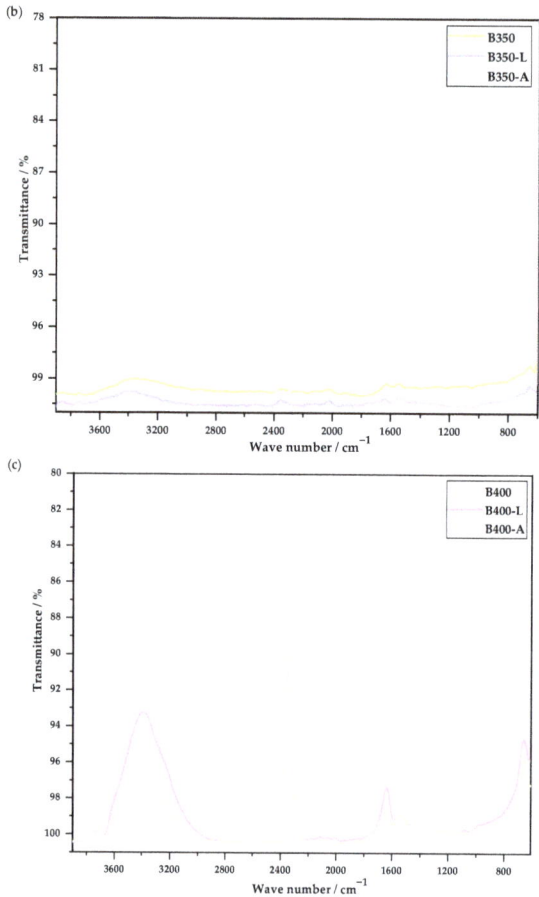

Figure 8. FTIR spectra: (**a**) biochars produced at 300 °C, (**b**) biochars produced at 350 °C and (**c**) biochars produced at 400 °C.

Except for B300-L and B350-A, it was possible to observe the presence of a peak at 3400 cm^{-1} that corresponded to the vibration of -OH elongation of the hydroxyl groups [52]. In biochars produced at temperatures of 350 and 400 °C, as the biochar underwent treatment and drying processes, the intensity of the peak decreased. According to Preston and Schmidt [53], the aromatic ring structure should be progressively formed with the increasing temperature. The high intensity and pronounced peak that we could observe at 2350–2340 cm^{-1} might occur due to the existence of atmospheric CO_2 and also due to the existence of some groups due to O = C = O stretching [54]. Another prominent intensity peak was 1600 cm^{-1}, which corresponds to the C = C stretching vibrations in aromatic compounds [55], and it was possible to verify a higher incidence as the biochar production temperature increased. In biochars B350-A and B400-A, it was possible to observe that a peak intensity appeared corresponding to the functional group and stretching C = C, which indicated that the projection of aromatic structures occurred [56]. The peak of 700–600 cm^{-1} corresponded to the C–Cl stretching vibration, and when washing and activation occurred, chlorine was absorbed or adsorbed in the case of activation [57].

3.2.6. Nitrogen Adsorption at 77 K

The removal of volatile matter produces a void within the biochar that leads to a large surface area, which is also dependent on the properties of the raw material used to produce the biochar. Biochar production based on a biomass develops pores and has been found to be very useful in land application, effluent purification and the removal of heavy metals and different chemicals [58–60]. A biochar with a high carbon content and a large surface area is also considered a potential electrode [61]. Figure 9 shows the results obtained for the surface area analysis of the produced biochars.

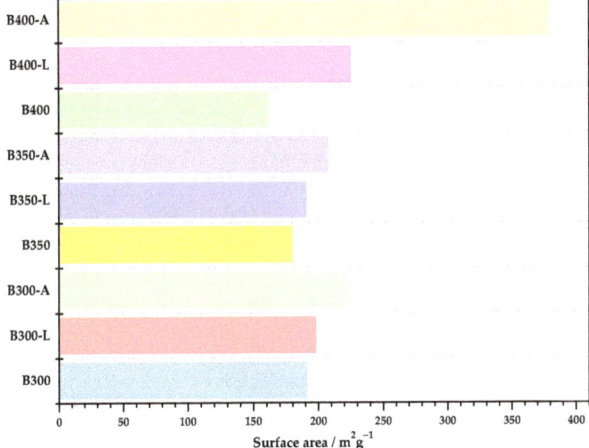

Figure 9. Surface area of all the biochar samples produced in this work.

Temperature is one of the main factors that affects the structure and physicochemical properties of biochar. Temperature affects the decomposition, formation and transformation of biomass and thus differs from the surface functionality of biochar [27].

In the tests carried out and presented in Figure 9, the increase in temperature did not significantly increase the surface area of biochars B300, B350 and B400. As the temperature increased, the surface area increased; however, it was possible to observe that, in the biochars produced without any type of treatment, there was a reduction in the surface area associated with a greater production of tars that were embedded on the biochar surface, and when they were washed, they increased the area due to the removal of the tars. When the biochars underwent the hot water washing process, the results showed an increase in surface area from 198 m^2g^{-1} to 226 m^2g^{-1} when the temperature was raised from 300 to 400 °C, respectively. Biochar samples B300-A and B350-A, when compared to the biochars without treatment, showed that the activation process did not significantly improve the area, with an increase corresponding to approximately 16%. Sample B400-A was the biochar that obtained the best surface area; when compared to sample B400, it had a 133% increase in surface area. These results indicated that the greater the removal of volatile matter, the better the surface area.

4. Conclusions

The characteristics of the biochars produced at temperatures of 300, 350 and 400 °C are strongly influenced by the production temperature and, subsequently, by washing and activation treatments. This was indicated by the different physicochemical properties that the biochars presented.

- The percentage of carbon present in the original feedstock and in the produced biochars were similar, differing mainly in the percentage of oxygen, which was lower, and in the ash, which increased as the temperature increased.

- The amount of volatile matter in the biochars was lower as the temperature of biochar production increased from 300 to 400 °C.
- The mass yield of biochars was not influenced by the temperature increase, ranging between 70 and 75%.
- The chlorine removal potential for biochars that were washed and activated was above 80%, demonstrating the efficiency of carbonization as a pretreatment for thermochemical processes to remove chlorinated compounds.
- In the FTIR analysis, it was possible to observe that there was a great difference between the spectra of the untreated biochars and the washed biochars, indicating the removal of compounds that were on the surface, such as chlorine. Biochar produced at 400 °C showed the lowest peaks after washing.
- In the analysis of the surfaces of the biochar samples, the differences between temperatures were more noticeable when the biochars were washed. When the activation process was carried out, the biochar samples produced at 300 and 350 °C were very similar, with the biochar produced at 400 °C having a higher surface area.
- Some results for the 350 °C biochars were not similar to the behaviors of the biochars produced at 300 and 400 °C, indicating that, when making the feedstock mixture, the amount of plastic and small metals may have been higher, thus making the carbonization process more difficult.

Author Contributions: Conceptualization, R.M.-P., M.M.G., P.B. and J.S.; Formal analysis, R.M.-P., A.A. and A.L.; Investigation, C.N.; Methodology, R.M.-P., A.A. and L.C.-C.; Project administration, P.B.; Resources, M.M.G., P.B. and J.S.; Supervision, J.S., M.M.G. and P.B.; Visualization, R.M.-P. and A.A.; Writing—original draft, R.M.-P. and C.N. and Writing—review and editing, R.M.-P., A.A., C.N., J.S., M.M.G. and P.B. All authors have read and agreed to the published version of the manuscript.

Funding: This work was supported by national funds through the FCT—Fundação para a Ciência e a Tecnologia within the R&D Units VALORIZA (UIDB/05064/2020) and MEtRICs (UIDB/04077/2020-2023) and project ALT20-03-0145-FEDER-039485 - SynDiesel: Fuels for diesel engines from dedicated waste and crop thermal gasification, co-financed by the ERDF through the Regional Operational Program of Alentejo.

Data Availability Statement: The data presented in this study are available on request from the corresponding author.

Conflicts of Interest: The authors declare no conflict of interest.

References

1. Liu, C.; Wang, H.; Karim, A.M.; Sun, J.; Wang, Y. Catalytic Fast Pyrolysis of Lignocellulosic Biomass. *Chem. Soc. Rev.* **2014**, *43*, 7594–7623. [CrossRef]
2. Reen, S.; Chyuan, H.; Wayne, K.; Loke, P.; Phang, S.; Chuan, T.; Nagarajan, D.; Lee, D. Sustainable Approaches for Algae Utilisation in Bioenergy Production. *Renew. Energy* **2018**, *129*, 838–852. [CrossRef]
3. Hoang, A.T.; Nizetic, S.; Ong, H.C.; Mofijur, M.; Ahmed, S.F.; Ashok, B.; Bui, V.T.V.; Chau, M.Q. Insight into the Recent Advances of Microwave Pretreatment Technologies for the Conversion of Lignocellulosic Biomass into Sustainable Biofuel. *Chemosphere* **2021**, *281*, 130878. [CrossRef] [PubMed]
4. Bhutto, A.W.; Qureshi, K.; Abro, R.; Harijan, K.; Zhao, Z.; Bazmi, A.A.; Abbas, T.; Yu, G. Progress in Production of Biomass-to-Liquid Biofuels to Decarbonize Transport Sector-Prospectus and Challenges. *RSC Adv.* **2016**, *6*, 32140–32170. [CrossRef]
5. Tuan, A.; Pham, V.V. 2-Methylfuran (MF) as a Potential Biofuel: A Thorough Review on the Production Pathway from Biomass, Combustion Progress, and Application in Engines. *Renew. Sustain. Energy Rev.* **2021**, *148*, 111265. [CrossRef]
6. Forti, V.; Balde, C.P.; Kuehr, R.; Bel, G. *The Global E-Waste Monitor 2020: Quantities, Flows and the Circular Economy Potential*; United Nations University/United Nations Institute for Training and Research, International Telecommunication Union, and International Solid Waste Association: Geneva, Switzerland, 2020; Volume 3, pp. 1–120.
7. Bhaskar, K.; Kumar, B. Electronic Waste Management and Sustainable Development Goals: Is There a Business Case for Linking the Two? *J. Indian Bus. Res.* **2019**, *11*, 120–137. [CrossRef]
8. Khetriwal, D.S.; Kraeuchi, P.; Widmer, R. Producer Responsibility for E-Waste Management: Key Issues for Consideration—Learning from the Swiss Experience. *J. Environ. Manag.* **2009**, *90*, 153–165. [CrossRef]
9. Widmer, R.; Oswald-Krapf, H.; Sinha-Khetriwal, D.; Schnellmann, M.; Böni, H. Global Perspectives on E-Waste. *Environ. Impact Assess. Rev.* **2005**, *25*, 436–458. [CrossRef]

10. Andrade, D.F.; Castro, J.P.; Garcia, J.A.; Machado, R.C.; Pereira-Filho, E.R.; Amarasiriwardena, D. Analytical and Reclamation Technologies for Identification and Recycling of Precious Materials from Waste Computer and Mobile Phones. *Chemosphere* **2022**, *286*, 131739. [CrossRef]
11. Perkins, D.N.; Brune Drisse, M.N.; Nxele, T.; Sly, P.D. E-Waste: A Global Hazard. *Ann. Glob. Health* **2014**, *80*, 286–295. [CrossRef]
12. Forti, V.; Baldé, K.; Kuehr, R. *E-Waste Statistics: Guidelines on Classifications, Reporting and Indicators*; United Nations University: Bonn, Germany, 2018; ISBN 9789280890662.
13. Ilankoon, I.M.S.K.; Ghorbani, Y.; Nan, M.; Herath, G.; Moyo, T. E-Waste in the International Context—A Review of Trade Flows, Regulations, Hazards, Waste Management Strategies and Technologies for Value Recovery. *Waste Manag.* **2018**, *82*, 258–275. [CrossRef] [PubMed]
14. Zeng, X.; Ali, S.H.; Tian, J.; Li, J. Mapping Anthropogenic Mineral Generation in China and Its Implications for a Circular Economy. *Nat. Commun.* **2020**, *11*, 1544. [CrossRef]
15. Wu, X.; Li, J.; Yao, L.; Xu, Z. Auto-Sorting Commonly Recovered Plastics from Waste Household Appliances and Electronics Using near-Infrared Spectroscopy. *J. Clean. Prod.* **2020**, *246*, 118732. [CrossRef]
16. Martinho, G.; Pires, A.; Saraiva, L.; Ribeiro, R. Composition of Plastics from Waste Electrical and Electronic Equipment (WEEE) by Direct Sampling. *Waste Manag.* **2012**, *32*, 1213–1217. [CrossRef] [PubMed]
17. Liu, X.; Lu, X.; Feng, Y.; Zhang, L.; Yuan, Z. Recycled WEEE Plastics in China: Generation Trend and Environmental Impacts. *Resour. Conserv. Recycl.* **2022**, *177*, 105978. [CrossRef]
18. Kumar, G.; Dharmaraja, J.; Arvindnarayan, S.; Shoban, S. A Comprehensive Review on Thermochemical, Biological, Biochemical and Hybrid Conversion Methods of Bio-Derived Lignocellulosic Molecules into Renewable Fuels. *Fuel* **2019**, *251*, 352–367. [CrossRef]
19. Chatterjee, C.; Pong, F.; Sen, A. Chemical Conversion Pathways for Carbohydrates. *Green Chem.* **2015**, *17*, 40–71. [CrossRef]
20. Gong, J.; Chen, X.; Tang, T. Recent Progress in Controlled Carbonization of (Waste) Polymers. *Prog. Polym. Sci.* **2019**, *94*, 1–32. [CrossRef]
21. Taherymoosavi, S.; Verheyen, V.; Munroe, P.; Joseph, S.; Reynolds, A. Characterization of Organic Compounds in Biochars Derived from Municipal Solid Waste. *Waste Manag.* **2017**, *67*, 131–142. [CrossRef]
22. Huang, J.; Qiao, Y.; Wei, X.; Zhou, J.; Yu, Y.; Xu, M. Effect of Torrefaction on Steam Gasification of Starchy Food Waste. *Fuel* **2019**, *253*, 1556–1564. [CrossRef]
23. Zheng, A.; Fan, Y.; Wei, G.; Zhao, K.; Huang, Z.; Zhao, Z.; Li, H. Chemical Looping Gasification of Torrefied Biomass Using NiFe2O4 as an Oxygen Carrier for Syngas Production and Tar Removal. *Energy Fuels* **2020**, *34*, 6008–6019. [CrossRef]
24. Laird, D.A. The Charcoal Vision: A Win-Win-Win Scenario for Simultaneously Producing Bioenergy, Permanently Sequestering Carbon, While Improving Soil and Water Quality. *Agron. J.* **2008**, *100*, 178–181. [CrossRef]
25. Brown, R.C.; Amonette, J.E.; National, N. Review of the Pyrolysis Platform for Coproducing Bio-Oil and Biochar. *Biofuels Bioprod. Biorefining* **2009**, *3*, 547–562.
26. Brewer, C.E.; Unger, R.; Schmidt-rohr, K.; Brown, R.C. Criteria to Select Biochars for Field Studies Based on Biochar Chemical Properties. *BioEnergy Res.* **2011**, *4*, 312–323. [CrossRef]
27. Kumar, A.; Saini, K.; Bhaskar, T. Bioresource Technology Hydochar and Biochar: Production, Physicochemical Properties and Techno- Economic Analysis. *Bioresour. Technol.* **2020**, *310*, 123442. [CrossRef]
28. Guo, S.; Dong, X.; Wu, T.; Zhu, C. Influence of Reaction Conditions and Feedstock on Hydrochar Properties. *Energy Convers. Manag.* **2016**, *123*, 95–103. [CrossRef]
29. Leng, L.; Huang, H. An Overview of the Effect of Pyrolysis Process Parameters on Biochar Stability. *Bioresour. Technol.* **2018**, *270*, 627–642. [CrossRef]
30. Dai, L.; Wang, Y.; Liu, Y.; He, C.; Ruan, R.; Yu, Z.; Jiang, L. A Review on Selective Production of Value-Added Chemicals via Catalytic Pyrolysis of Lignocellulosic Biomass. *Sci. Total Environ.* **2020**, *749*, 142386. [CrossRef]
31. Atienza-martínez, M.; Gea, G.; Plaza, D.; Lu, F. Pyrolysis of Cashew Nutshells: Characterization of Products and Energy Balance. *Energy* **2018**, *158*, 72–80. [CrossRef]
32. Mota-Panizio, R.; Hermoso-Orzáez, M.J.; Carmo-Calado, L.; Calado, H.; Goncalves, M.M.; Brito, P. Co-Carbonization of a Mixture of Waste Insulation Electric Cables (WIEC) and Lignocellulosic Waste, for the Removal of Chlorine: Biochar Properties and Their Behaviors. *Fuel* **2022**, *320*, 123932. [CrossRef]
33. Liu, Z. The Fate of Fluorine and Chlorine during Thermal Treatment of Coals. *Environ. Sci. Technol.* **2006**, *40*, 7886–7889.
34. Longo, A.; Nobre, C.; Sen, A.; Panizio, R.; Brito, P.; Gonçalves, M. Torrefaction Upgrading of Heterogenous Wastes Containing Cork and Chlorinated Polymers. *Environments* **2022**, *9*, 99. [CrossRef]
35. Yang, W.; Wang, H.; Zhang, M.; Zhu, J.; Zhou, J.; Wu, S. Fuel Properties and Combustion Kinetics of Hydrochar Prepared by Hydrothermal Carbonization of Bamboo. *Bioresour. Technol.* **2016**, *205*, 199–204. [CrossRef] [PubMed]
36. Xuan, Y.; Hua, Y.; Mubarak, N.M.; Kansedo, J.; Khalid, M.; Lokman, M.; Ghasemi, M. A Review on Biochar Production from Different Biomass Wastes by Recent Carbonization Technologies and Its Sustainable Applications. *J. Environ. Chem. Eng.* **2022**, *10*, 107017. [CrossRef]
37. Funke, A.; Ziegler, F.; Berlin, T.U. Hydrothermal Carbonization of Biomass: A Summary and Discussion of Chemical Mechanisms for Process Engineering. *Biofuels Bioprod. Biorefining* **2010**, *4*, 160–177. [CrossRef]

38. Zimmerman, A.R.; Hall, W.; Box, P.O. Abiotic and Microbial Oxidation of Laboratory-Produced Black Carbon (Biochar). *Environ. Sci. Technol.* **2010**, *44*, 1295–1301. [CrossRef]
39. Manyà, J.J.; Ortigosa, M.A.; Laguarta, S.; Manso, J.A. Experimental Study on the Effect of Pyrolysis Pressure, Peak Temperature, and Particle Size on the Potential Stability of Vine Shoots-Derived Biochar. *Fuel* **2014**, *133*, 163–172. [CrossRef]
40. Spokas, K.A. Review of the Stability of Biochar in Soils: Predictability of O: C Molar Ratios Review of the Stability of Biochar in Soils: Predictability of O:C Molar Ratios. *Carbon Manag.* **2014**, *3004*, 289–303. [CrossRef]
41. Budai, A.; Calucci, L.; Rasse, D.P.; Tau, L.; Pengerud, A.; Wiedemeier, D.; Abiven, S.; Forte, C. Effects of Pyrolysis Conditions on Miscanthus and Corncob Chars: Characterization by IR, Solid State NMR and BPCA Analysis. *J. Anal. Appl. Pyrolysis* **2017**, *128*, 335–345. [CrossRef]
42. Kuhlbusch, T.A.J. Method for Determining Black Carbon in Residues of Vegetation Fires. *Environ. Sci. Technol.* **1995**, *29*, 2695–2702. [CrossRef]
43. Yasuhara, A.; Amano, Y.; Shibamoto, T. Investigation of the Self-Heating and Spontaneous Ignition of Refuse-Derived Fuel (RDF) during Storage. *Waste Manag.* **2010**, *30*, 1161–1164. [CrossRef] [PubMed]
44. Sajjad, M.; Aamer, M.; Taha, S.; Taqvi, H. Pyrolysis, Kinetics Analysis, Thermodynamics Parameters and Reaction Mechanism of Typha Latifolia to Evaluate Its Bioenergy Potential. *Bioresour. Technol.* **2017**, *245*, 491–501. [CrossRef]
45. Zornoza, R.; Moreno-Barriga, F.; Acosta, J.A.; Muñoz, M.A.; Faz, A. Stability, Nutrient Availability and Hydrophobicity of Biochars Derived from Manure, Crop Residues, and Municipal Solid Waste for Their Use as Soil Amendments. *Chemosph. J.* **2016**, *144*, 122–130. [CrossRef] [PubMed]
46. Kan, T.; Strezov, V.; Evans, T. Effect of the Heating Rate on the Thermochemical Behavior and Biofuel Properties of Sewage Sludge Pyrolysis. *Eenrgy Fuels* **2016**, *30*, 1564–1570. [CrossRef]
47. Folgueras, M.B.; Alonso, M.; Díaz, R.M. Influence of Sewage Sludge Treatment on Pyrolysis and Combustion of Dry Sludge. *Energy* **2013**, *55*, 426–435. [CrossRef]
48. López, A.; De Marco, I.; Caballero, B.M.; Laresgoiti, M.F.; Adrados, A. Influence of Time and Temperature on Pyrolysis of Plastic Wastes in a Semi-Batch Reactor. *Chem. Eng. J.* **2011**, *173*, 62–71. [CrossRef]
49. Yang, H.; Yan, R.; Chen, H.; Lee, D.H.; Zheng, C. Characteristics of Hemicellulose, Cellulose and Lignin Pyrolysis. *Fuel* **2007**, *86*, 1781–1788. [CrossRef]
50. Zaker, A.; Chen, Z.; Zaheer-uddin, M.; Guo, J. Co-Pyrolysis of Sewage Sludge and Low-Density Polyethylene—A Thermogravimetric Study of Thermo-Kinetics and Thermodynamic Parameters. *J. Environ. Chem. Eng.* **2021**, *9*, 104554. [CrossRef]
51. Nobre, C.; Vilarinho, C.; Alves, O.; Mendes, B.; Gonçalves, M. Upgrading of Refuse Derived Fuel through Torrefaction and Carbonization: Evaluation of RDF Char Fuel Properties. *Energy* **2019**, *181*, 66–76. [CrossRef]
52. Movasaghi, Z.; Rehman, S.; Rehman, I. Fourier Transform Infrared (FTIR) Spectroscopy of Biological Tissues. *Appl. Spectrosc. Rev.* **2008**, *43*, 134–179. [CrossRef]
53. Preston, C.M.; Schmidt, M.W.I. Black (Pyrogenic) Carbon: A Synthesis of Current Knowledge and Uncertainties with Special Consideration of Boreal Regions. *Biogeosciences* **2006**, *3*, 397–420. [CrossRef]
54. IR Spectrum Table & Chart. Available online: https://www.sigmaaldrich.com/technical-documents/articles/biology/ir-spectrum-table.html#ir-spectrum-table-by-range (accessed on 5 May 2023).
55. Destainville, A.; Champion, E.; Laborde, E. Synthesis, Characterization and Thermal Behavior of Apatitic Tricalcium Phosphate. *Mater. Chem. Phys.* **2003**, *80*, 269–277. [CrossRef]
56. Tehreem, S.; Yousra, M.; Alamer, K.H.; Alsudays, I.M.; Sarwar, S.; Kamal, A.; Naeem, S. Analysis of the Role of Various Biochar in the Remediation of Heavy Metals in Contaminated Water and Its Kinetics Study. *J. Saudi Chem. Soc.* **2022**, *26*, 101518. [CrossRef]
57. Kakuta, Y. Study on Chlorine Removal from Mixture of Waste Plastics. *Waste Manag.* **2008**, *28*, 615–621. [CrossRef] [PubMed]
58. Galinato, S.P.; Yoder, J.K.; Granatstein, D. The Economic Value of Biochar in Crop Production and Carbon Sequestration. *Energy Policy* **2011**, *39*, 6344–6350. [CrossRef]
59. Di Natale, F.; Erto, A.; Lancia, A. Desorption of Arsenic from Exhaust Activated Carbons Used for Water Purification. *J. Hazard. Mater.* **2013**, *260*, 451–458. [CrossRef]
60. Mubarak, N.M.; Alicia, R.F.; Abdullah, E.C.; Sahu, J.N.; Haslija, A.B.A.; Tan, J. Statistical Optimization and Kinetic Studies on Removal of Zn 2 + Using Functionalized Carbon Nanotubes and Magnetic Biochar. *J. Environ. Chem. Eng.* **2013**, *1*, 486–495. [CrossRef]
61. Bernard, B.; He, X.; Wang, S.; Abomohra, A.E.; Hu, Y.; Wang, Q. Co-Pyrolysis of Biomass and Waste Plastics as a Thermochemical Conversion Technology for High-Grade Biofuel Production: Recent Progress and Future Directions Elsewhere Worldwide. *Energy Convers. Manag.* **2018**, *163*, 468–492. [CrossRef]

Disclaimer/Publisher's Note: The statements, opinions and data contained in all publications are solely those of the individual author(s) and contributor(s) and not of MDPI and/or the editor(s). MDPI and/or the editor(s) disclaim responsibility for any injury to people or property resulting from any ideas, methods, instructions or products referred to in the content.

From Waste to Resource: Utilizing Sweet Chestnut Waste to Produce Hydrothermal Carbon for Water Decontamination

Silvia Izquierdo [1,*], Nazaret Pacheco [1], Carlos J. Durán-Valle [2] and Ignacio M. López-Coca [3,*]

[1] Environmental and Sustainable Chemistry Research Group, School of Technology, University of Extremadura, 10003 Cáceres, Spain

[2] Environmental and Sustainable Chemistry Research Group, IACYS, Faculty of Sciences, University of Extremadura, 06006 Badajoz, Spain

[3] Environmental and Sustainable Chemistry Research Group, INTERRA, School of Technology, University of Extremadura, 10003 Cáceres, Spain

* Correspondence: sizquierdo@unex.es (S.I.); iglomar@unex.es (I.M.L.-C.)

Abstract: Carbonaceous materials are a highly appealing class of adsorbents, owing to their exceptional properties, such as high surface area and thermal and chemical stability. These materials have found successful applications in water purification. Sweet chestnut (*Castanea sativa*) cupules are disposed of as waste. Valorization of these residues is a step forward in terms of circular economy and sustainability. Meanwhile, per- and poly-fluoroalkyl substances (PFASs) pose significant concerns due to their persistence, bioaccumulation, and toxicity, emerging as contaminants of concern for human health and the environment. This study focuses on preparing carbonaceous material by hydrothermal carbonization from chestnut cupules, followed by their use as adsorbents for PFAS removal from polluted water. The cupule waste material was crushed, ground, sieved, and subjected to hydrothermal treatment at temperatures ranging from 180–200 °C to produce hydrothermal carbons. The adsorbents obtained were characterized by various techniques such as nitrogen adsorption isotherm, porosimetry, point of zero charge, Fourier-transform infrared, scanning electron microscopy, and thermal, elemental, and energy dispersive X-ray analyses. Surface area (S_{BET}) values of 42.3–53.2 m^2·g^{-1} were obtained; pH$_{PZC}$ ranged from 3.8 to 4.8. This study also determined the adsorption kinetics and isotherms for removing perfluorooctanoate-contaminated water. The equilibrium was established at 72 h and q_e = 1029.47 mg·g^{-1}. To summarize, this research successfully valorized a biomass residue by transforming it into hydrothermal carbon, which was then utilized as an adsorbent for water decontamination.

Keywords: carbonaceous material; biochar; hydrothermal carbon; *Castanea sativa*; biomass residues; residue valorization; per- and poly-fluoroalkyl substances (PFAS); water decontamination

1. Introduction

Carbonaceous materials constitute an attractive group of adsorbents due to their properties, such as large surface area and thermal and chemical stability. In addition, activated carbons with different pore sizes can be obtained, thus adapting them to the functions they are designed for. On the other hand, functionalizing the surface of activated carbons is possible, giving them specific chemical properties adapted to a particular purpose. It is also of great importance that carbonaceous materials are considered safe, accessible, and affordable [1]. Various types of biomass residues, due to their low cost and abundance, have been evaluated to obtain functional carbonaceous materials. Many methods provide materials whose physicochemical properties can be tuned at will, thus obtaining materials that have shown promising results in different fields, such as adsorption or energy storage [2]. For this reason, they are considered suitable candidates for obtaining sustainable materials. These materials have been successfully applied in catalysis, water purification, and soil remediation since they can remove heavy metals, agrochemicals, antibiotics, and other organic compounds. Excellent reviews on these aspects can be found in Soffian et al. [3], Adegoke et al. [4], Gęca et al. [5], Borchardt et al. [6], and Ma et al. [2]. One such type of carbonaceous material is biochar, which can be produced from biomass waste through hydrothermal carbonization, chemical carbonization, gasification, torrefaction, and pyrolysis [3,7–9].

Sweet chestnut (*Castanea sativa*) is a deciduous tree of the Fagaceae family, native to humid temperate regions of the Northern Hemisphere and widely distributed throughout Europe. It can reach 35 m in height and 2 m (exceptionally more) in diameter. Sweet chestnut trees live to an age of 600 years or even more. By autumn, the female flowers develop into spiny cupules containing 3–7 brownish nuts, which are shed in October [10–12]. After the chestnuts are harvested, the spiny dome is disposed of as waste and used as fuel for fireplaces.

On the other hand, per- and poly-fluoroalkyl substances (PFASs) are a group of synthetic chemicals that have been used in a wide range of industrial and consumer products, such as firefighting foams, non-stick coatings, and water-repellent fabrics [13,14]. Due to their persistence, bioaccumulation, and toxicity, PFASs are considered Contaminants of Emerging Concern (CECs), with potentially adverse effects on human health and the environment [15–21]. These substances have been detected in various environmental media, including water, soil, air, and biota, and have attracted increasing attention from regulators, scientists, and the public worldwide [22]. Among the different methods for treating PFAS-contaminated water, adsorption has been recognized as a promising technology, given its effectiveness, versatility, and ease of implementation [23,24]. Carbonaceous materials are widely used as adsorbents for PFAS removal due to their high surface area, microporous structure, and affinity for hydrophobic compounds. One of the classic and best-known carbonaceous materials is activated carbon (AC) [25–27]. It is usually obtained at high temperatures in a double carbonization and activation process. This material is characterized by a highly developed porous structure, a generally alkaline character, and a predominantly hydrophobic surface [28,29]. It is well known that these properties can be modified by adjusting the conditions of the activation processes or employing suitable chemical or thermal treatment [30,31].

In recent years, numerous studies have investigated using these adsorbents for PFAS adsorption in water, aiming to optimize the adsorbent properties and conditions, understand the adsorption mechanisms, and evaluate the performance and feasibility of the technology. The effectiveness of AC as an adsorbent for PFAS removal depends on its physicochemical properties, which can be tailored by adjusting the activation process and conditions. The surface area, pore size, surface chemistry, and surface charge of AC determine its adsorption capacity, selectivity, and kinetics for PFAS removal. The larger the surface area and the more mesopores present, the higher the adsorption capacity; this is due to the increased availability of adsorption sites and the enhanced accessibility of PFAS molecules to the internal surface of the carbonaceous material. The surface chemistry can also affect PFAS adsorption, as hydrophobic interactions between PFASs and the nonpolar surface are favored over electrostatic interactions with the polar surface of water molecules. Therefore, adsorbents with a higher content of surface functional groups, such as oxygen-containing or nitrogen-containing groups, may have a lower adsorption capacity for PFASs than non-functionalized ones due to the reduced hydrophobicity and the increased competition for adsorption sites.

In the last decade or so, a new type of carbonaceous material, hydrothermal carbons (HC), has attracted considerable interest from researchers in this field [32,33]. It is obtained by treating carbon-rich materials with water at a moderate temperature in a closed vessel so that endogenous water vapor pressure acts on this material [34]. Generally, some biomass is used as raw material. This type of material is characterized by low pore development, acidity, and a hydrophilic surface, with many surface groups. As in the previous case, these properties can be modified by thermal or chemical treatments. This has the advantage of being a more environmentally friendly manufacturing process, as it is carried out in hermetically sealed containers and at a lower temperature. This allows for easier management of possible waste and lower energy consumption. Additionally, despite its limited specific surface area, its high proportion of functional groups can give rise to adequate adsorption capacity. We have been working on the obtention and use of hydrothermal carbons from chestnut cupules for the last few years. Our preliminary results (unpublished) show that these carbons are suitable adsorbents in the decontamination of heavy metals, such as thorium, from water. Given these results, we thought it would be apt to check for the adsorption capacity of apolar organic molecules such as PFAS. Adsorption onto carbonaceous materials has been established as a promising technology for PFAS removal from water [20,35–42].

The objective of this work is to investigate the use of chestnut cupule waste as sustainable biomass raw material to obtain hydrothermal carbons—hence, adding value to it—and its use as adsorbent material for the removal from water of sodium perfluorooctanoate (PFONa), the anion of perfluorooctanoic acid, one of the most significant PFASs.

2. Materials and Methods

2.1. Materials

Sodium perfluorooctanoate 97%, CAS 335-95-5 (Figure 1), was purchased from Thermo Fisher. Milli-Q water was used as a solvent in all determinations. Sweet chestnut (*Castanea sativa*) cupules were collected from chestnut groves in Valle del Jerte, in Cáceres, Spain.

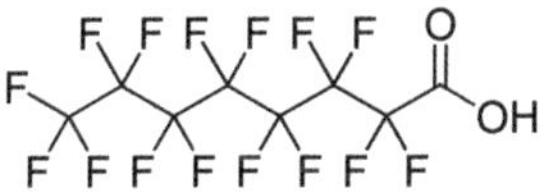

Figure 1. Structural formula of sodium perfluorooctanoate.

2.2. Methods

2.2.1. Synthesis of the Hydrothermal Carbons

The chestnut cupules were allowed to dry naturally. They were subsequently ground and sieved, collecting three fractions: large size (1–2 mm), medium size (0.5–1 mm), and small size (less than 0.5 mm). Then, they were subjected to hydrothermal treatment in a duralumin-coated Teflon™ reactor.

To obtain HC-A, in a hydrothermal reactor, 100 mL of distilled water and 15 g of small-size sieved cupule were added. The reactor was closed, placed in an oven previously programmed at 180 °C, and kept for 24 h. The reactor was then removed from the oven and, once cold, opened. The final product was filtered and washed with plenty of distilled water (ca. 200 mL) on a filter. Subsequently, it was dried in an oven at 110 °C for 8 h. The other carbons were prepared similarly, according to the data in Table 1.

Table 1. Operating conditions for the preparation of hydrothermal carbons.

Carbon	Raw Material		Oven Temperature (°C)
	Particle Size (mm)	Mass (g)	
HC-A	<0.5	15	180
HC-B	0.5–1	5	200
HC-C	1–2	15	200

2.2.2. Characterization of the Hydrothermal Carbons

Textural characterization was performed by N_2 adsorption isotherms at 77 K in a Quantachrome Quadrasorb Evo (Quantachrome Ltd., Hook, UK), and the specific surface area was calculated by applying the BET method [43]. Porosimetry was performed in a Quantachrome Poremaster 60 (Quantachrome, Hook, UK). Thermal analysis was carried out in an STA 449 F3 Jupiter thermobalance (Netzsch, Selb, Bavaria, Germany).

Fourier-transform infrared spectra were recorded in a Bruker Vertex70 spectrometer in the 400–4000 cm^{-1} range, with a DLaTGS detector and a 0.4 cm^{-1} spectral resolution. A KBr tablet was prepared with 1/250 dilution (Bruker, Billerica, Massachusetts, USA). Elemental analysis (C, H, N, S, O) was performed using a Leco CHMS-932 elemental analyzer (Leco, St. Joseph, MI, USA). C, H, N, and S were analyzed, and the difference was assigned to ash (measured from proximate analysis) and oxygen content. The point of zero charge (PZC) values were determined using the method proposed by Valente Nabais and Carrott [44]. To carry out this measurement, a 0.1 M solution of sodium nitrate was prepared, and 7% in weight of the carbon was added, followed by stirring for 48 h at 25 °C in a thermostatic bath. Subsequently, it was filtered, and the pH of the filtered solution was measured with a pH meter. This pH value corresponded to the point of zero charge. A field emission scanning electron microscope (SEM) Quanta 3D FEG (FEI Company, Hillsboro, OR, USA) was used to explore the surface morphological characteristics of all samples. The sample analysis was performed under high vacuum with a secondary electron. The surface of the samples was coated with gold due to their low conductivity. Energy dispersive X-ray analyses (EDX) were recorded with the same equipment.

2.2.3. Adsorption Experiments

The conjugate base sodium salt (PFONa), instead of the perfluorooctanoic acid, was used to carry out the adsorption experiments, since the acid has a K$_a$ ca. -1 and dissociates completely in water.

1. Adsorption kinetics

To determine the PFONa adsorption over time, the following procedure was used:
A 200 mg/L PFONa water solution was prepared by dissolving 10 mg of PFONa in a 50 mL volumetric flask. It was then transferred to an Erlenmeyer flask, and 50 mg of hydrothermal carbon (HC) was added, so the HC to PFONa ratio was 5:1. All flasks were thermostated at 25 °C and magnetically stirred at 400 rpm. Assays were performed in duplicate; 2 mL samples were taken at 2, 4, 24, 48, 72, and 96 h, then filtered with a 0.22 µm pore filter, and, finally, aliquots were analyzed for PFONa concentration.

2. Adsorption isotherm

Different PFONa solutions of 40, 70, 100, 130, 165, and 200 mg/L were prepared in 10 mL volumetric flasks. They were then transferred to separate Erlenmeyer flasks, and 10 mg of hydrothermal carbon was added to each one, so the HC to PFONa mass quotient was 25.0, 14.3, 10.0, 7.7, 6.1, and 5.0, respectively. All flasks were thermostated at 25 °C and magnetically stirred at 400 rpm. Assays were performed in duplicate; 2 mL samples were taken at 96 h, then filtered with a 0.22 µm pore filter, and, finally, aliquots were analyzed for PFONa concentration.

2.2.4. Analytical Determinations

The samples were analyzed using liquid chromatography (Agilent 1290 Infinity II LC (Agilent Technologies, Santa Clara, California, USA)) coupled to a triple quadrupole mass spectrometer (Agilent 6460 triple quad LC/MS (Agilent Technologies)). Aliquots of 3 µL were injected, and elution was carried out by operating in gradient mode at 0.4 mL/min with a reverse phase HPLC Zorbax Extend C18 column (3 µm, 100 mm × 1.8 mm) at 30 °C. As the mobile phase, 0.1% formic acid in ultrapure water (A) and 0.1% formic acid in acetonitrile (B) were used, as shown in Table 2.

Ionization was carried out in negative mode. The equipment parameters were gas temperature 225 °C, gas flow 10 L/min, nebulizer 45 psi, sheath gas temperature 350 °C, sheath gas flow 11 L/min, capillary voltage 3.6 kV, nozzle voltage 1500 V, and delta EMV 400. The MassHunter B.07.00 (Agilent) software was used for data acquisition and equipment control. Multiple Reaction Monitoring (MRM) was used, with the transitions sought to be 412.9 → 368.9 and 412.9 → 169, the former used in the quantification of the compound. The fragmentation and collision energy voltages were 86 V and 5 V, respectively.

Table 2. Mobile phase composition.

t, min	A, %	B, %
0	80	20
9	4	96
10	0	100
11	0	100
12	80	20
15	80	20

3. Results and Discussion

3.1. Characterization of the Hydrothermal Carbons

The nitrogen adsorption isotherms (Figure 2) corresponded to type III, according to the IUPAC classification [45,46]; therefore, the interaction between the N_2 molecules and the surface must have been weak. The surface obtained by the BET method produced values that were higher than those obtained by the DFT method but in the same order of magnitude (Table 3) [43,47]. Nevertheless, since the isotherms were of Type III, the BET model was not to be considered highly reliable. All three carbons showed a minimal surface area, low microporosity, and a considerably greater volume of large pores, i.e., mesopores and macropores (Figures 3–5). The HC-A coal had the least-developed porosity. The DFT method was used to study the distribution of micropores and mesopores. The results show that the three carbons presented a similar pore distribution. A high volume of narrow mesopores was detected, whereas no micropores were present. The HC-B carbon had a larger volume of narrow mesopores, while the HC-C showed more volume when the pores widened. The HC-A carbon showed a more significant development of wide porosity than the other two adsorbents. It should be noted that HC-B and HC-C carbons showed a more developed narrow mesoporosity than HC-A. However, the mesopore and macropore structure was very similar in the three materials.

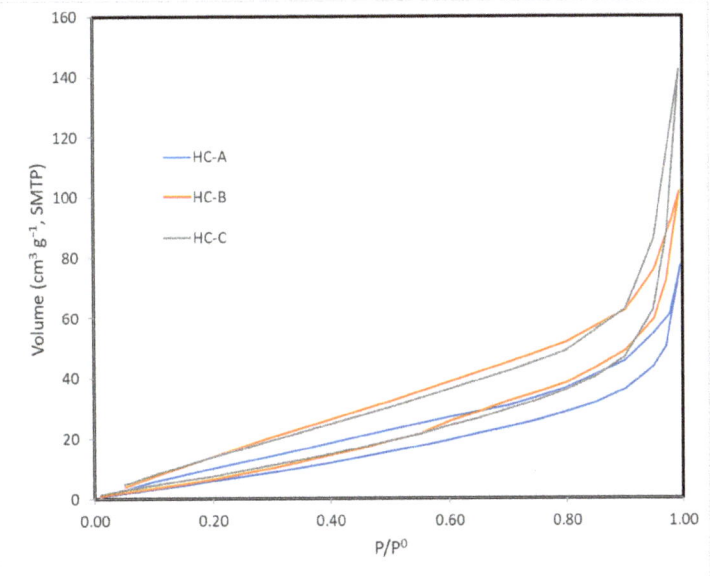

Figure 2. Isotherms of N_2 adsorption at 77 K.

Table 3. Porosity and specific surface area of carbons.

Material	BET [1]	DR [2]	Porosimetry	DFT [3]	
	S, $m^2\ g^{-1}$	V_{micro}, $cm^3\ g^{-1}$	V_{meso}, $cm^3\ g^{-1}$	S, $m^2\ g^{-1}$	V_{total}, $cm^3\ g^{-1}$
HC-A	42.3	0.007	0.078	27.68	0.085
HC-B	53.2	0.007	0.061	39.13	0.122
HC-B	53.2	0.007	0.061	39.13	0.122

[1] Brunauer, Emmett, and Teller model [43]. [2] Dubinin and Radushkevich model [48,49]. [3] Density Functional Theory [47].

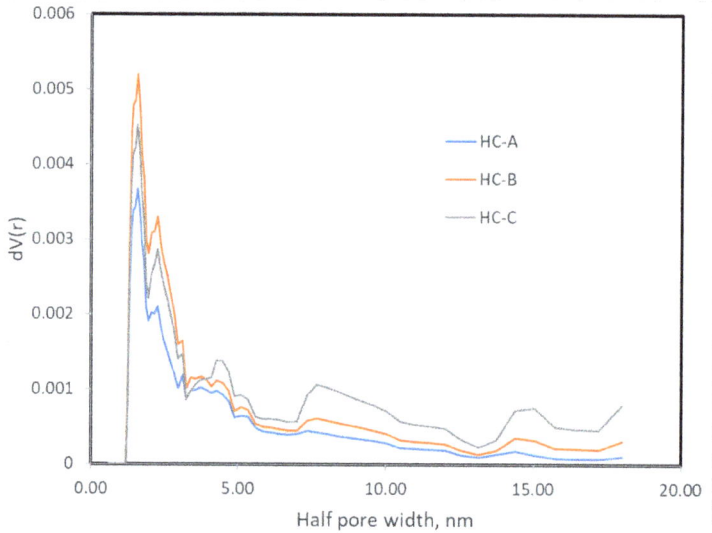

Figure 3. Distribution of porosity of the hydrothermal carbons.

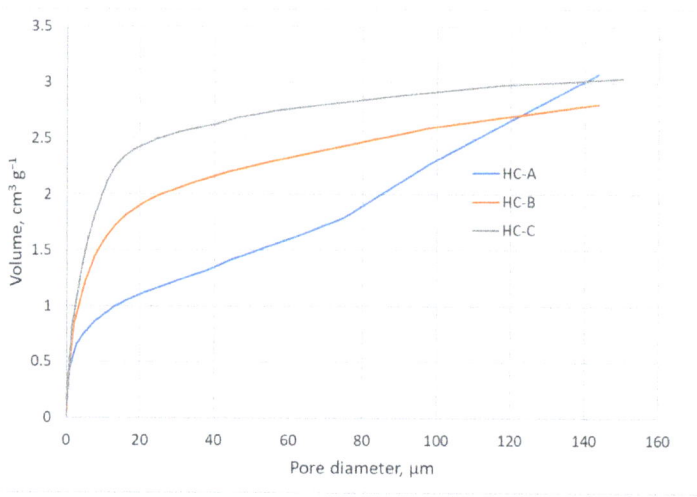

Figure 4. Cumulative pore volume of the hydrothermal carbons.

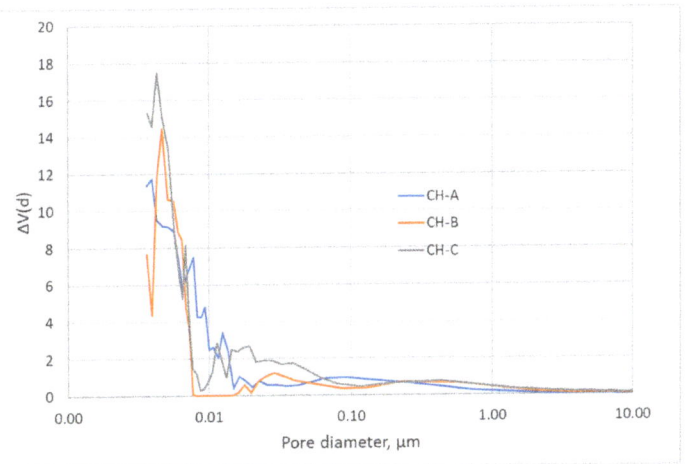

Figure 5. Pore width distribution of the hydrothermal carbons.

The DFT method was used to study the distribution of porosity in micro and mesopores, and the results are shown in Figure 3. The cumulative volume plot obtained from porosimetry is shown in Figure 4, and the width pore distribution plot, also obtained from porosimetry, is shown in Figure 5.

The measurements of the PZC show that all three materials were acidic (Table 4), as is usual in hydrothermal carbons. It must be considered that basic oxygenated functional groups (carbonyl) are more labile than those that provide acidic properties (alcohol and carboxylic), so it is to be expected that the former will degrade in a more significant proportion than the latter. In addition, as the degree of graphitization was low, as observed from the elemental composition with low carbon content, there was not a substantial contribution of basicity by the aromatic structures.

Table 4. Point of zero charge of the carbons.

Material	PZC
HC-A	4.0
HC-B	3.8
HC-C	4.8

The Fourier-transform infrared (FTIR) spectra obtained are shown in Figure 6. Numerous observable bands in the FTIR spectra belonging to many functional groups indicate that the degree of graphitization of these materials was not high. The most intense band at 3400 cm^{-1} corresponded to either alcohol or carboxylic O-H bonds. The band at 2935 cm^{-1} was assigned to C-H bonds, but its low wavenumber suggested they were mainly aliphatic bonds, since the aromatic ones give a signal at higher values. The 1710 cm^{-1} band was due to C-O double bonds that can be carbonyl or carboxylic, whereas the neighboring band of 1610 cm^{-1} was characteristic of C-C conjugated double bonds. Several bands between 1550 cm^{-1} and 1300 cm^{-1} can be attributed to the carbon backbone of these materials, for instance, those at 1515 cm^{-1} and 1460 cm^{-1}; however, the presence of O-H and C-O bands in this region made assigning them somewhat problematic. Between 1300 cm^{-1} and 100 cm^{-1}, the bands were mainly due to oxygenated or nitrogenous functional groups (C-O or C-N bonds) [50].

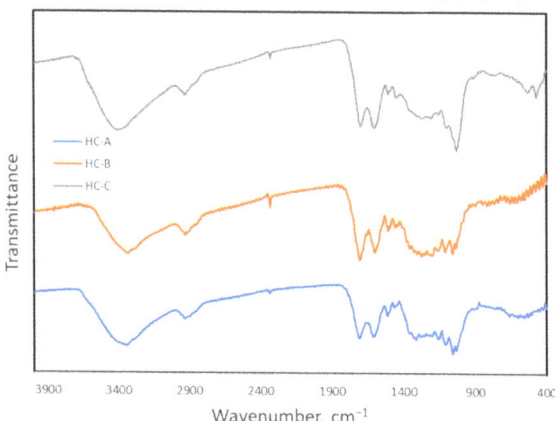

Figure 6. Fourier-transform infrared spectra of the hydrothermal carbons.

The scanning electron microscopy images of the samples in which hydrothermal carbonization had been carried out (Figures 7–9) showed a more defined structure than in the original material (Figure 10), in which the cells of the lignocellulosic material were more clearly observed. The treatment, therefore, removes matter from the original material, increasing the porosity. In some images, small spheres on the surface could be seen; it is known that obtaining hydrothermal coals from carbohydrates gives rise to spherical particles, so they could have come from dissolved substances of the chestnut cupule that underwent hydrothermal carbonization. Another option is that they were wax deposited on the surface of the biochar; given the similarity of composition in both cases, it is difficult to determine which of the processes gave rise to these structures.

Figure 7. SEM image of HC-A.

Figure 8. SEM image of HC-B.

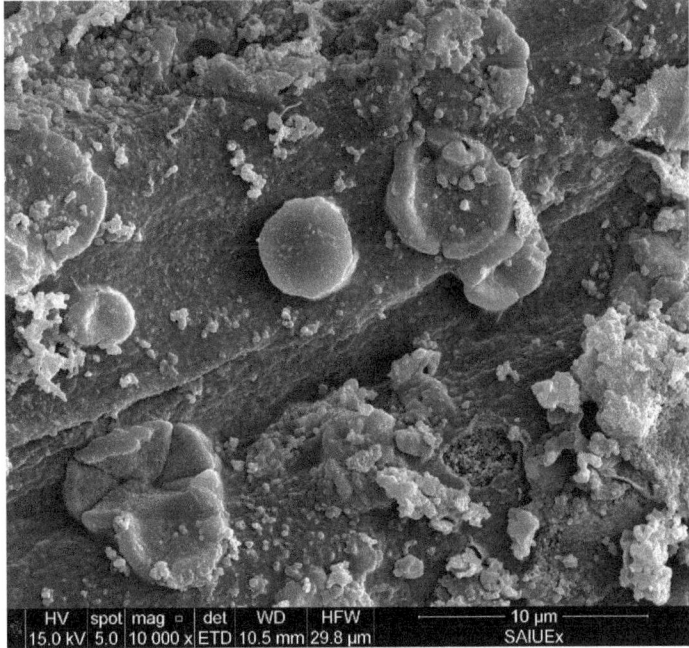

Figure 9. SEM image of HC-C.

Figure 10. SEM image of raw chestnut cupule.

The thermal analysis program can be found in the Supplementary Material (Table S1). Thermograms (Figure 11) show that the moisture of these materials was low, as well as the amount of residual ash, which was less than 1% in the three carbons. Regarding the volatile matter (from 50 to 80 min approximately), it followed the order HC-A > HC-B > HC-C, and the fixed carbon followed the opposite order. Therefore, they can be considered simply as carbonaceous materials with a low content of mineral matter.

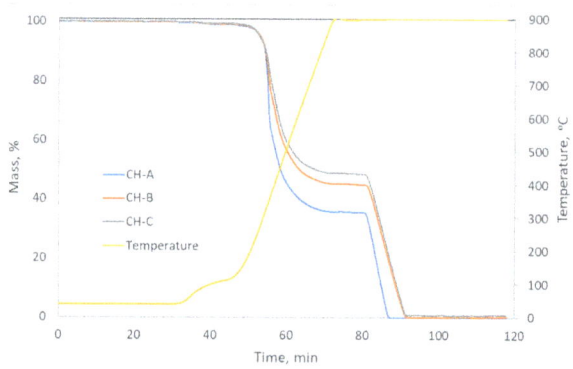

Figure 11. Thermograms of the hydrothermal carbons.

The results of the organic elemental analysis (C, H, N, S, and O) and the ash content (Table 5) show that the samples prepared at a higher temperature (HC-B and HC-C) presented more carbon, nitrogen, and ash, and less hydrogen and oxygen. This indicates that the carbonization process was more intense when the temperature was raised. The size did not seem to be crucial, though, since the smaller the particle size, the more intense the effect of pressurized water should have been, and this was not observed in the results. The data in the table agree with the fixed carbon content, which tended to increase with elemental carbon content, and the volatile matter, which tended to increase with oxygen and hydrogen content.

Table 5. Elemental analyses of the hydrothermal carbons.

Sample	C, %	H, %	N, %	S, %	O, %	Ash, %
HC-A	57.50	6.56	0.53	0.05	35.31	0.05
HC-B	61.70	5.93	0.58	0.05	31.55	0.19
HC-C	63.70	5.89	1.06	0.05	28.46	0.84

On the other hand, EDX analyses were performed on all three hydrothermal carbons and the raw chestnut cupule (Table 6). This determination, which studied a more superficial area than the elemental analysis, confirmed the same trend. All carbons had been enriched in carbon and depleted in oxygen. It is worth noticing that the HC-C sample contained the most significant mineral matter. This effect is larger than the one found in the elemental analysis, which indicates that the hydrothermal treatment affected the surface more than the interior of the material particles.

Table 6. EDX analyses of the hydrothermal carbons.

	Raw Material	HC-A	HC-B	HC-C
C	54.2	69.6	72.5	72.0
O	43.6	30.0	27.4	26.9
Mg	0.1	0.0	0.0	0.0
Al	0.1	0.1	0.0	0.0
Si	0.2	0.1	0.0	0.0
P	0.1	0.0	0.0	0.2
K	1.0	0.0	0.1	0.2
Ca	0.7	0.1	0.0	0.5
Fe	0.1	0.0	0.0	0.0
S	0.0	0.1	0.1	0.1

3.2. Adsorption Experiments

The kinetic study of adsorption processes makes it possible to relate the amount of adsorbate adsorbed per gram of adsorbent with the elapsed contact time between adsorbent and adsorbate. The variation in the excess PFONa concentration over time data is shown in Table 7. At 96 h, equilibrium can be considered established, having reached the maximum adsorption. Our preliminary studies show that HC-A and HC-B have adsorption of around 15%, while HC-C adsorbs 26%; therefore, further decontamination studies presented herein focused on the most promising adsorbent, HC-C.

Table 7. Adsorption kinetics results for HC-C.

Time (h)	PFONa [1]		
	Excess (mg/L)	Adsorbed (mg/L)	Adsorbed (%)
2	161	39	19
4	159	41	21
24	156	44	22
48	150	50	25
72	149	51	26
96	149	51	26

[1] Average value of duplicate runs.

The Lagergren adsorption kinetic model assumes that the adsorption rate is proportional to the number of unoccupied sites available for adsorption. However, when we applied it, we found it to be inappropriate for our data. On the other hand, the pseudo-second-order adsorption kinetic model assumes that the rate-limiting step is the chemical adsorption process, and it can be expressed mathematically, as shown in Equation (1):

$$\frac{t}{q_t} = \frac{1}{k \cdot q_e^2} + \frac{t}{q_e} \qquad (1)$$

where t is the time, q_t is the amount of solute adsorbed at time t, q_e is the amount of solute adsorbed at equilibrium, and k is the rate constant. The data are shown in Table 8, whereas the graph of t/q_t vs. t is plotted in Figure 12.

Table 8. Data for the pseudo-second-order kinetic model.

Time, t (h)	t/q_t (h·g·mg^{-1})
0	0
2	2.5907×10^{-03}
4	4.8603×10^{-03}
24	2.7027×10^{-02}
48	4.7619×10^{-02}
72	7.1146×10^{-02}
96	9.3842×10^{-02}

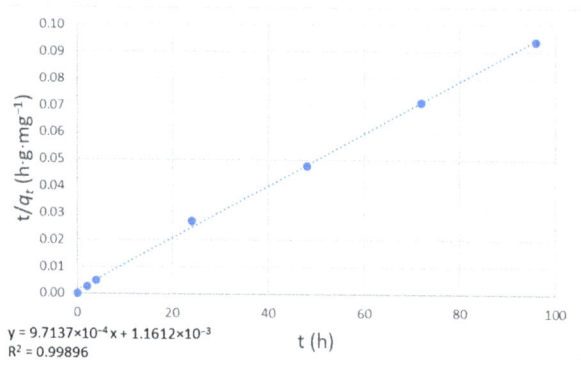

Figure 12. Plot for pseudo-second-order kinetics.

The data points follow a straight line, suggesting that the pseudo-second-order adsorption kinetic model applies to this system. The slope was 9.7137×10^{-4}, and the intercept was 1.1612×10^{-3}, so we can calculate the rate constant ($k = 8.126 \cdot 10^{-4}$ h^{-1}·mg^{-1}·g) and the amount of solute adsorbed at equilibrium ($q_e = 1029.47$ mg·g^{-1}).

Experiments were conducted with different HC-C to PFONa ratios to determine the carbon adsorption isotherm. The values obtained are shown in Table 9 and plotted in Figure 13.

Table 9. Adsorption isotherm results for HC-C [1].

c_0 (mg/L)	c_e (mg/L)	q_e (mg/g)
40	36.6	3.40
70	61.1	8.95
100	81.5	18.55
130	103.2	26.80
165	126.6	38.40
200	148.9	51.15

[1] Time = 96 h.

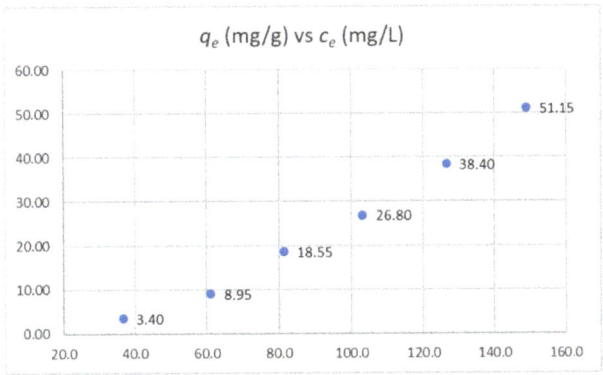

Figure 13. Adsorption isotherm for PFONa onto HC-C.

We found that the data obtained did not mathematically fit the Langmuir adsorption model, i.e., the adjustment of the model to the experimental data was inconsistent. On the other hand, the Freundlich model fit these results. Data for the Freundlich model are shown in Table 10.

Table 10. Data points for the Freundlich model.

Experiment	Log(c_e)	Log(q_e)
1	1.563481	0.531479
2	1.785686	0.951823
3	1.910891	1.268344
4	2.013680	1.428135
5	2.102434	1.584331
6	2.172749	1.708846

Using Equation (2) and the adjusted values, the Freundlich model constants can be calculated:

$$ln q_e = \frac{1}{n} ln c_e + ln K_F \qquad (2)$$

The calculated values of Freundlich isotherm constants were K_F = 12.3099; and $1/n$ = 1.9513 (r^2 = 0.9962).

The consistency with the Freundlich model indicates some degree of heterogeneity on the surface (not all active sites are identical) and the possibility that more than one layer can be adsorbed. Since the material used as adsorbent was amorphous, this relative heterogeneity can be expected. Moreover, the value of n indicates that the interaction between adsorbent and sorbate is mainly of the chemisorption type. Our results are promising, with a q_e = 1029.47 mg·g^{-1}; interestingly, they suggest that the maximum adsorption capacity had not been achieved at high concentrations. To put these results in perspective, we can refer to the excellent review by Liang et al.; in it, the authors compared the different adsorption methods for removing PFAS from water [37]. Analyzing the results obtained in water at pH 7, we found that, for granular activated carbon (GAC), a value of q_e = 2.34–102.7 mg·g^{-1} is reported; and for powdered activated carbon (PAC), q_e values of 16.9–500 mg·g^{-1} are found. On the other hand, for biochar derived from hardwood sawdust, the value obtained for q_e at pH 7.2 is 2.4 mg·g^{-1}, and for pinewood at the same pH, q_e is 2.1 mg·g^{-1}. Liang et al. found that ion exchange resins (q_e = 525–1500 mg·g^{-1}) and carbon nanotubes (q_e = 140–1110 mg·g^{-1}) show the highest adsorption capacities, yet they interestingly concluded that low-cost adsorbents with high adsorptive capacity are highly recommended in place of commercial AC due to their local availability, technical feasibility, and cost-effectiveness.

The mechanisms affecting PFAS adsorption have been reviewed by Liang et al. in an excellent review [37]. Diffusion, electrostatic interaction, hydrophobic interaction, ion exchange, and hydrogen bonds all can play a role in the adsorption process and are affected by the physicochemical characteristics of the adsorbent, such as pore size, pore distribution, and functional groups. In our case, HC-C is the carbon with the largest volume of mesopores; since PFONa is a large PFAS, it is essential for diffusion and adsorption that there are large numbers of mesopores. It is also the most graphitized hydrothermal carbon, i.e., the least polar; this also plays a vital role in the PFONa adsorption mechanism onto HC-C. Because of the hydrophobic nature of PFAS chains, the sorption of anionic PFAS can be dominated by hydrophobic interactions between the negatively-charged hydrophobic fluorinated tail and the surface of hydrophobic adsorbents, regardless of the electrostatic repulsion [51,52]. Particularly at high pH, hydrophobic interaction, which is not sensitive to changes in pH, should be the primary PFAS adsorption mechanism [53,54].

Given these results, and due to the advantages of the hydrothermal carbon herein reported, further work is underway to gain insight into the full potential of this adsorbent in water decontamination.

4. Conclusions

The preparation conditions of the hydrothermal carbons greatly influence the result; this allows for obtention of coals with tuned properties by optimizing the obtention process.

Carbonaceous material obtained by hydrothermal carbonization of sweet chestnut (*Castanea sativa*) cupule residues is a cost-effective, environmentally benign adsorbent, which effectively adds value to this biomass waste.

The carbons thus obtained show a limited surface area and a low microporosity while having a large volume of mesopores and macropores. All of them are acidic, with a low graphitization index, minimal moisture, and slight residual ash content; they are carbonaceous materials with a low mineral matter content.

It is possible to purify water contaminated with perfluorooctanoate by adsorption on this type of hydrothermal carbon. The adsorption process follows pseudo-second-order kinetics. Additionally, it fits the Freundlich model, showing that more than one layer can be adsorbed, which is expected in such an amorphous adsorbent.

Supplementary Materials: The following supporting information can be downloaded at: https://www.mdpi.com/article/10.3390/c9020057/s1, Table S1. Thermal Analysis Program; Figure S1. SEM Images of HC-A; Figure S2. SEM Images of HC-B; Figure S3. SEM Images of HC-C; Figure S4. SEM Images of raw chestnut cupule.

Author Contributions: Conceptualization, I.M.L.-C. and S.I.; methodology, I.M.L.-C. and C.J.D.-V.; investigation, S.I., N.P., C.J.D.-V. and I.M.L.-C.; writing—original draft preparation, S.I., C.J.D.-V. and I.M.L.-C.; writing—review and editing, S.I., N.P., C.J.D.-V. and I.M.L.-C.; funding acquisition, C.J.D.-V. All authors have read and agreed to the published version of the manuscript.

Funding: This research was funded by the Spanish Regional Government 'Junta de Extremadura' and European Regional Development Fund (ERDF), grants IB20026 and GR21107.

Data Availability Statement: Data may be obtained from the authors under reasonable request.

Conflicts of Interest: The authors declare no conflict of interest.

References

1. Tang, W.; Zhang, Y.; Zhong, Y.; Shen, T.; Wang, X.; Xia, X.; Tu, J. Natural biomass-derived carbons for electrochemical energy storage. *Mater. Res. Bull.* **2017**, *88*, 234–241. [CrossRef]
2. Ma, Q.; Yu, Y.; Sindoro, M.; Fane, A.G.; Wang, R.; Zhang, H. Carbon-Based Functional Materials Derived from Waste for Water Remediation and Energy Storage. *Adv. Mater.* **2017**, *29*, 1605361. [CrossRef] [PubMed]
3. Soffian, M.S.; Abdul Halim, F.Z.; Aziz, F.; Rahman, M.A.; Mohamed Amin, M.A.; Awang Chee, D.N. Carbon-based material derived from biomass waste for wastewater treatment. *Environ. Adv.* **2022**, *9*, 100259. [CrossRef]

4. Adegoke, K.A.; Akinnawo, S.O.; Ajala, O.A.; Adebusuyi, T.A.; Maxakato, N.W.; Bello, O.S. Progress and challenges in batch and optimization studies on the adsorptive removal of heavy metals using modified biomass-based adsorbents. *Bioresour. Technol. Rep.* **2022**, *19*, 101115. [CrossRef]
5. Gęca, M.; Wiśniewska, M.; Nowicki, P. Biochars and activated carbons as adsorbents of inorganic and organic compounds from multicomponent systems—A review. *Adv. Colloid Interface Sci.* **2022**, *305*, 102687. [CrossRef]
6. Borchardt, L.; Zhu, Q.-L.; Casco, M.E.; Berger, R.; Zhuang, X.; Kaskel, S.; Feng, X.; Xu, Q. Toward a molecular design of porous carbon materials. *Mater. Today* **2017**, *20*, 592–610. [CrossRef]
7. Enaime, G.; Baçaoui, A.; Yaacoubi, A.; Lübken, M. Biochar for Wastewater Treatment—Conversion Technologies and Applications. *Appl. Sci.* **2020**, *10*, 3492. [CrossRef]
8. Dzyazko, Y.S.; Palchik, O.V.; Ogenko, V.M.; Shtemberg, L.Y.; Bogomaz, V.I.; Protsenko, S.A.; Khomenko, V.G.; Makeeva, I.S.; Chernysh, O.V.; Dzyazko, O.G. Nanoporous Biochar for Removal of Toxic Organic Compounds from Water. In Proceedings of the 6th International Conference Nanotechnology and Nanomaterials (NANO2018), Kyiv, Ukraine, 27–30 August 2018; pp. 209–224.
9. Altundogan, H.S.; Bahar, N.; Mujde, B.; Tumen, F. The use of sulphuric acid-carbonization products of sugar beet pulp in Cr(VI) removal. *J. Hazard. Mater.* **2007**, *144*, 255–264. [CrossRef]
10. Kew, R.B.G. Castanea Sativa. 2023. Available online: https://powo.science.kew.org/taxon/urn:lsid:ipni.org:names:295349-1/general-information#descriptions (accessed on 3 February 2023).
11. Caudullo, G.; Welk, E.; San-Miguel-Ayanz, J. Chorological maps for the main European woody species. *Data Br.* **2017**, *12*, 662–666. [CrossRef]
12. Conedera, M.; Manetti, M.C.; Giudici, F.; Amorini, E. Distribution and economic potential of the Sweet chestnut (*Castanea sativa* Mill.) in Europe. *Ecol. Mediterr.* **2004**, *30*, 179–193. [CrossRef]
13. Schultz, M.M.; Barofsky, D.F.; Field, J.A. Fluorinated Alkyl Surfactants. *Environ. Eng. Sci.* **2003**, *20*, 487–501. [CrossRef]
14. Wang, Z.; DeWitt, J.C.; Higgins, C.P.; Cousins, I.T. A Never-Ending Story of Per- and Polyfluoroalkyl Substances (PFASs)? *Environ. Sci. Technol.* **2017**, *51*, 2508–2518. [CrossRef] [PubMed]
15. Andersson, E.M.; Scott, K.; Xu, Y.Y.; Li, Y.; Olsson, D.S.; Fletcher, T.; Jakobsson, K. High exposure to perfluorinated compounds in drinking water and thyroid disease. A cohort study from Ronneby, Sweden. *Environ. Res.* **2019**, *176*, 108540. [CrossRef] [PubMed]
16. Domingo, J.L.; Nadal, M. Human exposure to per- and polyfluoroalkyl substances (PFAS) through drinking water: A review of the recent scientific literature. *Environ. Res.* **2019**, *177*, 108648. [CrossRef] [PubMed]
17. Geissen, V.; Mol, H.; Klumpp, E.; Umlauf, G.; Nadal, M.; van der Ploeg, M.; van de Zee, S.E.A.T.M.; Ritsema, C.J. Emerging pollutants in the environment: A challenge for water resource management. *Int. Soil Water Conserv. Res.* **2015**, *3*, 57–65. [CrossRef]
18. Knutsen, H.K.; Alexander, J.; Barregård, L.; Bignami, M.; Brüschweiler, B.; Ceccatelli, S.; Cottrill, B.; Dinovi, M.; Edler, L.; Grasl-Kraupp, B.; et al. Risk to human health related to the presence of perfluorooctane sulfonic acid and perfluorooctanoic acid in food. *EFSA J.* **2018**, *16*, e05194. [CrossRef]
19. Gao, K.; Zhuang, T.; Liu, X.; Fu, J.; Zhang, J.; Fu, J.; Wang, L.; Zhang, A.; Liang, Y.; Song, M.; et al. Prenatal Exposure to Per- and Polyfluoroalkyl Substances (PFASs) and Association between the Placental Transfer Efficiencies and Dissociation Constant of Serum Proteins–PFAS Complexes. *Environ. Sci. Technol.* **2019**, *53*, 6529–6538. [CrossRef]
20. Park, M.; Wu, S.; Lopez, I.J.; Chang, J.Y.; Karanfil, T.; Snyder, S.A. Adsorption of perfluoroalkyl substances (PFAS) in groundwater by granular activated carbons: Roles of hydrophobicity of PFAS and carbon characteristics. *Water Res.* **2020**, *170*, 115364. [CrossRef]
21. Susmann, H.P.; Schaider, L.A.; Rodgers, K.M.; Rudel, R.A. Dietary Habits Related to Food Packaging and Population Exposure to PFASs. *Environ. Health Perspect.* **2019**, *127*, 107003. [CrossRef]
22. European Environment Agency. *Emerging Chemical Risks in Europe-PFAS?* Publications Office. 2019. Available online: https://data.europa.eu/doi/10.2800/486213 (accessed on 3 February 2023).
23. Taylor, S.; Terkildsen, M.; Stevenson, G.; de Araujo, J.; Yu, C.; Yates, A.; McIntosh, R.R.; Gray, R. Per and polyfluoroalkyl substances (PFAS) at high concentrations in neonatal Australian pinnipeds. *Sci. Total Environ.* **2021**, *786*, 147446. [CrossRef]
24. Li, F.; Duan, J.; Tian, S.; Ji, H.; Zhu, Y.; Wei, Z.; Zhao, D. Short-chain per- and polyfluoroalkyl substances in aquatic systems: Occurrence, impacts and treatment. *Chem. Eng. J.* **2020**, *380*, 122506. [CrossRef]
25. Wong, S.; Ngadi, N.; Inuwa, I.M.; Hassan, O. Recent advances in applications of activated carbon from biowaste for wastewater treatment: A short review. *J. Clean. Prod.* **2018**, *175*, 361–375. [CrossRef]
26. Abioye, A.M.; Ani, F.N. Recent development in the production of activated carbon electrodes from agricultural waste biomass for supercapacitors: A review. *Renew. Sustain. Energy Rev.* **2015**, *52*, 1282–1293. [CrossRef]
27. McNamara, J.D.; Franco, R.; Mimna, R.; Zappa, L. Comparison of Activated Carbons for Removal of Perfluorinated Compounds From Drinking Water. *J. Am. Water Works Assoc.* **2018**, *110*, E2–E14. [CrossRef]
28. Diamadopoulos, E.; Samaras, P.; Sakellaropoulos, G.P. The Effect of Activated Carbon Properties on the Adsorption of Toxic Substances. *Water Sci. Technol.* **1992**, *25*, 153–160. [CrossRef]
29. Jeguirim, M.; Belhachemi, M.; Limousy, L.; Bennici, S. Adsorption/reduction of nitrogen dioxide on activated carbons: Textural properties versus surface chemistry—A review. *Chem. Eng. J.* **2018**, *347*, 493–504. [CrossRef]
30. Durán-Valle, C.J.; Madrigal-Martínez, M.; Martínez-Gallego, M.; Fonseca, I.M.; Matos, I.; Botelho do Rego, A.M. Activated carbon as a catalyst for the synthesis of N-alkylimidazoles and imidazolium ionic liquids. *Catal. Today* **2012**, *187*, 108–114. [CrossRef]

31. Figueiredo, J.; Pereira, M.F.; Freitas, M.M.; Órfão, J.J. Modification of the surface chemistry of activated carbons. *Carbon* **1999**, *37*, 1379–1389. [CrossRef]
32. Zhao, L.; Baccile, N.; Gross, S.; Zhang, Y.; Wei, W.; Sun, Y.; Antonietti, M.; Titirici, M.-M. Sustainable nitrogen-doped carbonaceous materials from biomass derivatives. *Carbon* **2010**, *48*, 3778–3787. [CrossRef]
33. Titirici, M.-M.; White, R.J.; Falco, C.; Sevilla, M. Black perspectives for a green future: Hydrothermal carbons for environment protection and energy storage. *Energy Environ. Sci.* **2012**, *5*, 6796. [CrossRef]
34. Durán-Valle, C.J.; Botet-Jiménez, A.B.; Omenat-Morán, D. Hydrothermal Carbonisation: An Eco-Friendly Method for the Production of Carbon Adsorbents. In *Adsorption Processes for Water Treatment and Purification*; Springer International Publishing: Cham, Switzerland, 2017; pp. 77–108.
35. Kucharzyk, K.H.; Darlington, R.; Benotti, M.; Deeb, R.; Hawley, E. Novel treatment technologies for PFAS compounds: A critical review. *J. Environ. Manag.* **2017**, *204*, 757–764. [CrossRef] [PubMed]
36. Lei, X.; Lian, Q.; Zhang, X.; Karsili, T.K.; Holmes, W.; Chen, Y.; Zappi, M.E.; Gang, D.D. A review of PFAS adsorption from aqueous solutions: Current approaches, engineering applications, challenges, and opportunities. *Environ. Pollut.* **2023**, *321*, 121138. [CrossRef] [PubMed]
37. Zhang, D.Q.; Zhang, W.L.; Liang, Y.N. Adsorption of perfluoroalkyl and polyfluoroalkyl substances (PFASs) from aqueous solution—A review. *Sci. Total Environ.* **2019**, *694*, 133606. [CrossRef] [PubMed]
38. Gagliano, E.; Sgroi, M.; Falciglia, P.P.; Vagliasindi, F.G.A.; Roccaro, P. Removal of poly- and perfluoroalkyl substances (PFAS) from water by adsorption: Role of PFAS chain length, effect of organic matter and challenges in adsorbent regeneration. *Water Res.* **2020**, *171*, 115381. [CrossRef]
39. Zhang, K.; Sumita; Li, C.; Sun, C.; Marmier, N. A Review of the Treatment Process of Perfluorooctane Compounds in the Waters: Adsorption, Flocculation, and Advanced Oxidative Process. *Water* **2022**, *14*, 2692. [CrossRef]
40. Son, H.; Kim, T.; Yoom, H.-S.; Zhao, D.; An, B. The Adsorption Selectivity of Short and Long Per- and Polyfluoroalkyl Substances (PFASs) from Surface Water Using Powder-Activated Carbon. *Water* **2020**, *12*, 3287. [CrossRef]
41. Chen, R.; Huang, X.; Li, G.; Yu, Y.; Shi, B. Performance of in-service granular activated carbon for perfluoroalkyl substances removal under changing water quality conditions. *Sci. Total Environ.* **2022**, *848*, 157723. [CrossRef]
42. Du, Z.; Deng, S.; Bei, Y.; Huang, Q.; Wang, B.; Huang, J.; Yu, G. Adsorption behavior and mechanism of perfluorinated compounds on various adsorbents-A review. *J. Hazard. Mater.* **2014**, *274*, 443–454. [CrossRef]
43. Brunauer, S.; Emmett, P.H.; Teller, E. Adsorption of Gases in Multimolecular Layers. *J. Am. Chem. Soc.* **1938**, *60*, 309–319. [CrossRef]
44. Valente Nabais, J.M.; Carrott, P.J.M. Chemical Characterization of Activated Carbon Fibers and Activated Carbons. *J. Chem. Educ.* **2006**, *83*, 436. [CrossRef]
45. Thommes, M.; Kaneko, K.; Neimark, A.V.; Olivier, J.P.; Rodriguez-Reinoso, F.; Rouquerol, J.; Sing, K.S.W. Physisorption of gases, with special reference to the evaluation of surface area and pore size distribution (IUPAC Technical Report). *Pure Appl. Chem.* **2015**, *87*, 1051–1069. [CrossRef]
46. Bläker, C.; Muthmann, J.; Pasel, C.; Bathen, D. Characterization of Activated Carbon Adsorbents—State of the Art and Novel Approaches. *ChemBioEng Rev.* **2019**, *6*, 119–138. [CrossRef]
47. Lastoskie, C.; Gubbins, K.E.; Quirke, N. Pore size distribution analysis of microporous carbons: A density functional theory approach. *J. Phys. Chem.* **1993**, *97*, 4786–4796. [CrossRef]
48. Dubinin, M.M.; Radushkevich, L.V. The Equation of the Characteristic Curve of Activated Charcoal. *Proc. USSR Acad. Sci.* **1947**, *55*, 331.
49. Dubinin, M.M. The Potential Theory of Adsorption of Gases and Vapors for Adsorbents with Energetically Nonuniform Surfaces. *Chem. Rev.* **1960**, *60*, 235–241. [CrossRef]
50. Țucureanu, V.; Matei, A.; Avram, A.M. FTIR Spectroscopy for Carbon Family Study. *Crit. Rev. Anal. Chem.* **2016**, *46*, 502–520. [CrossRef]
51. Chen, X.; Xia, X.; Wang, X.; Qiao, J.; Chen, H. A comparative study on sorption of perfluorooctane sulfonate (PFOS) by chars, ash and carbon nanotubes. *Chemosphere* **2011**, *83*, 1313–1319. [CrossRef] [PubMed]
52. Yu, Q.; Zhang, R.; Deng, S.; Huang, J.; Yu, G. Sorption of perfluorooctane sulfonate and perfluorooctanoate on activated carbons and resin: Kinetic and isotherm study. *Water Res.* **2009**, *43*, 1150–1158. [CrossRef]
53. Gao, Y.; Deng, S.; Du, Z.; Liu, K.; Yu, G. Adsorptive removal of emerging polyfluoroalky substances F-53B and PFOS by anion-exchange resin: A comparative study. *J. Hazard. Mater.* **2017**, *323*, 550–557. [CrossRef]
54. Pan, G.; Jia, C.; Zhao, D.; You, C.; Chen, H.; Jiang, G. Effect of cationic and anionic surfactants on the sorption and desorption of perfluorooctane sulfonate (PFOS) on natural sediments. *Environ. Pollut.* **2009**, *157*, 325–330. [CrossRef]

Disclaimer/Publisher's Note: The statements, opinions and data contained in all publications are solely those of the individual author(s) and contributor(s) and not of MDPI and/or the editor(s). MDPI and/or the editor(s) disclaim responsibility for any injury to people or property resulting from any ideas, methods, instructions or products referred to in the content.

Article

Arsenic, Iron, and Manganese Adsorption in Single and Trinary Heavy Metal Solution Systems by Bamboo-Derived Biochars

Anawat Pinisakul [1], Nattakarn Kruatong [2], Soydoa Vinitnantharat [2,3,*], Ponwarin Wilamas [4], Rattikan Neamchan [3], Nareerat Sukkhee [3], David Werner [5] and Saichol Sanghaisuk [6]

1. Chemistry for Green Society and Healthy Living Research Unit (ChGSH), Department of Chemistry, Faculty of Science, King Mongkut's University of Technology Thonburi, Bangkok 10140, Thailand
2. Environmental Technology Program, School of Energy, Environment and Materials, King Mongkut's University of Technology Thonburi, Bangkok 10140, Thailand
3. Environmental and Energy Management for Community and Circular Economy (EEC&C) Research Group, King Mongkut's University of Technology Thonburi, Bangkok 10140, Thailand
4. Department of Biological Science, Faculty of Science, Ubon Ratchathani University, Ubon Ratchathani 34190, Thailand
5. School of Engineering, Newcastle University, Newcastle upon Tyne NE1 7RU, UK
6. Department of Pollution Control, Ministry of Natural Resources and Environment, Bangkok 10400, Thailand
* Correspondence: soydoa.vin@mail.kmutt.ac.th

Citation: Pinisakul, A.; Kruatong, N.; Vinitnantharat, S.; Wilamas, P.; Neamchan, R.; Sukkhee, N.; Werner, D.; Sanghaisuk, S. Arsenic, Iron, and Manganese Adsorption in Single and Trinary Heavy Metal Solution Systems by Bamboo-Derived Biochars. *C* **2023**, *9*, 40. https://doi.org/10.3390/c9020040

Academic Editor: Dimitrios Kalderis

Received: 28 January 2023
Revised: 1 April 2023
Accepted: 12 April 2023
Published: 16 April 2023

Copyright: © 2023 by the authors. Licensee MDPI, Basel, Switzerland. This article is an open access article distributed under the terms and conditions of the Creative Commons Attribution (CC BY) license (https:// creativecommons.org/licenses/by/ 4.0/).

Abstract: Currently, heavy metal-contaminated groundwater is an environmental concern. This study investigated the use of bamboo biochar, chitosan-impregnated biochar, and iron-impregnated biochar for arsenic, iron, and manganese removal from groundwater. Isotherms of arsenic, iron, and manganese adsorption by bamboo derived biochar were compared with those of commercial activated carbon in simulated groundwater composed of single and trinary heavy metal solutions. The binding of heavy metals by virgin and loaded bamboo biochar and activated carbon was also investigated by sequential extraction. Chitosan and iron-impregnated biochar had enhanced arsenic adsorption, but these sorbents turned the pH of solution acidic, while it was alkaline for activated carbon. Adsorption equilibrium times of arsenic and iron were faster for single than trinary heavy metal systems because less ion competition occurred at active sites. The Langmuir model fitted the adsorption data well. The maximum adsorption capacities of arsenic, iron, and manganese by bamboo biochar in trinary heavy metal system were 2.2568, 0.6393, and 1.3541 mg g^{-1}, respectively. The main mechanism for arsenic removal was precipitation with iron. Bamboo biochar bound iron in organic and sulfide fractions and manganese with iron-oxide. Bamboo biochar can replace activated carbon as a more efficient and sustainable carbonaceous sorbent material for removal of mixed heavy metals from groundwater within acceptable pH ranges.

Keywords: adsorption; fractionation; heavy metal removal; isotherm; modified biochar

1. Introduction

Iron (Fe) and manganese (Mn) are ubiquitous in soil and normally found in surface and groundwater from rock weathering. In some regions of Asian countries, heavy metal contamination in water resources was associated with mining, manufacturing, and rock weathering [1,2]. Arsenic (As) is one of the heavy metals causing concern, and about 180 million people are at risk of arsenic poisoning [2]. In addition, environmental impacts will differ between single-metal and mixed-metal pollution [1]. Excessive arsenic, iron, and manganese concentrations were found in groundwater in the rural areas of developing countries where groundwater is the main water resource for drinking water. Groundwater reportedly contained As, Fe, and Mn at maximum concentrations of 0.09, 3.68, and 0.38 mg/L in Jashor, Bangladesh [3]; 0.112, 46.3, and 6.16 mg L^{-1} in Shuangliao, China [4]; and 0.416, 68, and 1.9 mg L^{-1} in Lampang, Thailand [5]. Iron and manganese are necessary for human health as iron relates to a wide variety of metabolic processes, including oxygen transport, deoxyribonucleic acid synthesis, and electron transport [6]. Manganese is also essential for development, metabolism, and the antioxidant system [7] serving as a cofactor of several critical enzymes [8]. However, prolonged consumption of high amounts of these heavy metals results in severe health impacts such as organ dysfunction including cell death, fibrosis, and carcinogenesis from iron toxicity [9]; psychiatric symptoms including emotional liability, mania, compulsive or aggressive behavior, irritability, reduced response speed from manganese toxicity [10]; and disturbance in the nervous system, while carcinogenic effects on numerous organs such as lung, urinary tract, and skin result from arsenic toxicity [11]. The maximum concentrations of As, Fe, and Mn for drinking water recommended by the WHO (2017) [12] are 0.01, 0.3, and 0.4 mg L^{-1}, respectively. As the excessive presence of As and Mn in water resources is a serious concern to public health, they should be removed to the allowable concentrations by water treatment.

The adsorption process is widely used in water treatment due to its ease of operation and cost-effectiveness. Adsorption is a mass transfer process in which the pollutant from the liquid phase transfer to the solid phase or adsorbent. The porous structure, surface area, and functional groups of adsorbents play an important role in heavy metal removal. Heavy metals tend to adsorb onto the oppositely charged adsorbents. Among the numerous studied adsorbents, biochar has proven to be effective for the removal of heavy metals from water and wastewater due to its negative charge from oxygen functional groups and also other related mechanisms such as complexation, physical sorption, reduction of metal species, electrostatic interactions, and precipitation [13,14]. Thus, biochar is increasingly considered as alternative that can replace commercial activated carbon. Biochar can also be modified on its surfaces with chemicals to improve its adsorption properties. Chitosan impregnation of biochar was reported for the removal of inorganic and organic pollutants including heavy metals [15]. Chitosan is widely used because it is a biodegradable and renewable polymer that possesses both cationic charges (from amino groups-NH$_2$) and anionic charges (from hydroxyl groups-OH). Loc et al. [16] reported the removal of an organic dye by chitosan-modified biochar through electrostatic interaction and complexation. Another simple modification of biochar surfaces is iron impregnation for the removal of acid red dye [17], phosphate [18], and arsenic [19,20]. Sun et al. [21] concluded that the main mechanisms for arsenic removal by modified biochars were electrostatic interaction, complexation, and precipitation.

Biochar composition is highly heterogeneous, containing not only the main elements of carbonaceous adsorbents (carbon, hydrogen, and oxygen) but also nutrients (nitrogen, phosphorus, and potassium) and some heavy metals. It was reported that biochar produced from coconut residues and rice straw contains Fe, Zn, and Al [22]. Wang et al. [23] researched the amounts of heavy metals in chicken manure biochar and water-washed swine manure biochar indicating high concentrations of arsenic, chromium, and manganese; however, the proportions of labile fractions were decreased with increased pyrolysis temperature during thermal conversion of biomass. Thus, if biochar is used as adsorbent for water treatment it may release nutrients and heavy metals. It was reported that the release of ions from biochar into deionized water followed the order of $Cl^- > K^+ > Na^+ > PO_4^{3-} > SO_4^{2-} > Ca^{2+} > NO_3^- > Mg^{2+} > NH_4^+ = NO_2^-$ and chitosan-impregnated biochar released less ions than unmodified biochar [24]. Previous research mostly evaluated the removal of heavy metals by biochar and modified biochar in single solute systems. Few studies have addressed the removal of mixed heavy metals and considered the fraction of heavy metals in biochar.

Bamboo is abundantly available in tropical and subtropical countries. To increase its stability in the environment, it is necessary to convert raw biomass to biochar. Bamboo may be a promising adsorbent as bamboo biomass has been previously transformed into biochar, activated carbon, and aerogel [25]. Previous research reported the main structural components of bamboo biomass were cellulose and hemicellulose of 47.5 and 15.3%, respectively [26]. A review on bamboo-based biochar indicated the high surface area and mesoporous structure of bamboo biochar enables it to adsorb antibiotics (fluoroquinolone, sulfamethoxazole, and sulphapyridine), nutrients (ammonium ion, nitrate ion, and phosphate ion), heavy metals (Cd, Cr, and U), dyes (Congo red and acid black 172), and 2,4-dichlorophenol [25]. Hernandez-Mena et al. [26] reported bamboo biochar could adsorb heavy metals with strong adsorption intensity, but less is known about the removal of mixed heavy metals, such as As, Fe, and Mn in a trinary system [26]. In addition, the effects of surface modification on adsorption capacities for a trinary system has also hardly been investigated. Thus, the present study aimed to (i) study the performance of bamboo biochar without and with modification by iron and chitosan impregnation, in comparison with commercial activated carbon, in terms of the adsorption of As, Fe, and Mn in single and trinary systems and (ii) to investigate the labile and stable fractions of heavy metals bound onto bamboo-derived biochars.

2. Materials and Methods

2.1. Simulated Water Solution

Tap water was used to provide background ions for all experiments. It was analyzed for cations and anions by Ion Chromatography (IC, 761 Compact IC, Metrohm, Switzerland) as shown in Table 1. Before being used, tap water was spiked to achieve the desired concentration of each heavy metal in the preparation of the single or trinary heavy metal solution systems. The Fe(II), Mn(II), and As(V) ions were prepared from $FeSO_4 \cdot 7H_2O$ (Ajax Finechem Pty Ltd., Taren Point, Australia), $MnSO_4 \cdot H_2O$ (Ajax Finechem Pty Ltd., Taren Point, Australia), and $Na_2HAsO_4 \cdot 7H_2O$ (Sigma-Aldrich, St. Louis, MO, USA) salts in analytical grade, respectively.

Table 1. Background ions in simulated groundwater solution.

Ions	Na^+	NH_4^+	K^+	Ca^{2+}	Mg^{2+}	Cl^-	NO_3^-	SO_4^{2-}	PO_4^{3-}
(mg L^{-1})	11.95	0.03	4.66	57.05	9.52	12.75	4.03	34.89	ND

Note: ND = not detectable.

2.2. Bamboo-Derived Biochars and Characterization

Bamboo biochar (B) was derived from Ubon Ratchathani province, Thailand. Bamboo stem (*Bambusa beecheyana*) at the age of 2–3 years were cut into lengths of 30 cm and then the cuttings were placed in a traditional earth mound kiln (hemispherical shape with diameter of 2.5 m and height of 2.5 m). Carbonization was completed within 6 days with the temperatures range 400–600 °C. Then, the kiln was left to cool for 6 days, and biochar was taken out from the kiln. It was crushed to a uniform size in the range of 2.0–3.0 mm. All experiments used same batch of biochar production.

Chitosan-impregnated bamboo biochar (BC) was prepared following Vinitnantharat et al. [27] by placing 5 g crushed B in 100 mL chitosan solution in a 250 mL Erlenmeyer flask. Then, the suspension was agitated in a shaker at 100 rpm for 24 h at room temperature (30 °C). The chitosan solution was prepared by dissolving 1 g chitosan powder in 100 mL of 1% ($v\ v^{-1}$) acetic acid. BC was separated from the mixture and rinsed with tap water until the pH of the wash water reached 7.0 ± 0.5. BC was rinsed with deionized water (200 mL) and air dried for 2 days before storage in a bottle.

Iron-impregnated bamboo biochar (BFe) was prepared by adapting the method of Kalaruban et al. [28]. A total of 20 g of crushed B was mixed with 1.35 % ($w\ v^{-1}$) of $FeCl_3 \cdot 6H_2O$ solution in a 250 mL Erlenmeyer flask. Then, the suspension was agitated in a shaker at 100 rpm for 24 h at room temperature (30 °C). BFe was separated from the mixture by filtering through a plastic sieve, then it was rinsed with tap water until pH was 7.0 ± 0.5 and rinsed with deionized water (200 mL). BFe was left to dry at room temperature and then kept in a bottle.

Commercial activated carbon (AC) in the size of 2.0–3.0 mm was obtained from Charcoal Home Co., Ltd., Bangkok, Thailand. It was made from bamboo (*Dendrocalamus asper*) by carbonization in a brick kiln at high temperature (1000 °C). It was used as received.

Adsorbents were characterized by the physical adsorption of nitrogen gas according to the method of Brunauer, Emmet, and Teller (BET, Quantachrome, Quandrasorb evo, Anton Paar GmbH, Graz, Austria); the surface features and morphology were studied by a Scanning Electron Microscope and Energy Dispersive X-ray Spectrometer (SEM-EDS, Jeo., JSM-7610F Plus, Japan); surface functional groups were characterized by Fourier transform infrared spectroscopy (FTIR, Thermo Scientific Nicolet 6700, Waltham, MA, USA); and chemical composition was studied by X-ray fluorescence (XRF, Rigaku, ZSX Primus II, Japan).

Iodine number was analyzed to determine the iodine adsorption following the ASTM D4607 method. The pH of each biochar was measured after shaking 5 g of biochar in 50 mL of deionized water (pH_{DI}) at 100 rpm at 30 °C and the pH at point of zero charge (pH_{pzc}) was analyzed following the method of Khawkomol et al. [22]. The pH value was measured by a pH meter (pH3210, WTW, Weilheim, Germany).

2.3. Adsorption Equilibration Time Study

The adsorption equilibration time was performed for single and trinary heavy metal solution systems. The single heavy metal system was performed using 0.5 g of adsorbents with 50 mL of solution containing 0.05 mg L^{-1} As(V), or 7.0 mg L^{-1} Fe(II), or 1.0 mg L^{-1} Mn(II) in 100 mL plastic bottles. As for trinary heavy metal system, the concentrations of As(V), Fe(II), and Mn(II) were 0.02, 7.0, and 1.0 mg L^{-1}, respectively. The solution pH was adjusted to 7.00 ± 0.50, except that it was 4.40 ± 0.50 for adsorption of single Fe(II) solution to avoid iron precipitation. The mixtures were agitated in an incubator shaker at 100 rpm and temperature of 28 ± 0.5 °C. After contact time at 2, 5, 12, 24, 36, 48, 72, 84, 96, 108, 120, 132, 144, and 156 h, the mixture was passed through a 0.45 μm nylon membrane and the filtrate was analyzed for As(V) using an inductively coupled plasma-optical emission spectrometer (ICP-OES, Perkin Elmer Optima 8000), and for Fe(II) and Mn(II) using an atomic absorption spectrometer (AAS, Shimadzu Europa AA-6300, Japan). All experiments were performed in duplicate. The amount of Fe, Mn, and As adsorbed onto the adsorbent, q (mg g^{-1}), was calculated using Equation (1):

$$q_t = [(C_0 - (C_t - C_c)) \cdot V]/m \tag{1}$$

where C_0 is the initial concentration of heavy metal ion (mg L^{-1}), C_t is the concentration of heavy metal ions at time t (mg L^{-1}), C_c is the concentration of heavy metal ions released by adsorbent into simulated water solution for the same condition of heavy metal ions adsorption at time t (mg L^{-1}), V is the volume of solution (L), and m is the mass of adsorbent (g).

2.4. Isotherm Study

The batch adsorption experiments consisted of single and tri-heavy metals solutions for determining the adsorption capacities of each biochar type. Adsorption studies were performed in 100 mL bottles shaken in an incubator shaker at 100 rpm and temperature of 28 ± 0.5 °C until reaching equilibrium.

For single heavy metal isotherm experiments, a biochar dosage of 0.5 g was contacted with 50 mL simulate water solution containing different concentration of As(V) 0.01–0.10 mg L^{-1} or Fe(II) 1.00–10.00 mg L^{-1} or Mn(II) 1.00–18.00 mg L^{-1}. The pH of the solution was adjusted to 7.00 ± 0.50, except it was 4.40 ± 0.50 for Fe(II) adsorption. In the case of the trinary heavy metal system, different masses of biochars (0.2–1.0 g) were contacted with fixed concentrations of 0.02 mg L^{-1} As(V), 7.0 mg L^{-1} Fe(II), and 1.0 mg L^{-1} Mn(II) in 50 mL simulated water solution. The initial and final equilibrium concentrations were measured for each heavy metal and used for the construction of isotherm curves. The pH of the solution at equilibrium (pH_E) was also measured.

Two adsorption models, Freundlich and Langmuir, were used to fit the experimental data. The Freundlich model is an empirical model suitable for nonideal adsorption on heterogeneous surfaces as expressed in Equation (2).

$$q_e = K_F C_e^{1/n}, (0 < 1/n < 1) \tag{2}$$

where q_e is the amount of heavy metal adsorbed at equilibrium (mg g^{-1}), C_e is the heavy metal concentration at equilibrium (mg L^{-1}), K_F is the Freundlich constant related to maximum adsorption capacity ((mg g^{-1})(L mg^{-1})$^{1/n}$), and $1/n$ is the Freundlich constant relate to the intensity (dimensionless).

The Langmuir adsorption model has been used to describe the uniform energies of adsorption onto the surface and no transmigration of adsorbate in the plane of the surface. The Langmuir model is given by Equation (3).

$$\frac{C_e}{q_e} = \frac{1}{K_L q_m} + \frac{C_e}{q_m} \tag{3}$$

q_m defines the maximum amount of heavy metal per unit weight of adsorbent to form a complete monolayer coverage on the surface (mg g^{-1}), and K_L is the Langmuir constant (L mg^{-1}).

2.5. Fractionation Analysis

The distribution pattern of heavy metals bound on the specific functional groups of selected biochars was investigated by using biochar before and after being loaded with trinary heavy metal solution (biochar dosage 1% $w\ v^{-1}$). Two grams of biochar were sequential extracted into 6 fractions: F1 Water soluble (Soluble); F2 Exchangeable (Exch); F3 Bound to carbonate (CO$_3$); F4 Bound to Fe/Mn oxides (Fe/MnO); F5 Bound to organic matter and sulfide (Org/S); and F6 Residual. The extraction method was adapted from Kashem et al. [29] as shown in Table 2. All extracts were centrifuged for 10 min at 3000 rpm to separate the residue and supernatant. The supernatant was filtered and adjusted to pH < 2 for analysis of As(V), Fe(II), and Mn(II).

Table 2. Sequential extraction procedures.

Fraction	Extractants	Extraction Conditions
F1 Soluble	20 mL of deionized water	Shaking at room temperature for 1 h
F2 Exch	20 mL of 1 M NH$_4$OAc, pH 7	Shaking at room temperature for 2 h
F3 CO$_3$	20 mL of 1 M NH$_4$OAc, pH 5	Shaking at room temperature for 2 h
F4 Fe/MnO	20 mL of 0.04 M NH$_2$OH.HCl in 25% acetic acid ($v\ v^{-1}$), pH 3	Placing in the water bath at 80 °C for 6 h
F5 Org/S	15 mL of 30% H$_2$O$_2$ ($v\ v^{-1}$), after being cooled add 5 mL of 3.2 M NH$_4$OAc in 20% HNO$_3$ ($v\ v^{-1}$)	Placing in the water bath at 80 °C for 5.5 h Shaking at room temperature for 0.5 h
F6 Residual	20 mL of 7 M HNO$_3$	Placing in the water bath at 80 °C for 6 h

The percentage of a heavy metal in each fraction (Fi) was calculated following Equation (4)

$$\%\text{Fraction}(i) = \frac{100 \times Fi}{(F1 + F2 + F3 + F4 + F5 + F6)} \tag{4}$$

3. Results

3.1. Biochar Characteristics

Table 3 shows the chemical and physical properties of each biochar. Results from XRF revealed that there were Fe and Mn in all bamboo-derived biochars and activated carbon, but arsenic was not detected. The highest content of metal from XRF analysis of B and AC was potassium (K). Chongtham et al. [30] reported that potassium was the main metal in bamboo shoot ranging from 41.90–66.60% whereas Fe and Mn were in the range from 0.047–0.10% and 0.012–0.097%, respectively. The oxides of calcium (Ca) and K could be hydrated to $Ca(OH)_2$ and KOH, respectively, and produce alkaline solutions. The pH_{DI} of AC was higher than B because it contains more base cations. Impregnation of chitosan and iron changed the elemental composition and decreased pH_{DI}. BC was prepared by coating chitosan on B, for which chitosan was dissolved in acetic acid solution. Thus, metals in B were released particularly for easily soluble metal such as potassium, resulting in higher relative contents of other metal contents than B. In addition, the percentages of Fe, Ca, silica (Si), magnesium (Mg), and phosphorus (P) were increased. Fe and chloride (Cl) contents were high in BFe because of the impregnation by $FeCl_3$. The pH_{DI} of BFe was lowest compared to the other adsorbents as the hydrolysis of Fe^{3+} formed $FeOH^{2+}$ and released hydrogen ions (H^+). BC and BFe had pH_{DI} lower than pH_{pzc}, hence their surfaces exhibited a net positive charge. BET surface area (S_{BET}) and pore volume (V_p) of BC and BFe were lower than B indicating that impregnation was achieved. However, pore diameters (D_p) of BC and BFe were closed to the original biochar (B). Iodine number measures the elementary iodine (I_2) which has a statistical radius of 2.46 Å [31]. Ferric chloride was used for impregnation and the Fe^{3+} with a radius of 0.65 Å [32] could fill up the micropores. The D_p of AC was less than B according to the pyrolysis temperature of AC was 1000 °C higher than those of B. Sahool et al. [33] reported the evolution of micropores in the biochar leads to a decrease in pore diameter at high pyrolysis temperature. This result agrees with the remarkably high iodine number of AC indicating its microporous structure.

Table 3. Chemical and physical characteristics of bamboo-derived biochar.

	Metal Composition (%)										S_{BET} ($m^2 g^{-1}$)	V_p ($cm^3 g^{-1}$)	D_p (Å)	pH_{DI}	pH_{pzc}	Iodine Number ($mg\ g^{-1}$)
	Mn	Fe	Ca	Si	S	P	Cl	Mg	As	K						
B	0.70	0.32	7.78	7.93	2.87	7.06	5.62	2.69	ND	64.5	191.9	0.10	21.64	8.66	7.63	75.30
BC	1.68	0.82	13.7	14.5	6.44	10.9	ND	3.41	ND	47.7	66.6	0.04	21.96	4.31	4.56	72.09
BFe	0.59	19.6	3.17	25.9	3.08	6.04	24.3	1.39	ND	13.2	127.7	0.07	22.06	2.59	3.61	45.54
AC	0.88	2.21	9.77	9.62	1.04	2.41	4.44	3.60	ND	65.7	136.0	0.11	19.52	9.35	8.34	267.38

The SEM images of biochar samples confirmed the high porosity of biochar with longitudinal pores originating from the vascular bundles of bamboo, whereas AC showed cracked surfaces (Figure 1). Chitosan was mostly impregnated on the surfaces, whereas iron was impregnated in the mesopores and micropores of biochar. AC had irregular surfaces and porous structure. In addition, there were some particles on the surfaces. Result from EDS revealed that carbon (C) content was high in B, BC, and AC, ranging from 76.24 to 88.51%, followed by oxygen (O), ranging from 7.84 to 25.37%. After chitosan impregnation of biochar, the lower percentage of carbon content and higher oxygen content suggested the presence of chitosan on the BC surface. As for BFe, the C and O contents were 33.46% and 25.37%, respectively. The O/C ratio of BFe was highest at 0.76, followed by BC, B, and AC, at 0.16, 0.08, and 0.09, respectively, indicating the increased oxygen functional groups on BFe surfaces. The low O/C ratio of B and AC was due to the loss in oxygen functional groups such as hydroxy, carboxyl, and carbonyl groups in the thermo-conversion process. At pyrolysis temperature of 300–400 °C, cellulose and hemicellulose are decomposed and lignin is decomposed above 400 °C [34]. However, Mn was found at the outer surfaces of

BC and BFe at 6.08% and 2.39%, respectively. This is due to the acid solution of BC and BFe which enhanced Mn mobility allowing its transport from the inner layers to outer surfaces. The stability of biochar can be estimated from the O/C ratio and the O/C of BFe was greater than 0.6, indicating a half-life of less than 100 years whereas the O/C of the rest of the biochars were less than 0.2 implying a half-life of more than 1000 years [35].

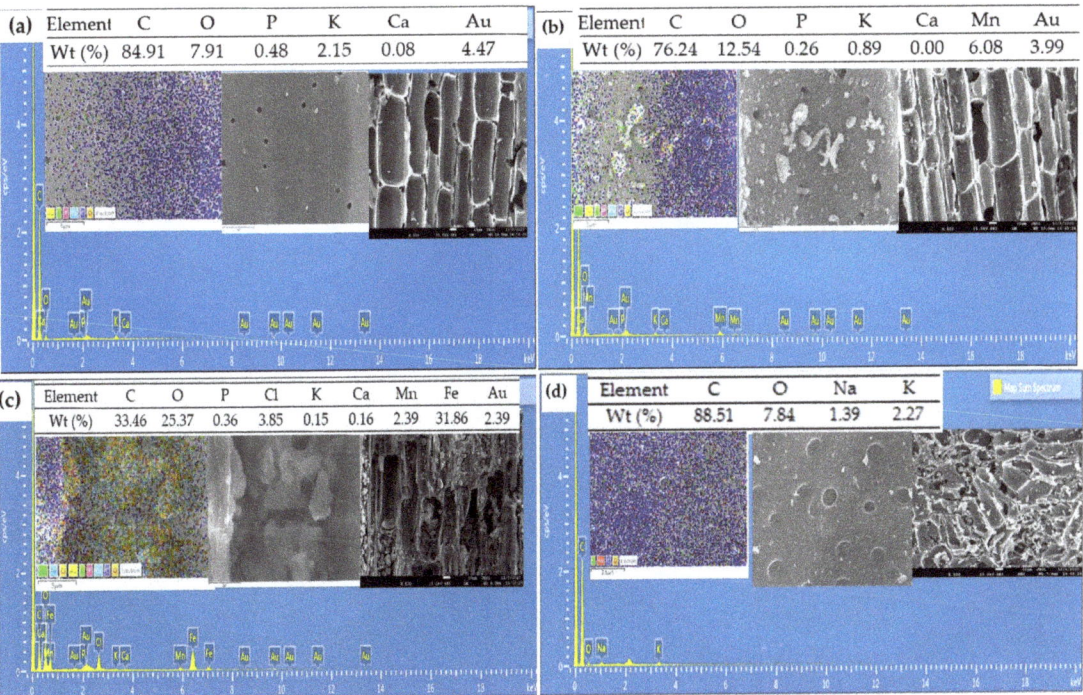

Figure 1. SEM-EDS images of bamboo-derived biochars. (**a**) B (**b**) BC (**c**) BFe (**d**) AC.

3.2. Adsorption Equilibration Time

Adsorption equilibration times of As and Fe in the single heavy metal system were faster than in the trinary system because heavy metal ions in the trinary system competed for the same active sites of biochar. BC and BFe reached equilibrium for As adsorption in the single heavy metal system at 5 and 2 h, respectively, which was faster than B and AC that reached equilibrium at 12 h (Figure 2a). Dissolved As(V) is negatively charged at neutral pH so it reacted with the positively charged surfaces of BC and BFe. Likewise, Fe (II) adsorption also reached equilibrium within 3 h for B and AC because of the opposite charge of adsorbent and adsorbate (Figure 2b). It took 12 and 48 h for BC and BFe, respectively, to reach Fe adsorption equilibrium in the single heavy metal system. Adsorption of Mn(II) required more time than As and Fe adsorption in the range of 72–108 h. In addition, the equilibration times of Mn adsorption in the single system were the same as in the trinary system (Figure 2c).

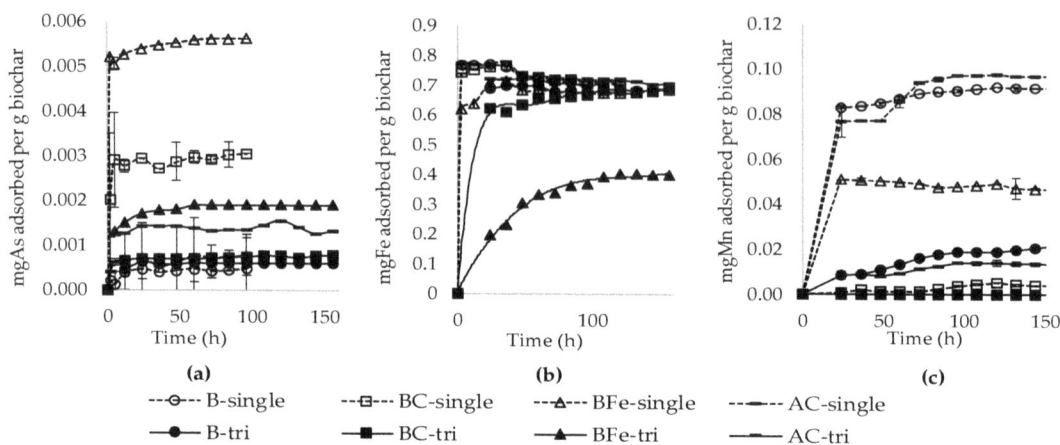

Figure 2. Equilibration time of (**a**) As(V), (**b**) Fe(II), and (**c**) Mn(II) adsorption in single and trinary heavy metal systems by different biochars and activated carbon.

3.3. Single Heavy Metal Solution Adsorption

Freundlich and Langmuir adsorption isotherms are widely used to explain the adsorption on heterogeneous and homogeneous of adsorbent surfaces, respectively. The corresponding parameters, correlation coefficients (R^2), and pH_E are listed in Table 4. The R^2 values calculated for each model indicated that the adsorption data by B and BC were more suitable described by the Langmuir model than by the Freundlich model. The Langmuir model also provided the best fit for Mn adsorption by BFe, and for As and Fe adsorption by AC. In addition, kinetic studies from previous work also revealed that the adsorption of As(V), Fe(II), and Mn (II) by all adsorbents were following pseudo-second order kinetics. This indicates that adsorption could be achieved via chemical processes [36]. The q_m from the Langmuir model revealed that chitosan impregnation enhanced all heavy metals adsorption, whereas iron impregnation enhanced only As adsorption. The highest As adsorption was 0.019 mg g^{-1} by BFe, and the highest Fe and Mn adsorption were 303.03 and 5.2029 mg g^{-1} by BC. BFe and BC significantly removed As and Fe, respectively. It shows that q_m values of BFe to remove As were higher than q_m values of B, BC and AC of 3.36, 2.03 and 10.83 times, respectively. The q_m values of BC were higher than q_m values of B, BFe, and AC of 1.76, 7.12, and 22.70 times for removal of Fe, respectively. Other work on iron-modified biochar reported high q_m for As(V) adsorption of 0.926 mg g^{-1} by palm leave biochar [37] and 0.19 mg g^{-1} by parsley leaf biochar [38]; however, those As(V) concentrations were higher in the ranges pf 0.5–30 mg L^{-1} and 0.05–2.00 mg g^{-1}, respectively, compared to the present study of 0.01–0.1 mg L^{-1}.

Table 4. Isotherm parameters and correlation coefficients for As, Fe, and Mn adsorption in the single heavy metal system.

Adsorbent	Heavy Metal	Freundlich			Langmuir			pH_E
		K_F (mg g^{-1}) (L mg^{-1})$^{1/n}$	$1/n$	R^2	q_m (mg g^{-1})	K_L (L mg^{-1})	R^2	
B	As (V)	0.0044	0.8435	0.9215	0.0058	0.7986	0.9630	7.98–8.78
	Fe(II)	353.25	0.9353	0.9450	172.41	0.3621	0.9482	8.00–8.33
	Mn (II)	1.1410	0.7727	0.9911	5.0226	3.3451	0.9914	8.19–8.59
BC	As (V)	0.0594	0.7384	0.8924	0.0096	0.0334	0.9162	5.00–5.33
	Fe(II)	274.21	0.9513	0.9959	303.03	0.9091	0.9966	6.13–6.34
	Mn (II)	0.0260	1.0336	0.9230	5.2029	181.92	0.9914	4.96–5.94
BFe	As (V)	0.1018	0.5052	0.9050	0.0195	0.0087	0.8639	6.05–6.98
	Fe(II)	97.55	0.8309	0.7684	42.55	0.2128	0.8971	6.33–6.37
	Mn (II)	0.0515	0.7436	0.9449	0.2170	2.6296	0.8137	5.03–6.96
AC	As (V)	0.0038	0.6713	0.8499	0.0018	0.1268	0.8889	9.32–9.50
	Fe(II)	25.23	0.2672	0.8541	13.35	0.0080	0.8153	8.95–8.99
	Mn (II)	2.5796	0.3375	0.9222	4.8355	0.4265	0.9317	8.89–9.24

BC gave high adsorption capacities because chitosan has free amino and hydroxyl groups for interaction with negative and positive charges of heavy metals, respectively. The pK_a of the amino group in chitosan ranges from 6.2–7.0 [39]. Hence, at a solution pH below 6.2 the amino group is protonated. Reportedly, the extent of protonation is 9, 50, 91, and 99 % at pH 7.3, 6.3, 5.3, and 4.3, respectively [40]. The pH of solution also affects the adsorbent surfaces and the specification of heavy metals. The speciation of As at different pHs is shown in Equations (5)–(7) [41].

$$H_3AsO_4 \rightleftharpoons H^+ + H_2AsO_4^- \quad pK_1 = 2.3 \quad (5)$$

$$H_2AsO_4^- \rightleftharpoons H^+ + HAsO_4^{2-} \quad pK_2 = 6.8 \quad (6)$$

$$HAsO_4^{2-} \rightleftharpoons H^+ + AsO_4^{3-} \quad pK_3 = 11.6 \quad (7)$$

The adsorption experiments for As adsorption were performed at the initial pH of 7.00 ± 0.50. Use of BC as adsorbent could decrease the pH values and the pH_E for As adsorption were in the range of 5.00–5.33. Thus, the main speciation of As was $H_2AsO_4^-$, whilst the amino and hydroxyl groups of BC were protonated. These protonated groups at acidic pH result in electrostatic attractions with negatively charge arsenate species. At the solution pH below pH_{pzc}, anions favor to adsorb on the adsorbent surfaces. The pH_{pzc} values of B and AC were 7.63 and 8.34, respectively, enabling adsorption of $HAsO_4^-$ species. As for BFe, the main mechanisms for arsenate removal were the surface complexation with iron oxyhydroxide and partial inclusion into the crystalline iron oxides [20,28].

The initial pH for Fe adsorption is 4.40 ± 0.50, which is lower than pH_{pzc} of B, BC, and AC, indicating these adsorbents exhibit positive charge. Thus, the repulsion of Fe^{2+} could occur. BFe exhibits negative charge resulting in electrostatic attraction. However, the pH values were increased after contacting with adsorbents. This was due to the release of base cations and hydroxide ions from adsorbents into solution. The pH_E values of B and AC were alkaline (8.00–8.99), whereas the pH_E values of BC and BFe ranged from 6.13–6.37. In the agitation during adsorption experiments, Fe^{2+} could be oxidized to $Fe(OH)_3$. Hove et al. [41] expressed that the oxidation rate at pH 9.0 was higher than at pH 6.0. The results agree with the study of contact time showing B and AC reached equilibration times faster than BC and BFe.

Adsorption of Mn was performed at neutral pH; thus, the main species was Mn^{2+}. The pH_{pzc} values of all adsorbents were lower than pH_E indicating favorable conditions to adsorb Mn^{2+}. The formation of manganese oxide (MnO_2) from aeration at pH below 9.0 is a slow process and may not be achieved [42]. Because of this, Mn is hardly removed via adsorption.

It should be noted that R^2 values of the Freundlich model were closed to those of the Langmuir model, therefore, heavy metals possibly adsorbed in the inner pores. The radius of $H_2AsO_4^-$, Fe^{2+} and Mn^{2+} are 2.27, 0.77, and 0.83 Å [32], respectively which are smaller than pore diameters of the adsorbents. In addition, multilayer adsorption such as surface complexation may have occurred as the simulated groundwater contains cations and anions.

The pH_E values of BC were lower than 6.0 for As and Mn adsorption which was outside the allowable range of pH 6.5–9.2 of drinking water standards in Thailand [43]. Thus, it may not be suitable for groundwater treatment as low pH could corrode and induce the release of other ions from the adsorbent. Hence, B, BFe, and AC were selected for the trinary adsorption experiments.

3.4. Trinary Heavy Metal Solution Adsorption

Arsenic was completely removed from the trinary heavy metal system by adsorption; therefore, the data could not be fitted by the adsorption models. Saikia et al. [44] also reported that iron oxide coated sand and other adsorbents removed As(III) 100 % from water at concentration ranges 0.6–1.0 mg L^{-1} in the presence of Fe(II) and Mn(II). The initial pH of solution was 7.00 ± 0.50. After being agitated, Fe^{2+} was oxidized to Fe^{3+} and then it could react with As(V) to ferric arsenate (i.e., $FeAsO_4$). Table 5 shows that the Langmuir model provides the best fit for Fe and Mn adsorption by B and BFe, and for Fe adsorption by AC.

Table 5. Isotherm parameters and correlation coefficients for Fe and Mn adsorption in trinary system.

Adsorbent	Heavy Metal	Freundlich			Langmuir			pH_E
		K_F (mg g^{-1}) (L mg^{-1})$^{1/n}$	1/n (L g^{-1})	R^2	q_m (mg/g^{-1})	K_L (L mg^{-1})	R^2	
B	Fe(II)	2.6710	0.6808	0.8022	2.2568	0.3056	0.8360	6.82–8.44
	Mn(II)	0.2559	0.6979	0.9020	0.6393	1.5398	0.9155	
Bfe	Fe(II)	36.874	0.7530	0.8686	26.1780	0.3010	0.9008	5.58–6.80
	Mn(II)	0.2715	0.8946	0.8388	3.6643	12.342	0.8732	
AC	Fe(II)	3.6370	0.8387	0.8632	3.1756	0.5062	0.8868	6.80–9.47
	Mn(II)	0.2142	0.8680	0.9031	1.3541	5.3369	0.8823	

The q_m values show that adsorption of Fe and Mn is in the order of BFe > AC > B. The q_m values from trinary heavy metal adsorption experiments were less than q_m values from single heavy metal adsorption, except for Mn adsorption by BFe. The pH_E values of BFe were far higher than pH_{pzc} (3.61), because of the negative charge of BFe surfaces. AC had higher adsorption than B because its porous structure enabled adsorption of heavy metals in the inner pores. In addition, Mn can be oxidized to MnO_2 at pH 9. To examine the main functional groups for heavy metal adsorption, the FTIR spectra of virgin and loaded B, BFe, and AC were compared in Figure 3.

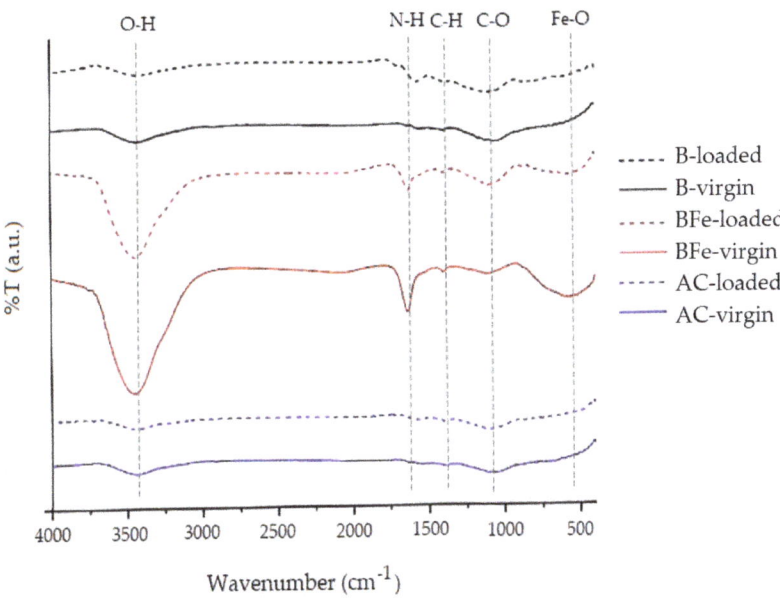

Figure 3. Infrared spectra of virgin and loaded B, BFe, and AC.

The spectral bands obtained at 3200–3600 cm^{-1} were attributed to the O-H stretching vibration of alcohols [45]. The band ranges 1550–1640 cm^{-1} were denoted as N-H stretching vibration of amides [45], and ranges 1400–1450 cm^{-1} was the C-H bending vibration of alkanes [45]. The spectral bands at 1000–1300 cm^{-1} were attributed C-O stretching vibration of ethers. The peaks of virgin adsorbents were similar to the peaks of loaded adsorbents, but the peak intensities of loaded B and AC were smaller than those of virgin B and AC, especially for the O-H group. This indicates the binding of Fe(II) and Mn(II) by this group. Virgin and loaded BFe were found to have additional peaks at 582 and 597 cm^{-1}, respectively, which represented the Fe-O bond from iron impregnation [46]. The intensity of characteristic bands of AC was low comparing to the bands of B and BFe because the high pyrolysis temperature (1000 °C) for AC preparation could degrade cellulose, hemicellulose, and lignin. The presence of O-H, N-H and C-O functional groups confirms their availability for adsorption via different mechanisms.

3.5. Fraction of Heavy Metal Adsorption onto Bamboo Biochar and Activated Carbon

Table 6 presents the distribution pattern of As, Fe, and Mn in virgin and loaded B and AC. As was not found in virgin B, but small amounts were found in F1 and F4 of virgin AC. After loading with mixed heavy metals, As contents were found in F1. Fe was found in all fractions, except in F3 in both virgin B and AC, and was predominantly in F6 for virgin B at 76.2%, and in virgin AC at 47.6%. Fe in the soluble fraction may have reacted with As and precipitated from the solution, thus As was found in the soluble form after being loaded onto the adsorbents. Fe content was decreased after adsorption indicating Fe also reacted with other anionic ions such as sulfate ions. The main fraction of Fe after adsorption was in F5 for B and in F6 for AC. Mn fractions in virgin B and AC were in the following order: F6 > F4 > F5~F2 > F1. It previously reported that the highest Mn fraction was found in the residual fraction (F6) in goat manure biochar and bound to organic matter (F5) in swine manure biochar [47]. Once B and AC were loaded with heavy metals, the metals were mainly found in the fractions F4 and F3. Since F1, F2, and F3 are weakly binding, it was found that Mn is more present in mobile fractions than Fe. The mobile fractions of Mn after being adsorbed were 44.15 and 41.03% for B and AC, respectively.

Table 6. Comparison of As, Fe, and Mn contents in each fraction of virgin and loaded B and AC.

Metal	Adsorbents			Fractions (mg kg^{-1})						
			F1 Soluble	F2 Exch	F3 CO$_3$	F4 Fe/MnO	F5 Org/S	F6 Residual	Total
As	B	Virgin	0.00 ± 0.00	0.00 ± 0.00	0.00 ± 0.00	0.00 ± 0.00	0.00 ± 0.00	0.00 ± 0.00	0.00 ± 0.00
		Loaded	0.02 ± 0.00	0.00 ± 0.00	0.00 ± 0.00	0.00 ± 0.00	0.00 ± 0.00	0.00 ± 0.00	0.02 ± 0.00
	AC	Virgin	0.04 ± 0.00	0.00 ± 0.00	0.00 ± 0.00	0.00 ± 0.00	0.00 ± 0.00	0.00 ± 0.00	0.04 ± 0.00
		Loaded	0.01 ± 0.01	0.00 ± 0.00	0.00 ± 0.00	0.00 ± 0.00	0.00 ± 0.00	0.00 ± 0.00	0.01 ± 0.01
Fe	B	Virgin	0.23 ± 0.01	1.20 ± 0.42	0.00 ± 0.00	2.03 ± 0.95	9.45 ± 3.45	41.47 ± 3.99	54.40 ± 3.99
		Loaded	0.41 ± 0.19	1.36 ± 0.26	2.44 ± 0.30	9.35 ± 0.13	14.75 ± 2.73	7.98 ± 2.18	36.29 ± 2.73
	AC	Virgin	0.20 ± 0.06	0.34 ± 0.16	0.00 ± 0.00	94.64 ± 7.41	9.61 ± 1.65	95.21 ± 18.16	200.01 ± 18.16
		Loaded	0.41 ± 0.06	0.77 ± 0.00	0.62 ± 0.23	54.46 ± 4.07	3.97 ± 1.44	72.73 ± 12.32	132.96 ± 12.32
Mn	B	Virgin	0.67 ± 0.02	3.55 ± 0.28	8.32 ± 0.06	11.84 ± 3.79	9.15 ± 0.37	14.00 ± 1.69	47.53 ± 3.79
		Loaded	0.34 ± 0.04	15.83 ± 1.36	22.91 ± 0.88	24.24 ± 0.04	12.57 ± 0.64	12.64 ± 0.24	88.54 ± 1.36
	AC	Virgin	0.00 ± 0.00	0.14 ± 0.09	0.72 ± 0.01	2.22 ± 0.20	0.91 ± 0.14	2.98 ± 0.39	6.97 ± 0.39
		Loaded	0.00 ± 0.00	1.88 ± 0.36	3.82 ± 0.29	4.65 ± 0.48	1.30 ± 0.09	2.26 ± 0.25	13.92 ± 0.36

4. Discussion

Bamboo biochar can replace commercial activated carbon for heavy metal removal from groundwater. Bamboo biochar and activated carbon had the same functional groups of O-H, N-H, C-H and C-O. The pH of solution and heavy metal speciation influenced the adsorption of As(V), Fe(II), and Mn(II) both in single and trinary heavy metal systems. The pH values of treated water after heavy metal adsorption by bamboo biochar were in the range suitable for drinking water production, whereas AC produced alkaline water, and BC and BFe produced acidic water. However, the maximum adsorption of As(V), Fe(II), and Mn(II) were improved for the chitosan or iron impregnated biochars. BC and BFe can thus be used for removal of As, Fe, and Mn contamination in groundwater, but pH needs readjustment to the acceptable values for drinking water. In addition, treated water should be analyzed for other heavy metals which may be released from biochar. Further work should scrutinize the chemicals used for impregnation and the release of other heavy metals after adsorption. BC and BFe may also be suitable for removal of high concentration of heavy metals such as from mining wastewater.

Author Contributions: Conceptualization, S.V.; Funding acquisition, S.V.; Investigation, A.P., N.K., P.W., R.N. and N.S.; Writing—original draft, S.V.; Writing—review and editing, A.P., S.V., D.W. and S.S. All authors have read and agreed to the published version of the manuscript.

Funding: This research was funded by the National Research Council of Thailand, grant number NRCT.MHESI(A)(PS)/122/2563.

Data Availability Statement: Not applicable.

Acknowledgments: Additional supports for biochar characterization were provided by the School of Energy, Environment and Materials and from the study of Kruatong N. through Petchra Pra Jom Klao Ph.D. Research Scholarship (No. 54/2564), King Mongkut's University of Technology Thonburi.

Conflicts of Interest: The authors declare no conflict of interest.

References

1. Zhou, Q.; Yang, N.; Li, Y.; Ren, B.; Ding, X.; Bian, H.; Yao, X. Total Concentrations and Sources of Heavy Metal Pollution in Global River and Lake Water Bodies from 1972 to 2017. *Glob. Ecol. Conserv.* **2020**, *22*, e00925. [CrossRef]
2. Shaji, E.; Santosh, M.; Sarath, K.V.; Prakash, P.; Deepchand, V.; Divya, B.V. Arsenic Contamination of Groundwater: A Global Synopsis with Focus on the Indian Peninsula. *Geosci. Front.* **2021**, *12*, 101079. [CrossRef]
3. Chakraborty, T.K.; Chandra Ghosh, G.; Hossain, M.R.; Islam, M.S.; Habib, A.; Zaman, S.; Bosu, H.; Nice, S.; Haldar, M.; Khan, A.S. Human Health Risk and Receptor Model-Oriented Sources of Heavy Metal Pollution in Commonly Consume Vegetable and Fish Species of High Ganges River Floodplain Agro-Ecological Area, Bangladesh. *Heliyon* **2022**, *8*, e11172. [CrossRef]
4. Zhang, Z.; Xiao, C.; Adeyeye, O.; Yang, W.; Liang, X. Source and Mobilization Mechanism of Iron, Manganese and Arsenic in Groundwater of Shuangliao City, Northeast China. *Water* **2020**, *12*, 534. [CrossRef]

5. Santha, N.; Sangkajan, S.; Saenton, S. Arsenic Contamination in Groundwater and Potential Health Risk in Western Lampang Basin, Northern Thailand. *Water* **2022**, *14*, 465. [CrossRef]
6. Abbaspour, N.; Hurrell, R.; Kelishadi, R. Review on Iron and Its Importance for Human Health. *J. Res. Med. Sci.* **2014**, *19*, 164–174. [PubMed]
7. Li, L.; Yang, X. The Essential Element Manganese, Oxidative Stress, and Metabolic Diseases: Links and Interactions. *Oxidative Med. Cell. Longev.* **2018**, *2018*, 7580707. [CrossRef] [PubMed]
8. Ye, Q.; Park, J.E.; Gugnani, K.; Betharia, S.; Pino-Figueroa, A.; Kim, J. Influence of Iron Metabolism on Manganese Transport and Toxicity. *Metallomics* **2017**, *9*, 1028–1046. [CrossRef] [PubMed]
9. Kohgo, Y.; Ikuta, K.; Ohtake, T.; Torimoto, Y.; Kato, J. Body Iron Metabolism and Pathophysiology of Iron Overload. *Int. J. Hematol.* **2008**, *88*, 7–15. [CrossRef]
10. Avila, D.S.; Puntel, R.L.; Aschner, M. Manganese in Health and Disease. In *Interrelations between Essential Metal Ions and Human Diseases*; Metal Ions in Life Sciences; Sigel, A., Sigel, H., Sigel, R.K.O., Eds.; Springer: Dordrecht, The Netherland, 2013; Volume 13, pp. 199–227.
11. Fatoki, J.O.; Badmus, J.A. Arsenic as an Environmental and Human Health Antagonist: A Review of Its Toxicity and Disease Initiation. *J. Hazard. Mater.* **2022**, *5*, 100352. [CrossRef]
12. WHO. Progress on Drinking Water, Sanitation and Hygiene: 2017 Update and SDG Baselines. World Health Organization (WHO), Geneva and the United Nations Children's Fund (UNICEF), New York. Available online: https://www.who.int/publications-detail-redirect/9789241549950 (accessed on 27 December 2022).
13. Inyang, M.I.; Gao, B.; Yao, Y.; Xue, Y.; Zimmerman, A.; Mosa, A.; Pullammanappallil, P.; Ok, Y.S.; Cao, X. A Review of Biochar as a Low-Cost Adsorbent for Aqueous Heavy Metal Removal. *Crit. Rev. Environ. Sci. Technol.* **2016**, *46*, 406–433. [CrossRef]
14. Li, H.; Dong, X.; da Silva, E.B.; de Oliveira, L.M.; Chen, Y.; Ma, L.Q. Mechanisms of Metal Sorption by Biochars: Biochar Characteristics and Modifications. *Chemosphere* **2017**, *178*, 466–478. [CrossRef]
15. Gao, N.; Du, W.; Zhang, M.; Ling, G.; Zhang, P. Chitosan-Modified Biochar: Preparation, Modifications, Mechanisms and Applications. *Int. J. Biol. Macromol.* **2022**, *209*, 31–49. [CrossRef] [PubMed]
16. Loc, N.X.; Tuyen, P.T.T.; Mai, L.C.; Phuong, D.T.M. Chitosan-Modified Biochar and Unmodified Biochar for Methyl Orange: Adsorption Characteristics and Mechanism Exploration. *Toxics* **2022**, *10*, 500. [CrossRef] [PubMed]
17. Rubeena, K.K.; Hari Prasad Reddy, P.; Laiju, A.R.; Nidheesh, P.V. Iron Impregnated Biochars as Heterogeneous Fenton Catalyst for the Degradation of Acid Red 1 Dye. *J. Environ. Manag.* **2018**, *226*, 320–328. [CrossRef] [PubMed]
18. Dalahmeh, S.S.; Stenström, Y.; Jebrane, M.; Hylander, L.D.; Daniel, G.; Heinmaa, I. Efficiency of Iron- and Calcium-Impregnated Biochar in Adsorbing Phosphate from Wastewater in Onsite Wastewater Treatment Systems. *Front. Environ. Sci.* **2020**, *8*, 538539. [CrossRef]
19. He, R.; Peng, Z.; Lyu, H.; Huang, H.; Nan, Q.; Tang, J. Synthesis and Characterization of an Iron-Impregnated Biochar for Aqueous Arsenic Removal. *Sci. Total Environ.* **2018**, *612*, 1177–1186. [CrossRef] [PubMed]
20. Fan, J.; Xu, X.; Ni, Q.; Lin, Q.; Fang, J.; Chen, Q.; Shen, X.; Lou, L. Enhanced As (V) Removal from Aqueous Solution by Biochar Prepared from Iron-Impregnated Corn Straw. *J. Chem.* **2018**, *2018*, 5137694. [CrossRef]
21. Sun, Y.; Yu, F.; Han, C.; Houda, C.; Hao, M.; Wang, Q. Research Progress on Adsorption of Arsenic from Water by Modified Biochar and Its Mechanism: A Review. *Water* **2022**, *14*, 1691. [CrossRef]
22. Khawkomol, S.; Neamchan, R.; Thongsamer, T.; Vinitnantharat, S.; Panpradit, B.; Sohsalam, P.; Werner, D.; Mrozik, W. Potential of Biochar Derived from Agricultural Residues for Sustainable Management. *Sustainability* **2021**, *13*, 8147. [CrossRef]
23. Wang, A.; Zou, D.; Zeng, X.; Chen, B.; Zheng, X.; Li, L.; Zhang, L.; Xiao, Z.; Wang, H. Speciation and Environmental Risk of Heavy Metals in Biochars Produced by Pyrolysis of Chicken Manure and Water-Washed Swine Manure. *Sci. Rep.* **2021**, *11*, 11994. [CrossRef] [PubMed]
24. Thongsamer, T.; Vinitnantharat, S.; Pinisakul, A.; Werner, D. Chitosan Impregnation of Coconut Husk Biochar Pellets Improves Their Nutrient Removal from Eutrophic Surface Water. *Sustain. Environ. Res.* **2022**, *32*, 39. [CrossRef]
25. Lamaming, J.; Saalah, S.; Rajin, M.; Ismail, N.M.; Yaser, A.Z. A Review on Bamboo as an Adsorbent for Removal of Pollutants for Wastewater Treatment. *Int. J. Chem. Eng. Res.* **2022**, *2022*, 7218759. [CrossRef]
26. Hernandez-Mena, L.; Pecora, A.; Beraldo, A. Slow Pyrolysis of Bamboo Biomass: Analysis of Biochar Properties. *Chem. Eng. Trans.* **2014**, *37*, 115–120.
27. Vinitnantharat, S.; Rattanasirisophon, W.; Ishibashi, Y. Modification of Granular Activated Carbon Surface by Chitosan Coating for Geosmin Removal: Sorption Performances. *Water Sci. Technol.* **2017**, *55*, 145–152. [CrossRef]
28. Kalaruban, M.; Loganathan, P.; Nguyen, T.V.; Nur, T.; Johir, M.A.H.; Nguyen, T.H.; Trinh, M.V. Iron-impregnated Granular Activated Carbon for Arsenic Removal: Application to Practical Column Filters. *J. Environ. Manag.* **2019**, *239*, 235–243. [CrossRef]
29. Kashem, M.A.; Singh, B.R.; Kondo, T.; Imamul Huq, S.M.; Kawai, S. Comparison of Extractability of Cd, Cu, Pb and Zn with Sequential Extraction in Contaminated and Non-cantaminated Soils. *Int. J. Environ. Sci. Tech.* **2007**, *4*, 169–176. [CrossRef]
30. Chongtham, N.; Bisht, M.S.; Santosh, O.; Bajwa, H.K.; Indira, A. Mineral Elements in Bamboo Shoots and Potential Role in Food Fortification. *J. Food Compos. Anal.* **2021**, *95*, 103662. [CrossRef]
31. Mianowski, A.; Owczarek, M.; Marecka, A. Surface Area of Activated Carbon Determined by the Iodine Adsorption Number. *Energy Sources Part A* **2007**, *29*, 839–850. [CrossRef]

32. Simoes, M.C.; Hughes, K.J.; Ingham, D.B.; Ma, L.; Pourkashanian, M. Estimation of the Thermochemical Radii and Ionic Volumes of Complex Ions. *Inorg. Chem.* **2017**, *56*, 7566–7573. [CrossRef]
33. Sahool, S.S.; Vijay, V.K.; Chandra, R.; Kumar, H. Production and Characterization of Biochar Produced from Slow Pyrolysis of Pigeon Pea Stalk and Bamboo. *J. Clean. Prod.* **2021**, *3*, 100101.
34. Chen, D.; Yu, X.; Song, C.; Pang, X.; Huang, J.; Li, Y. Effect of Pyrolysis Temperature on the Chemical Oxidation Stability of Bamboo Biochar. *Bioresour. Technol.* **2016**, *218*, 1303–1306. [CrossRef] [PubMed]
35. Spokas, K.A. Review of the Stability of Biochar in Soils: Predictability of O:C Molar Ratios. *Carbon Manag.* **2010**, *1*, 289–303. [CrossRef]
36. Kruatong, N.; Vinintnantharat, S.; Pinisakul, A.; Wilamas, A.; Sukkhe, N. Use of Biochar Impregnated with Iron and Chitosan for Heavy Metal Removal: Sorption Performances. In Proceedings of the 47th International Congress on Science, Technology and Technology-Based Innovation: Sciences for SDGs: Challenges and Solutions, Kasetsart University, Kamphaeng Saen Campus, Nakhon Pathom, Thailand, 5–7 October 2021; pp. 614–620.
37. Kirmizakis, P.; Tawabini, B.; Siddiq, O.M.; Kalderis, D.; Ntarlagiannis, D.; Soupios, P. Adsorption of Arsenic on Fe-modified Biochar and Monitoring Using Spectral Induced Polarization. *Water* **2022**, *14*, 563. [CrossRef]
38. Jiménez-Cedillo, M.J.; Olguín, M.T.; Fall, C.; Colin-Cruz, A. As(III) and As(V) Sorption on Iron-modified Non-pyrolyzed and Pyrolyzed Biomass from *Petroselinum crispum* (Parsley). *J. Environ. Manag.* **2013**, *117*, 242–252. [CrossRef]
39. Lodhi, G.; Kim, Y.S.; Hwang, J.W.; Kim, S.K.; Jeon, Y.J.; Je, J.Y.; Ahn, C.B.; Moon, S.H.; Jeon, B.T.; Park, P.J. Chitooligosaccharide and Its Derivatives: Preparation and Biological Applications. *Biomed. Res. Int.* **2014**, *2014*, 654913. [CrossRef]
40. Nomanbhay, S.M.; Palanisamv, K. Removal of Heavy Metal from Industrial Wastewater Using Chitosan Coated Oil Palm Shell Charcoal. *Electron. J. Biotechnol.* **2005**, *8*, 43–53. [CrossRef]
41. Hove, M.; van Hille, R.P.; Lewis, A.E. Mechanisms of Formation of Iron Precipitates from Ferrous Solutions at High and Low pH. *Chem. Eng. Sci.* **2008**, *63*, 1626–1635. [CrossRef]
42. Aziz, H.A.; Smith, P.G. The Influence of pH and Coarse Media on Manganese Precipitation from Water. *Wat. Res.* **1992**, *26*, 853–855. [CrossRef]
43. LIRT. Groundwater Quality Standard for Drinking Purpose, Notification of the Ministry of Industry, No. 12, BE 2542 (1999). Legislative Institutional Repository of Thailand. Available online: https://dl.parliament.go.th/handle/20.500.13072/230880 (accessed on 27 March 2023).
44. Saikia, A.; Agnihotri, G.; Raul, K.P.; Banerjee, S.; Dwivedi, K.S. Integrated Approach to Remove Iron, Arsenic and Manganese from Water Using Manganese Greensand and Other Adsorbent. *Indian J. Environ. Sci.* **2020**, *16*, 105.
45. Naik, D.K.; Monika, K.; Prabhakar, S.; Parthasarathy, R.; Satyavathi, B. Pyrolysis of Sorghum Bagasse Biomass into Bio-char and Bio-oil Products. *J. Therm. Anal. Calorim.* **2017**, *127*, 1277–1289. [CrossRef]
46. Liang, H.; Zhu, C.; Ji, S.; Kannan, P.; Chen, F. Magnetic Fe_2O_3/biochar Composite Prepared in a Molten Salt Medium for Antibiotic Removal in Water. *Biochar* **2022**, *4*, 3. [CrossRef]
47. Zeng, X.; Xiao, Z.; Zhang, G.; Wang, A.; Li, Z.; Liu, Y.; Wang, H.; Zeng, Q.; Liang, Y.; Zou, D. Speciation and Bioavailability of Heavy Metals in Pyrolytic Biochar of Swine and Goat manures. *J. Anal. Appl. Anal.* **2018**, *132*, 82–93. [CrossRef]

Disclaimer/Publisher's Note: The statements, opinions and data contained in all publications are solely those of the individual author(s) and contributor(s) and not of MDPI and/or the editor(s). MDPI and/or the editor(s) disclaim responsibility for any injury to people or property resulting from any ideas, methods, instructions or products referred to in the content.

Article

Replacing Lime with Rice Husk Ash to Reduce Carbon Footprint of Bituminous Mixtures

Raja Mistry [1], Tapash Kumar Roy [1,*], Sand Aldagari [2] and Elham H. Fini [2,*]

[1] Department of Civil Engineering, Indian Institute of Engineering Science and Technology, Shibpur, Howrah 711 103, West Bengal, India
[2] School of Sustainable Engineering and the Built Environment, Arizona State University, 660 S. College Avenue, Tempe, AZ 85287, USA
* Correspondence: tapash@civil.iiests.ac.in (T.K.R.); efini@asu.edu (E.H.F.)

Abstract: There have been several emphasized pathways toward a reduction in carbon footprint in the built environment such as recycling, technologies with lower energy consumption, and alternative materials. Among alternative materials, bio-based materials and nature inspired solutions have been well-received. This study examines the merits of using rice husk ash as a replacement for lime; lime has a high carbon footprint mainly associated with the decomposition of calcium carbonate to calcium oxide to form lime. Lime is commonly used in bituminous composites for roadway construction to mitigate their susceptibility to moisture damage. Replacing lime with a low-carbon alternative could allow a reduction in CO_2 equivalent of bituminous composites. This paper studies the merits of using rice husk ash (RHA) as a substitute for conventional hydrated lime (HL) in bituminous composites. It should be noted that rice industries burn rice husks in a boiler as fuel, generating a substantial volume of RHA. The disposal of this ash has major environmental impacts associated with the contamination of air and water. Here, we study physical and chemical characteristics of both HL and RHA for use in bitumen mixtures. This was followed by examining the extent of dispersion of each filler in bitumen via optical microscopy to ensure their uniform dispersion. The properties of the mixtures were further studied using the Marshall mix design method. It was found that a 25.67% increase in Marshall stability and a 5.95% decrease in optimum binder content were achieved when HL was replaced by RHA at 4% filler concentration. In addition, mixtures containing RHA exhibited higher resistance to cracking and permanent deformation compared to mixtures containing HL. Additionally, 4% RHA in the mix showed stripping resistance similar to the conventional mix with HL. The mixture with 4% RHA had a lower carbon footprint with enhanced economic and environmental impacts compared to the conventional mix with HL. The study results provide insights pertaining to the merits of bio-based materials to reduce the carbon footprint of pavements.

Keywords: sustainability; carbon footprint; rice husk ash; lime; silica; asphalt; concrete; roadways

Citation: Mistry, R.; Roy, T.K.; Aldagari, S.; Fini, E.H. Replacing Lime with Rice Husk Ash to Reduce Carbon Footprint of Bituminous Mixtures. C **2023**, 9, 37. https://doi.org/10.3390/c9020037

Academic Editors: Indra Neel Pulidindi, Pankaj Sharma and Aharon Gedanken

Received: 26 December 2022
Revised: 8 March 2023
Accepted: 10 March 2023
Published: 27 March 2023

Copyright: © 2023 by the authors. Licensee MDPI, Basel, Switzerland. This article is an open access article distributed under the terms and conditions of the Creative Commons Attribution (CC BY) license (https:// creativecommons.org/licenses/by/ 4.0/).

1. Introduction

Generally, a bituminous mix is composed of aggregate, mastic (bitumen mixed with filler), and air voids [1]. Fillers are fine material that mostly passes through a 75-μm sieve and occupies 2–10% of the total mixture's aggregate weight [2]. The critical roles of filler are to steady the mix by filling the holes in the aggregate skeleton and to enhance the viscoelastic properties of the mastic that glues the aggregate together [3]. It is commonly acknowledged that the performance of bituminous mixes is greatly affected by the overall characteristics of the filler [3].

Conventionally, stone dust produced during the process of stone crushing was used as filler in a bituminous mix. Presently, cement and hydrated lime (HL) are used as fillers in place of stone dust, since they deliver better binding and antistripping properties in the mix [4]. In comparison with cement, HL acts as an active filler and enhances adhesion by altering the aggregate's surface chemistry and reacting with acid components in bitumen [5]. These conventionally used fillers are acquired by mining, and their constant use in bituminous mixes has led to their shortage in different parts of the world. Regular extraction for control fillers also causes the loss of riparian vegetation, damage in water-retentive strata, lowering of the groundwater level, and disturbance of the remaining ecosystem [6]. There are additional negative impacts on the environment: every ton of cement production emits 0.73–0.99 tons of CO_2, and every ton of lime production emits 0.2–0.45 tons of CO_2 [7].

The use of available waste as an eco-friendly replacement for conventional fillers will improve sustainability in building roads and create a way to lessen the carbon footprint throughout the world. However, before applying such wastes, it is essential to check the technical and economic viability of the resulting bituminous mix.

Rice is a primary food source for more than 100 countries and is consumed as an essential food by nearly half of the world's population. In 2014, the global rice harvesting area was 162.72 million ha, and 741.48 million tons of rice were produced. Asia alone produced more than 90% of the rice; India contributed 21.20% of the total output [8]. In the production of milled rice, huge quantities of rice husk are generated as by-products. For instance, India produced 104.8 million tons of rice in 2015 [9]. Theoretically, in 2015, India generated 20.96 million tons of rice husk, which caused a problem for the rice-grinding industry. Due to the inherent properties of rice husk (hard surface, poor nutritional value, high silica content, fair resistance to bacterial decomposition), it is left unused or burned in a boiler to generate steam or electricity. Additionally, the burning of rice husk produces another source of pollutants called rice husk ash (RHA). The volume of this waste generated every year (5.58 million tons in 2019) is so large that the safe disposal of this waste powder has become a severe problem for the rice-milling industry [10]. Usually, RHA is discarded in nearby landfills, resulting in severe environmental pollution and contamination of water resources [10].

To overcome these problems, investigators have made many attempts to use RHA as added building materials to reduce the contamination effect and total construction cost. For instance, RHA has been used as a pozzolanic material in the construction industry [11]. From previous studies, it is known that RHA can be grinded finer and become more active because of the metastable state of the surface structure [12]. As a result, the activated fine nature of RHA makes it a potential filler/modifier in other materials, such as a bituminous binder or mix [12]. However, very few studies deal with RHA's use in a bituminous mix. The application of RHA was examined as a partial replacement for the standard filler, i.e., limestone powder [4]. One study investigated the applicability of RHA combined with brick powder and waste glass dust as a filler in a bituminous mix; in that study, RHA did not show any noteworthy performance compared to the other fillers [13]. The result of another study showed that RHA could boost the visco-elastic properties of a bituminous binder. Furthermore, RHA upgraded the elastic response of the modified binder [12]. Yet, nearly all these research works are limited to only a few properties of the bituminous mixes, or the studies applied RHA in an arbitrary percentage. Various studies also established that RHA could be used as filler in a bituminous mix, but tests of RHA's effectiveness against stripping acquired conflicting results [13]. Additionally, up to now, no study has investigated how modification with RHA affects the physical and morphological properties of the resultant bituminous mastic. Thus, further investigation is needed on the usefulness of RHA as a filler in a bituminous mix.

Objectives

This research has four objectives:
1. Analyze the suitability of HL and RHA individually as fillers by evaluating their essential properties.
2. Examine the effect of each filler on the physical and morphological behavior of the resultant bituminous mastics.

3. Design dense bituminous macadam mixes using each filler at four percentages (2%, 4%, 6%, and 8%) and compare their mechanical and durability properties at the corresponding optimum bitumen content.
4. Analyze the economic viability of each filler by comparing the costs of the prepared mixes.

2. Materials and Methods

2.1. Materials

Aggregates of the basalt type were gathered from a Pakur quarry (Birbhum district, West Bengal, India) and used in this study. Table 1 provides a description of the aggregates' attributes. To suit the continuous aggregate gradation of dense bituminous macadam (DBM) grade-II established by Indian standard specification [2], the collected coarse and fine aggregates were sieved. The results are shown in Figure 1.

Table 1. Evaluated properties of studied aggregates.

Property	Values	Specified Limits [2]
Aggregate impact value	17%	max 27%
Los Angeles abrasion value	20%	max 35%
Water absorption value	1.3%	max 2%
Specific gravity		
■ Coarse aggregate	2.86	2.5–3.0
■ Fine aggregate	2.71	
Combined Flakiness and Elongation Index	26.3%	max 35%

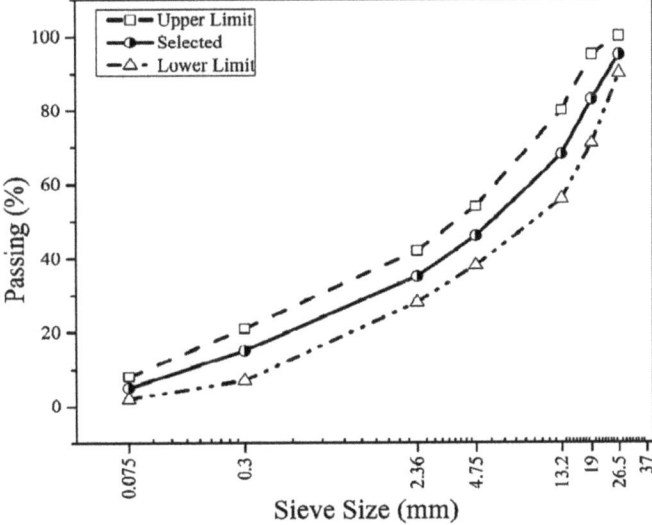

Figure 1. Adopted gradation of DBM grading II mix.

A VG30 grade of bitumen was used as a binder. The bitumen came from Haldia Petrochemicals (Purba Medinipur district, West Bengal, India) and is typically used in the hot-mix asphalt industry in India. The binder had a penetration value of 57, a kinematic viscosity (135 °C cSt) of 368, an SG of 1.045, and a softening point of 49 °C.

HL collected from a local market of Shibpur (Howrah district, West Bengal, India) was used as the conventional filler. RHA was collected from a rice mill in the Purba Burdwan district of West Bengal, India. Oven-dried fillers were used in this investigation after sieving through a 0.075 mm sieve.

2.2. Test for Filler Characterization

The physical and chemical features of each filler were estimated according to associated specifications. Specific gravities of fillers were determined using a pycnometer as per ASTM D854 [14]. The particle size distribution test was conducted as per ASTM D422 [15], and the curves were plotted according to the guidelines. The fineness modulus (F_M) and uniformity coefficient (U_C) were used to examine the curves. The specific surface area (SSA) was evaluated by Blaine's air permeability test using IS 4031 Part 2 [16]. Scanning electron microscopy (SEM) imaging methods were used to observe that the fillers' particle shape and surface texture conformed to ASTM E 986–04 [17].

The quantity of harmful clays in the fillers was estimated by the methylene blue values (MBV) according to ASTM C837-09 [18]. X-ray diffraction (XRD) analysis was used to identify the prevalent minerals present in the fillers; the XRD was conducted using a Rigaku D/Max-IIIC diffractometer operated with CuKα radiation over a range of 5–40° (2θ = 10° to 80°) at a rate of 0.2° min^{-1}. X-ray fluorescence spectroscopy was used to determine the chemical composition of each filler as per ASTM E1621-13 [19]. The organic content of each filler was quantified using a Loss-On-Ignition (LOI) test.

2.3. Designing and Testing of Bituminous Mastics and Mixes

2.3.1. Preparation of Mastic Samples

The mastic samples were fabricated by adding a suitable amount of fillers to preheated bitumen in order to maintain a filler/bitumen (f/b) ratio of 0.5–1.4 with a rate of 0.3 rises by weight; each sample was then mixed for 1 h. A shear blender with blending conditions of 2800 rpm at 160 °C was applied to attain homogeneity, as detailed in earlier studies [20]. Afterward, 100 gm of each mastic sample was stored in an airtight container for physical and morphological characterization.

2.3.2. Determination of the Physical Properties of Mastic

The conventional bitumen tests of penetration and softening point were performed to assess the physical properties of the prepared mastics. Those two tests may be considered a significant indicator of the stiffening effect of the fillers in mastic. Again, a conventional approach was taken from the Shell bitumen handbook to forecast the thermal susceptibility of the mastic samples in respect of the penetration index [21]. Such an index for mastics is based on their penetration value and softening point, as shown in Equation (1). A lower value of a binder's penetration index indicates greater temperature susceptibility, whereas a higher value indicates more resistance to low-temperature cracking [22].

$$\text{Penetration Index (PI)} = \frac{1952 - 500 \log(Pen_{25}) - 20SP}{50 \log(Pen_{25}) - SP - 120} \tag{1}$$

where Pen_{25} = Penetration value of the mastic at 25 °C, and SP = Softening point (°C) of the mastic.

2.3.3. Morphological Analysis by Optical Microscopy Test

A Ziuss optical microscope was used to study the extent of dispersion of fillers in the binder. A droplet of warm mastic was crammed between two microscope slides that were then bound by adhesive tape at both ends and left to dry for a day in a dirt-free Petri dish (Figure 2a). Then, they were studied using an optical microscope (Figure 2b) with a 20× magnification power; the images were captured by Axiovision software. The photos of 16 numbered slides (2 replicates × 2 types of filler × 4 f/b ratios) were captured and considered for analysis.

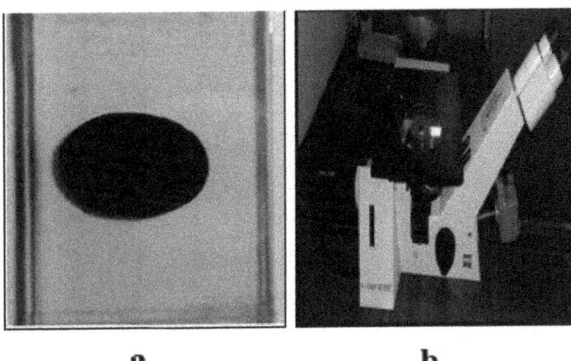

Figure 2. (a) Microscope slide having bituminous mastic (b) Optical microscope.

2.3.4. Marshall and Volumetric Properties

The Marshall mix design method was used to evaluate the optimum bitumen content (OBC) of all mixes according to MS-2 [23] guidelines. For each mix, about 1200 g of aggregate was sieved at the chosen gradation (Table 2) with five bitumen contents (4.5–6.5%) and four filler contents (2%, 4%, 6%, and 8% by the weight of aggregate). For each mix, 15 numbered samples (three for each bitumen content) were fabricated, and their volumetric properties, Marshall stability, and flow were calculated (ASTM D6927 [24]). The OBC of each mixture was considered as a binder percentage in the compacted samples with 4% air voids [25]. The air voids in the compacted specimens were measured according to ASTM D3203-17 [26]. The increase in filler content in the mix was achieved by reducing the fine aggregate fraction to maintain the required grading. In this study, the mix having 2% HL as filler was considered the conventional/control mix.

Table 2. Characteristics of Fillers.

Property	HL	RHA
Specific Gravity	2.15	2.00
F_M	3.18	2.86
U_C	2.33	2.92
Specific surface area (m^2/kg)	431.5	565.8
MBV (gm/kg)	0.97	1.27
Particle shape and texture (SEM)	Angular particles with uneven texture	Irregular porous particles with very rough texture
Chemical composition		
SiO_2	3.23	86.64
CaO	72.42	1.88
Al_2O_3	0.41	1.66
Fe_2O_3	0.31	1.06
MgO	0.46	0.97
K_2O	0.13	0.40
SO_3	1.22	0.12
Loss on ignition (LOI)	21.69	6.15

2.3.5. Cracking Resistance

The resistance to cracking of each mix was evaluated by measuring the average indirect tensile strength (ITS) of compacted samples by the ASTM D 6931-12 guidelines [27]. The cracking resistance of bituminous mixes was extensively examined with the help of the indirect tensile strength test [6,20]. According to the specification, testing was executed at 25 °C, and a compressive load was applied on compacted Marshall specimens (at 4% air voids) diametrically with steel strips at a continuous rate of 50.8 mm/min. Twenty-four samples were fabricated, and average ITS values for each mix were compared. The ITS was calculated using Equation (2):

$$ITS = \frac{2000 P_{max}}{\pi D T} \qquad (2)$$

where ITS = indirect tensile strength (kPa); P_{max} = peak load (N); D = the diameter of the samples (mm); and T = the thickness of the samples (mm).

2.3.6. Resistance to Moisture Damage

In this study, the resistance to moisture damage of the bituminous mixes was determined following AASHTO T283 [28]. After calculating the OBC, 48 Marshall samples were fabricated. Before casting each mix, the appropriate minimum specimen weight to reach the air voids of 7 ± 1.0% was achieved by trial and error. Each mixture was divided into two subgroups: dry and saturated. A partial vacuum was applied to the saturated specimens to achieve 55% to 80% saturation. Next, the samples were placed in a deep freezer at −18 ± 2 °C for 16 h. Then, the specimens were positioned under a 60 ± 1 °C water bath for 24 ± 1 h. Before the ITS test, the saturated specimens were located in an additional water bath at 25 ± 0.5 °C for two hours. The ITS of dry samples and the ITS of water-conditioned samples were measured at 25 °C. The ability of the bituminous mix to withstand the harmful effect of moisture was represented as the ratio of the ITS of the saturated specimens to the ITS of the dry samples, which is known as the tensile strength ratio (TSR). The TSR (expressed as a percent) was calculated using Equation (3).

$$TSR = \frac{ITS_{saturated}}{ITS_{dry}} \times 100\% \qquad (3)$$

2.3.7. Creep Resistance

As described in BS598-111 [29], the uniaxial static creep recovery test was used to measure the resistance to permanent deformation of bituminous mixes with different fillers. This test was performed by applying a static load to a specimen and measuring the resulting deformation with time. The reversible part of the total deformation was also recorded by removing the entire load and measuring deformation after a recovery time equal to the loading time. A set-up was fabricated at the transportation Engineering Laboratory, IIEST, Shibpur. This test was performed using the Marshall loading frame, a dial gauge with an accuracy of at least 0.001 mm, and a video camera for recording the dial gauge reading and time. In this test, a 10 kg load was applied to provide axial stress of 100 kg for 1 h on the sample, followed by unloading and a recovery time of 1 h, all at a test temperature of 27 °C.

3. Results and Discussion

3.1. Filler Characterization

The characteristics of the considered fillers are shown in Table 2 and Figures 3–5. The specific gravity of RHA (2.00) was marginally less than that of HL (2.15). RHA was a finer filler than HL; the F_M of RHA (2.86) was less than the F_M of HL (3.18). Earlier studies [3,30] stated that fillers that are finer in nature show a stronger physicochemical interaction with bitumen, thus providing higher stiffness and better moisture resistance. The U_C values of RHA (2.92) and HL (2.33) indicated that both were uniformly graded fillers. As expected, the specific surface area of RHA (566.8 m^2/kg) was larger than the specific surface area of HL (432.5 m^2/kg).

Figure 4 shows the SEM images of the fillers. As shown in Figure 4a, HL usually has a rough texture with variations in sizes, and it has a proclivity toward agglomeration. Figure 4b shows that RHA has larger particles of irregular shape and is poriferous in nature. In the SEM image of RHA, a unique honeycomb structure can be seen. RHA particles range in size from 10 to 100 microns. Experience indicates that the cellular structure of RHA has been gradually broken-down during grinding, which can account for the RHA's higher specific surface area compared to HL [12]. This is particularly evident in ashes derived from the uncontrolled RHA combustion processes, which can produce high carbon content and wide distributions of particle size [12].

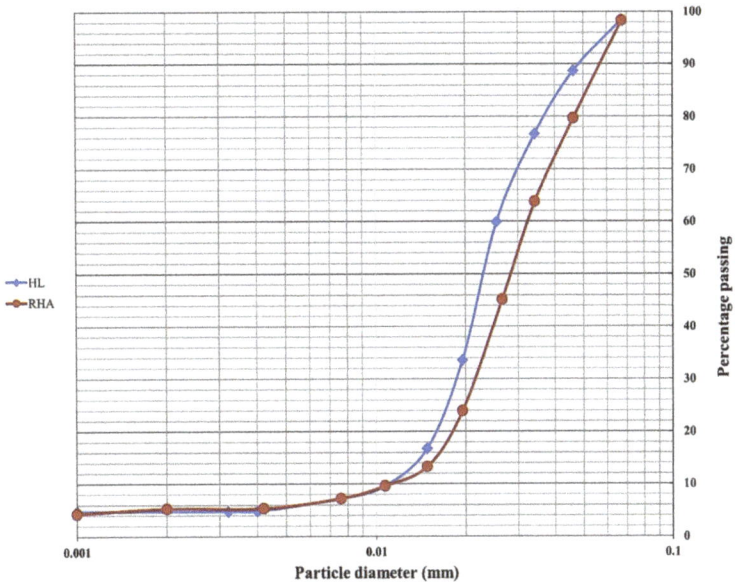

Figure 3. Particle size distribution curves of fillers.

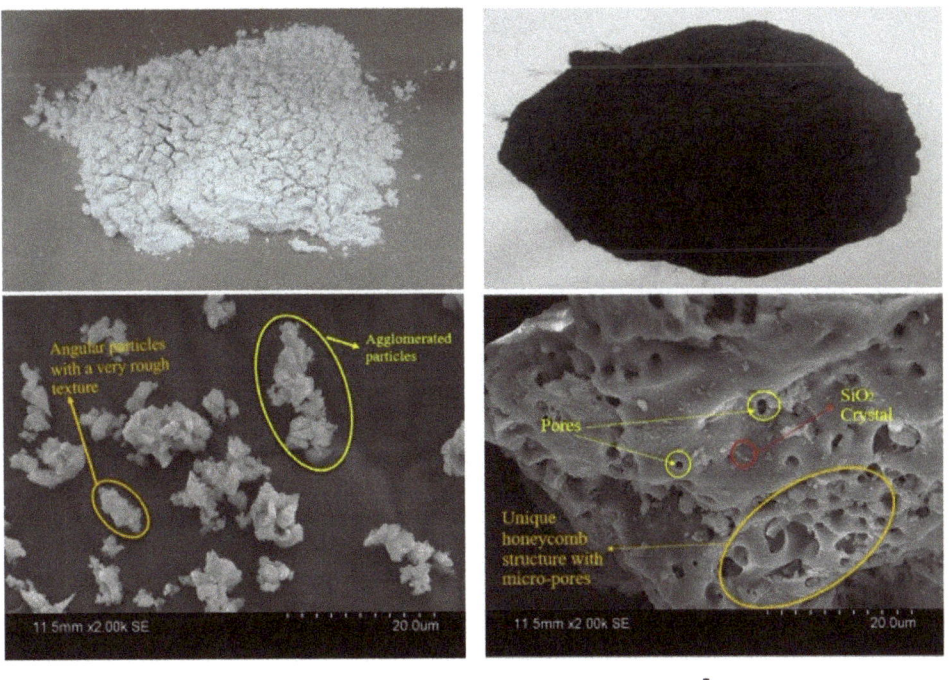

Figure 4. Physical appearance and SEM image of (**a**) HL and (**b**) RHA.

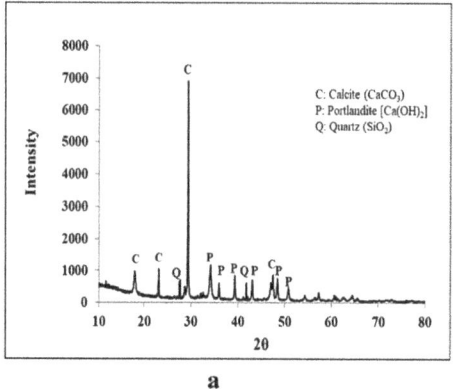 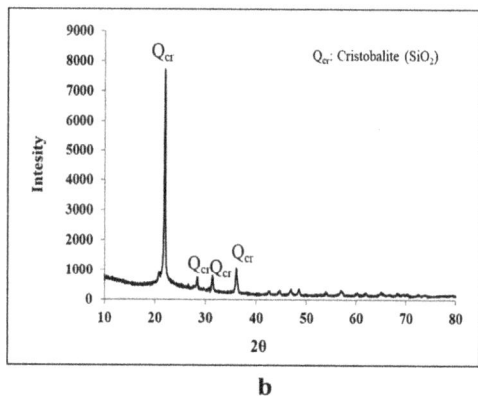

Figure 5. X-ray diffraction spectrum of (**a**) HL and (**b**) RHA.

A filler containing a high quantity of detrimental clay may swell when interacting with water and act as a partition between binder and aggregates, weakening the adhesion at the bitumen–aggregate interface [6]. There is no specific range for methylene blue value (MBV) given in India's specifications. In Portugal's specification, the highest allowable limit for MBV is 10 g/kg. The MBV values for HL (0.97 g/kg) and RHA (1.27 g/kg) were much lower than the permissible limit. XRD diffractograms of the studied fillers are shown in Figure 5. The HL composition had a prevalence of calcite and portlandite. These compounds are popularly known as anti-stripping agents and are generally found in HL and ordinary Portland cement. A small percentage of calcite was also found in RHA. Silica in the form of cristobalite and quartz was observed in all fillers but in different quantities. Silica in the form of cristobalite was observed in RHA. As shown in Table 2, the chemical characteristics of HL show a high proportion of CaO (approximately 72%), along with an LOI of 21.69%. RHA shows a composition similar to the F pozzolans [31], since the sum of Fe_2O_3, Al_2O_3, and SiO_2 is more than 70%, and the LOI is 6.15%. Although there is no permissible limit for the LOI value, it would be more feasible to use fillers with a relatively low LOI in order to reduce the possible preoccupation of binders by unburned carbon particles [32].

3.2. Mastic Evaluation

3.2.1. Physical Properties of Mastics

The change in physical properties of mastics prepared with various proportions of reference fillers is shown in Figures 6 and 7. The penetration and the softening point are inversely co-related. The penetration value shows the hardness and consistency of the bituminous binder, and the softening point indicates the temperature at which the binder changes its phase. Figure 6 shows a continuous decreasing tendency in penetration value of the mastic blend, and Figure 7 shows an increasing drift in the softening point of the mastic blend. Compared to HL at the same f/b ratio of the mastic samples, RHA shows a lower penetration value and a higher softening point. This might be due to the lower density and higher specific surface area of RHA compared to HL. It is well-known that the inclusion of filler with fine particles makes a bituminous binder stiffer and improves the performance of the resultant mixtures under permanent deformation [33]. Further, a bituminous binder with a high softening point can resist deformation due to a temperature increase. The test results for softening point show that the mastic having RHA is an excellent solution for a hot climate such as India's.

Figure 8 shows that the inclusion of RHA in binder increases the penetration index with an increase in the f/b ratio. At an f/b ratio of 0.8, the penetration index of bituminous mastic having either HL or RHA as filler lies within the specified range of −1 to +1 [34]. This indicates that the thermal susceptibility of the bitumen was improved by the addition of HL or RHA fillers. Therefore, it is possible to use those waste fillers for road construction in a hot climate; this is in line with observations from previous studies [12].

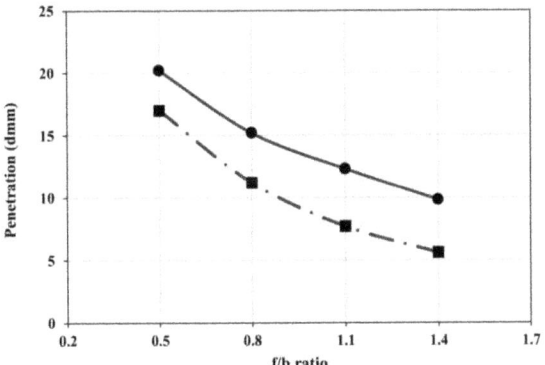

Figure 6. Variation in the penetration values of the mastic samples.

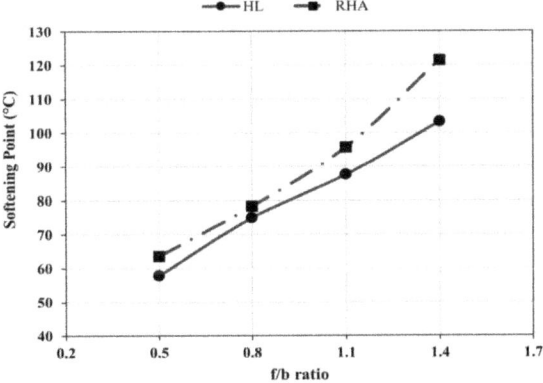

Figure 7. Variation in the softening point value of the mastic samples.

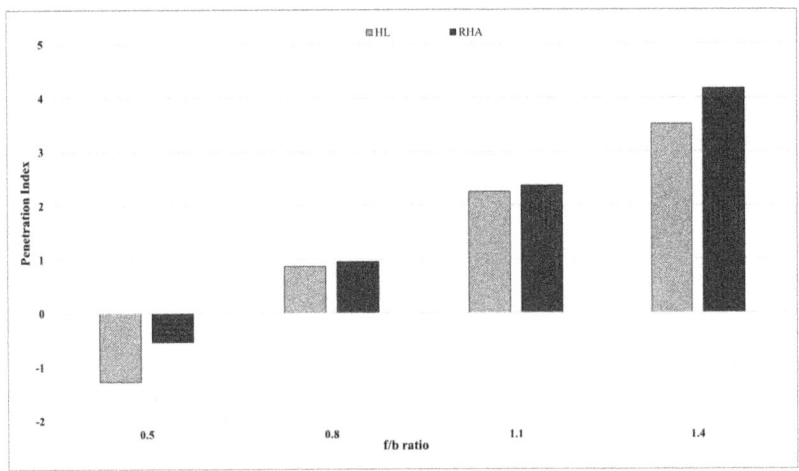

Figure 8. Evaluation of thermal susceptibility of bituminous mastic by penetration index.

3.2.2. Morphological Analysis of Mastic

The mastic is the delicate phase of a bituminous mix; the affinity between filler and bitumen is vital for good performance of the bituminous mix [1]. The morphology of mastic made with HL or RHA at f/b ratios of 0.5, 0.8, 1.1, and 1.4 is shown in Figure 9 for HL and Figure 10 for RHA. The uniform dispersion of HL molecules in bitumen is shown at the f/b ratio of 0.5 (Figure 9a). HL is very active chemically and has a relatively low molecular weight, and it has a high relative concentration of reactive chemical functionality. HL also has a comparatively strong base and reacts with acid components such as carboxylic acid and 2-quinolone-type. Removal of these acid components from the bitumen matrix enhances the compatibility in the bitumen–filler system; this may account for the excellent stripping performance of the conventional mix. For HL at f/b ratios of 0.8, 1.1, and 1.4, as shown in Figure 9b–d, the HL particles tend to be clustered in the bitumen. Additionally, the absorption of bitumen's oil fraction by calcium hydroxide in HL increases the rate of particle agglomeration, leading to asphaltene precipitation in the mastic. For RHA, accumulation was more likely to occur at the f/b ratio of 0.5; Figure 10a shows the heterogeneous dispersion of RHA particles in bitumen. RHA molecules had their best bonding arrangement in the mastic at the f/b ratio of 0.8; Figure 10b shows the homogeneous dispersion of RHA and adequate filling of the pores through absorption of the acid molecules of bitumen by siliceous nanoparticles of RHA, providing a steadier adhesive system [33]. This result parallels the mix performance, which suggests that RHA modification at the f/b ratio of 0.8 may significantly enhance the stiffness and resistance to moisture damage of the RHA-modified mix.

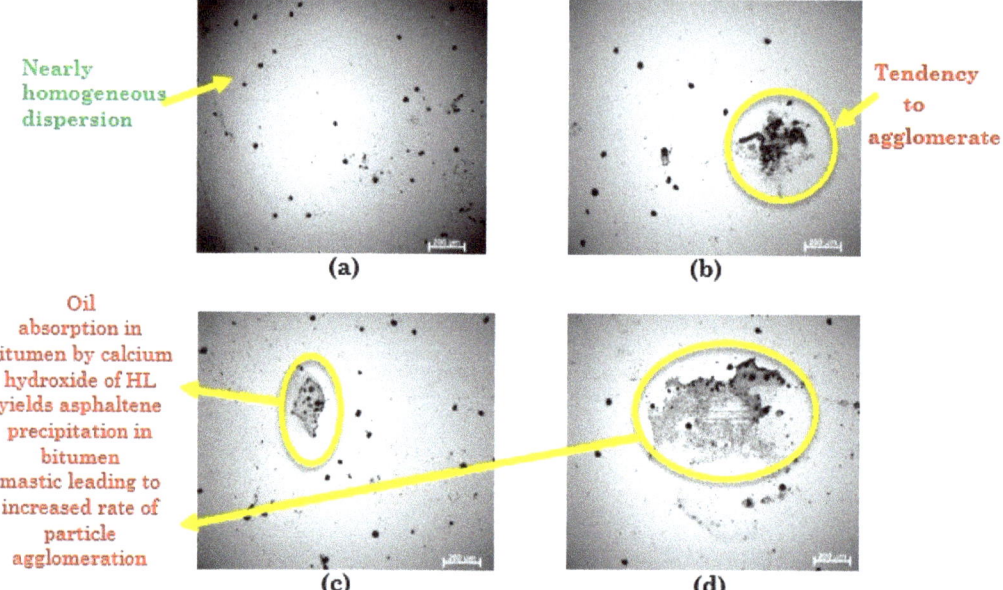

Figure 9. Optical micrograph of mastics with HL at f/b ratio of (**a**) 0.5, (**b**) 0.8, (**c**) 1.1, and (**d**) 1.4.

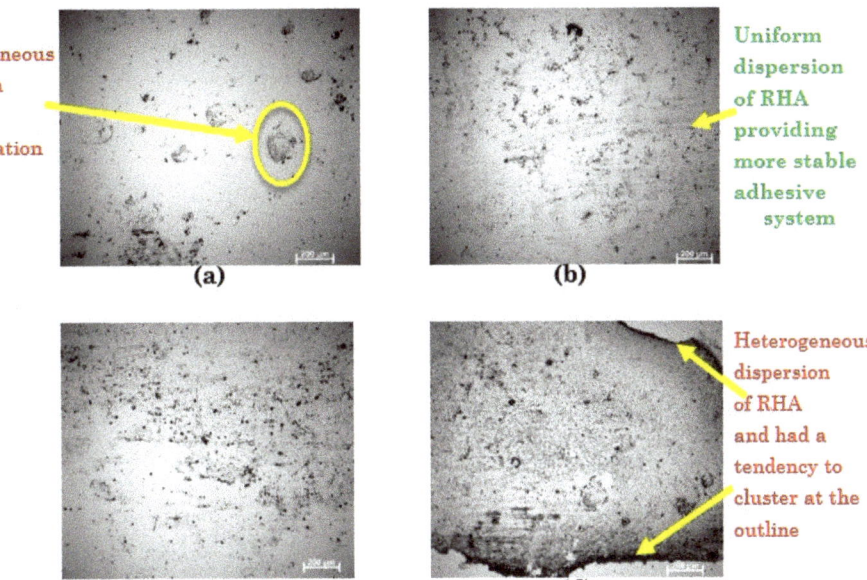

Figure 10. Optical micrograph of mastics with RHA at f/b ratio of (**a**) 0.5, (**b**) 0.8, (**c**) 1.1, and (**d**) 1.4.

3.3. Performance of Bituminous Mixes
3.3.1. Marshall and Volumetric Properties

The change in average Marshall properties (air voids, voids in mineral aggregates (VMA), voids filled with bitumen (VFB), stability, and flow) of the mixes with OBCs are shown in Table 3. The conventional mix (HL at 2% filler content) has an OBC of 5.2%; the OBC at 4% filler content is lower at 5.17%, and then reaches 5.6% at 8% filler content. In contrast, the RHA-modified mixes display lower OBC in comparison with HL for filler content levels of 2%, 4%, and 8%. It should be noted that RHA shows nearly 5.9% lower OBC than the conventional mix when 4% filler is used. A similar trend was observed in earlier studies [12,30,33]. Compared to HL, RHA has larger particles, as shown in Figure 4. Nonetheless, RHA has a larger specific surface area than HL. This can be attributed to the mesoporous structure of RHA particles. However, the pores in RHA are not accessible to binder molecules [12], thus the OBC of RHA is lower. The range of pore diameters in RHA is 0.002 µm to 0.12 µm [35], while the average bitumen film thickness is typically 2–8 µm [36].

Table 3. Average Marshall and Volumetric Properties of the Mixes.

Filler Type	Amount of Filler (%)	OBC (%)	Marshall Stability (kN)	Flow (mm)	VMA (%)	VFB (%)
HL	2	5.2	13.85	3.82	16.18	74.57
	4	5.17	15.75	3.77	16.07	72.59
	6	5.35	14.75	3.79	16.32	74.79
	8	5.60	14.18	3.96	16.82	74.85
RHA	2	5.15	14.5	3.65	15.93	70.44
	4	4.91	17.25	3.48	15.66	70.83
	6	5.38	16.62	3.73	16.21	73.34
	8	5.42	15.93	3.82	16.5	73.87
Requirements (MORTH 2013)		-	9.00 (min)	2–4	13.00 (min)	65–75

At 4% filler content, the peak stability of the HL mixture is 15.75 kN, while that of the RHA mixture is 17.25 kN. The superiority in stability values of RHA mixtures relative to HL mixtures can be attributed to the irregular particle shape of RHA giving rise to interlocking actions (see Figure 4b). This, in turn, may increase the shear strength and stiffness of the modified binder, which can increase resistance to the plastic flow of the mix.

All mixes have flow values within the desired range (Table 3). The VMA of all the mixes is well above the minimum requirement. RHA mixes have the lowest VMA values. The HL mixes have the greater VMA values, despite HL having small-sized particles. This result may be attributed to the clustered HL particles. All mixes have VFB values in the acceptable range as specified in the Indian Code (Table 3). RHA mixes have lower VFBs than the HL mixes and can be used in areas having hotter environments because of a low bleeding prospect [37].

3.3.2. Cracking Resistance

Figure 11 shows the ITS values of the mixtures with various percentages of HL and RHA. As shown in Figure 11, the ITS increases with the rise in filler content and comes to a peak at 4% content. The conventional mix (HL at 2% filler) has an ITS value of 646 kPa, but the ITS of HL mix at 4% filler is 23.06% higher at 795 kPa compared to the conventional mix. Further, the RHA mix at 2% filler has an ITS value of 720 kPa; however, the ITS of RHA mix at 4% filler is 50.15% higher at 970 kPa compared the conventional mix. The higher ITS values of HL mix and RHA mix at 4% filler may be attributed to the fine nature of HL and RHA as indicated by their high specific surface areas [10,30]. However, further increments in filler content (6% and 8%) decrease the values of the said properties. This may be owing to the higher stiffness of modified mixes at increasing filler proportion. Since the HMA with high stiffness is prone to be more brittle and has less ability to bear tensile strength. The improvement in ITS value of RHA mixes may be attributed to the multilayer structure and micro-pores in RHA (see Figure 4b), which successfully enhance the cohesion and adhesion of the mastic, thereby helping to increase the cracking resistance of the mixture.

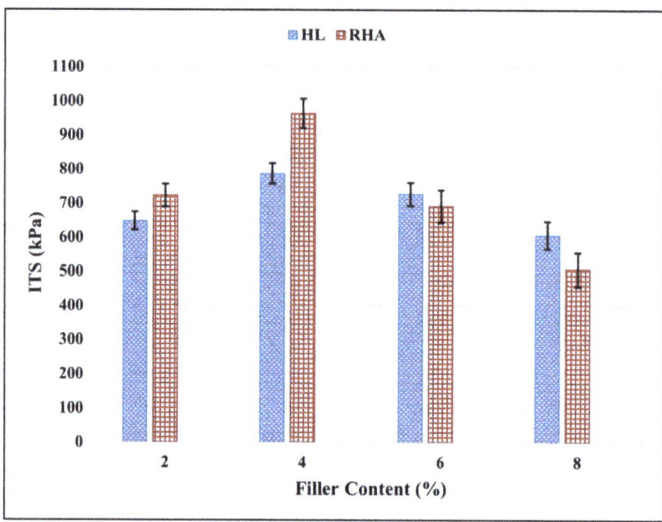

Figure 11. ITS values of HL mixes and RHA mixes with different filler contents.

3.3.3. Resistance to Moisture Damage

Bituminous mixes with higher TSR values have better resistance to moisture damage. Figure 12 shows the TSR values of all mixes. According to the Indian Code, each bituminous mix should have a TSR of 80% or more [2]. All the mixtures met this requirement at 2%, 4%, and 6% filler. As expected, the conventional mix (HL at 2% filler) displays the maximum TSR value of 98.13% among all mixes. HL in this mix may enhance the mixture's resistance to moisture damage by neutralizing the acidic compounds in bitumen [38]. Neutralizing the acidic compounds in bitumen helps prevent moisture from weakening the interfacial bond between bitumen and aggregate [39].

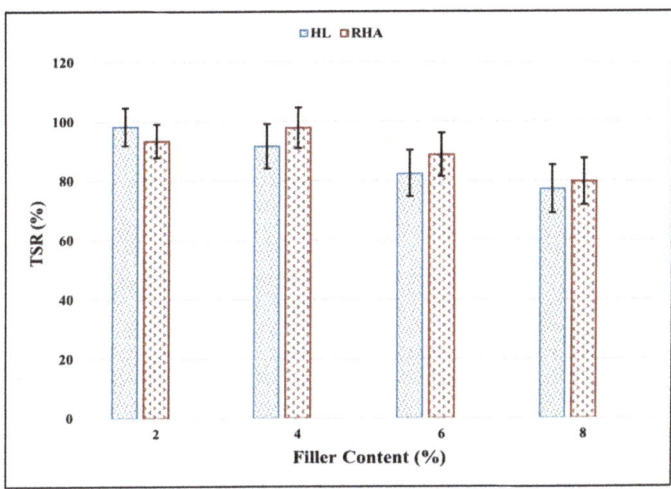

Figure 12. TSR values of HL mixes and RHA mixes with different filler contents.

Nonetheless, at 4%, the RHA mix has a TSR value of 97.74%, which is almost equal to the 98.13% TSR of the control mix (HL at 2%). The acid components in bitumen (carboxylic acid and 2-quinolone-type) are responsible for weakening the ability to resist stripping at the bitumen–aggregate interface, especially for siliceous aggregates and the gathering of a layer of brush-like acid crystals at the bitumen–silica interface [39]. Due to the unique nature of RHA, it can act as a sink for acid compounds and prevent their crystallization at the surface of silica particles; this, in turn, helps prevent moisture damage. [39]. It has been reported that fillers with a high specific surface area help to prevent accumulation and crystallization of the acids at the bitumen–stone interface [40].

The size and surface chemistry of siliceous particles play a major role in their performance in asphalt; functionalized, surface-treated silica has been used to improve asphalt's durability and moisture resistance [40,41]. In the case of RHA, the presence of carboxyl groups on the surface can promote interaction with bitumen compounds. It has been shown that modifying bitumen with nano clay (which includes silica platelets) promotes asphalt's resistance to moisture damage [42]. Additionally, Fini et al. showed that active mineral powders can be used as modifiers in bitumen to control the migration of alkane acids to the interface of bitumen and aggregate, thereby improving the resistance to moisture damage [38].

3.3.4. Creep Resistance

The test results for static creep recovery and permanent deformation are shown for HL mixes in Figure 13 and for RHA mixes in Figure 14. (Each reported test result is an average of three samples.) In the present study, creep and permanent deformation (due to sustained 10 kg load) for all the mixtures shows that the creep performance improved with increased filler content. This trend could be observed for both fillers at all percentages of filler. This is expected, since an increase in filler percentage increases the stiffness, thereby providing more resistance to deformation. At filler levels of 2%, 4%, 6%, and 8%, the respective permanent deformation of HL mixes was found to be 0.20, 0.13, 0.10, and 0.06 mm, and the respective permanent deformation of RHA mixes was found to be 0.16, 0.12, 0.07, and

0.04 mm. As shown in Figure 4b, RHA has a rough interior and exterior layer. Further, the RHA micro-particles contain three-dimensional micro-sheets and columnar SiO_2 crystals, which might be attracted to base bitumen. In contrast, HL particles are more regular and uniform than RHA. Compared to HL, RHA could form a more stable and viscous three-dimensional meshed bonding system with bitumen, increasing the mixture's resistance to permanent deformation [12].

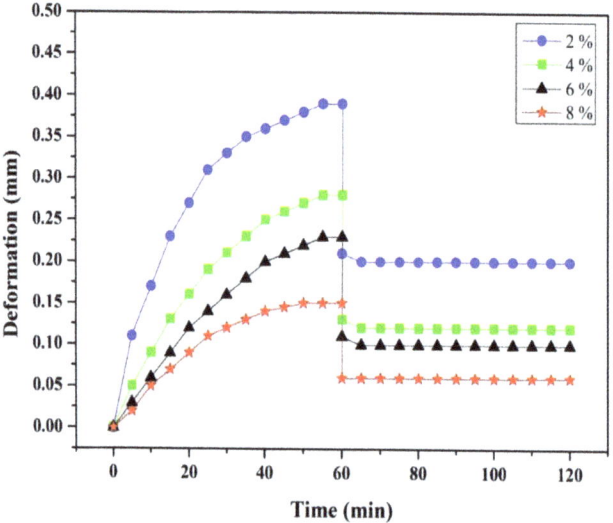

Figure 13. Creep performance of mixes containing different percentages of HL.

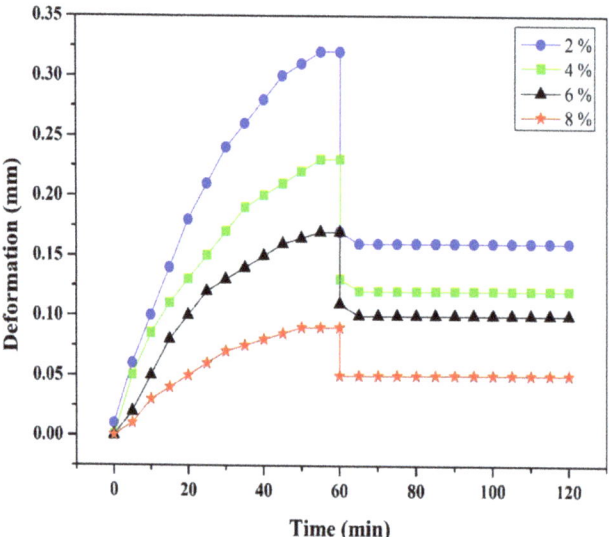

Figure 14. Creep performance of mixes containing different percentages of RHA.

3.4. Cost Analysis

The economic benefit from using RHA as filler was evaluated after comparing the material cost for the production of a 1.0 m^3 of DBM grading-II (50 to 75 mm thickness) layer. Based on the performance of all the studied mixes, mixes up to 4% filler were considered in the cost analysis. The quantities of materials were calculated as per the mix design used in this study. Table 4 shows the present unit cost of each ingredient (coarse aggregate, fine aggregate, hydrated lime (HL), and bitumen) for production of dense bituminous macadam mixes according to the schedule of rates of the Public Works Department, West Bengal, India [43]. Since RHA is a waste material, RHA is freely available, apart from its transportation cost. Thus, the carrying cost for RHA was considered to be the same as for cement. RHA does not require any processing, since it mostly passed through a 75 μm sieve. RHA's processing (labor) cost in the worst situation can be taken as 0.5% of the total cost needed to produce 1.0 m^3 of DBM mix.

Table 4. Comparison of Cost of Selected Mixes.

Materials	WBPWD Rates	Quantity in 1 m^3 of Dense Bituminous Macadam Grade-II (50 to 75 mm) Thickness			
		2% HL	4% HL	2% RHA	4% RHA
Coarse Aggregate (m^3)	1536.35/m^3	0.952	0.948	0.952	0.948
Fine Aggregate (m^3)	1167.10/m^3	0.483	0.452	0.483	0.452
Hydrated Lime (kg)	7.50/kg	44	88	0	0
Rice Husk Ash (kg)	0	0	0	44	88
Bitumen (kg)	35.18/kg	114.62	113.96	113.08	107.8
Cost in (INR/m^3)		6389	6659	6005	5777
Transportation cost (INR/MT)	84/MT	0	0	3.696	7.392
0.5% for processing (INR/m^3)		0	0	300	289
Total Cost (INR/m^3)		6389	6659	6305	6066
Percentage saving in cost with respect to conventional mix (2% HL) (%)		0	−4.22	1.31	5.06

The results of the cost analysis clearly show that the mixes composed of 2% RHA or 4% RHA filler are economical compared to the conventional mix (2% HL). The significant parameter in the cost reduction is the lesser amount of bitumen consumed by RHA mixes. The cost comparison is also made by considering the increasing quantity of alternative filler (mixes with 2% RHA or 4% RHA) and observing that the cost of the mixes gradually decreases with the increase in filler content. Thus, the significant observation is that the total material cost of constructing 1.0 m^3 of DBM layer using 4% RHA mix is nearly 5.06% lower than that of the conventional mix (2% HL) after satisfying all standard specifications. Apart from these monetary benefits, using RHA as a filler has environmental benefits such as saving a significant amount of valuable land from being used as landfill for discarded RHA.

4. Conclusions

This study examined the merits of replacing hydrated lime (HL) used in bituminous composites with rice husk ash (RHA) to reduce their carbon footprint and promote both construction sustainability and resource conservation. Based on the study results, these are the conclusions:

- Mixtures containing RHA (up to 4%) were shown to have enhanced Marshall stabilities and volumetric characteristics compared to those containing HL.
- The mixture containing 4% RHA had the lowest optimum binder content; this can be attributed to RHA's physiochemical characteristics and particle size distribution.
- Compared to the control mixture containing HL, the mixture containing RHA had a nearly 47% higher resistance to cracking and 40% higher resistance to permanent deformation. This was attributed to the RHA micro-particles containing three-dimensional micro sheets and columnar SiO_2 crystals, forming a stable and viscous three-dimensional network with bitumen compared to the control mixture.

- Resistance to moisture damage in mixtures containing RHA was found to be slightly better than that of mixtures containing HL. This can be attributed to the active silica in RHA and physiochemical characteristics of RHA, which can adsorb acidic compounds of bitumen, as shown by prior studies [38].
- Mixtures containing RHA had lower carbon footprints than mixtures containing HL. In terms of production cost, mixtures containing RHA cost 5% less than the control mixture containing HL.

The study results showed that replacing HL by RHA had positive effects on the performance of bituminous composites while reducing their carbon footprint. Accordingly, the study outcomes promote construction sustainability and resource conservation.

Author Contributions: R.M. wrote original draft and performed data curation; T.K.R. assisted data interpretation and analysis. S.A. assisted with writing and revision of the manuscript. E.H.F. provided guidance with paper development, data interpretation, review & editing. All authors have read and agreed to the published version of the manuscript.

Funding: This research received no external funding.

Data Availability Statement: All data used in this study has been included in the manuscript.

Conflicts of Interest: The authors declare no conflict of interest.

References

1. Mistry, R.; Karmakar, S.; Kumar Roy, T. Experimental evaluation of rice husk ash and fly ash as alternative fillers in hot-mix asphalt. *Road Mater. Pavement Des.* **2019**, *20*, 979–990. [CrossRef]
2. MORTH (Ministry of Road Transport and Highways). *Specifications for Road and Bridge Works (Fifth Revision)*; Indian Road Congress: New Delhi, India, 2013.
3. Kandhal, P.S.; Lynn, C.Y.; Parker, F. *Characterization Tests for Mineral Fillers Related to Performance of Asphalt Paving Mixes*; NCAT Rep. No. 98-2; Auburn University: Auburn, AL, USA, 1998.
4. Sargın, Ş.; Saltan, M.; Morova, N.; Serin, S.; Terzi, S. Evaluation of rice husk ash as filler in hot mix asphalt concrete. *Constr. Build. Mater.* **2013**, *48*, 390–397. [CrossRef]
5. Lesueur, D.; Petit, J.; Ritter, H.J. The mechanisms of hydrated lime modification of asphalt mixtures: A state-of-the-art review. *Road Mater. Pavement Des.* **2013**, *14*, 1–16. [CrossRef]
6. Choudhary, J.; Kumar, B.; Gupta, A. Application of waste materials as fillers in bituminous mixes. *Waste Manag.* **2018**, *78*, 417–425. [CrossRef] [PubMed]
7. EEA (European Environment Agency). *Air Pollutant Emission Inventory Guidebook*; Technical Report No 9; European Environment Agency: Copenhagen, Denmark, 2009.
8. FAOSTAT. 2014. Available online: http://www.fao.org/faostat/en/#data/QC/visualize (accessed on 16 June 2020).
9. National Food Security Mission. 2016. Available online: https://nfsm.gov.in/StatusPaper/Rice2016.pdf (accessed on 23 June 2020).
10. Mistry, R.; Kumar, T. Utilization of rice husk ash in hot mix asphalt concrete as mineral filler replacement. *J. Indian Road Cong.* **2015**, *76*, 13–17.
11. Mehta, P.K. *Testing and Correlation of Fly Ash Properties with Respect to Pozzolanic Behavior*; University of California: Berkeley, CA, USA, 1984.
12. Arabani, M.; Tahami, S.A. Assessment of mechanical properties of rice husk ash modified asphalt mixture. *Constr. Build. Mater.* **2017**, *149*, 350–358. [CrossRef]
13. Arabani, M.; Tahami, S.A.; Taghipoor, M. Laboratory investigation of hot mix asphalt containing waste materials. *Road Mater. Pavement Des.* **2017**, *18*, 713–729. [CrossRef]
14. *ASTM D854-14*; Standard Test Methods for Specific Gravity of Soil Solids by Water Pycnometer. ASTM: West Conshohocken, PA, USA, 2014.
15. *ASTM D422-63*; Standard Test Method for Particle-Size Analysis of Soils. ASTM: West Conshohocken, PA, USA, 2007.
16. *IS 4031 (Part 2)*; Methods of Physical Tests for Hydraulic Cement (Determination of Fineness by Blaine Air Permeability Test). IS (Indian Standard): New Delhi, India, 1978.
17. *ASTM E986-04*; Standard Practice for Scanning Electron Microscope Beam Size Characterization. ASTM: West Conshohocken, PA, USA, 2017.
18. *ASTM C837-09*; Standard Test Method for Methylene Blue Index of Clay. ASTM: West Conshohocken, PA, USA, 2009.
19. *ASTM E1621-13*; Standard Guide for Elemental Analysis by Wavelength Dispersive X-ray Fluorescence Spectrometry. ASTM: West Conshohocken, PA, USA, 2010.

20. Mistry, R.; Roy, T.K. Performance evaluation of bituminous mix and mastic containing rice husk ash and fly ash as filler. *Constr. Build. Mater.* **2021**, *268*, 121187. [CrossRef]
21. Zhang, J.; Wang, J.; Wu, Y.; Wang, Y.; Wang, Y. Evaluation of the improved properties of SBR/weathered coal modified bitumen containing carbon black. *Constr. Build. Mater.* **2009**, *23*, 2678–2687. [CrossRef]
22. Price, A.; Yeargin, R.; Fini, E.; Abu-Lebdeh, T. Investigating effects of introduction of corncob ash into Portland cements concrete: Mechanical and thermal properties. *Am. J. Eng. Appl. Sci.* **2014**, *7*, 133–144. [CrossRef]
23. Asphalt Institute. *Mix Design Methods for Asphalt Concrete and Other Hot-Mix Types: Manual Series No. 2 (MS-2)*, 6th ed.; Asphalt Institute: Lexington, KY, USA, 1997.
24. *ASTM D6927-15*; Standard Test Method for Marshall Stability and Flow of Asphalt Mixtures. ASTM: West Conshohocken, PA, USA, 2015.
25. *ASTM D3203–17*; Standard Test Method for Percent Air Voids in, Compacted Asphalt Mixtures. ASTM: West Conshohocken, PA, USA, 2017.
26. *ASTM D6931-12*; Indirect Tensile (IDT) Strength for Bituminous Mixtures. ASTM: West Conshohocken, PA, USA, 2012.
27. *T283*; Resistance of Compacted Bituminous Mixture to Moisture Induced Damage. AASHTO: Washington, DC, USA, 1989.
28. *BS 598-111*; Method for Determination of Resistance to Permanent Deformation of Bituminous Mixtures Subjected to Uniaxial Loading. British Standards Institution: London, UK, 1995.
29. Choudhary, J.; Kumar, B.; Gupta, A.C. Feasible utilization of waste limestone sludge as filler in bituminous concrete. *Constr. Build. Mater.* **2020**, *239*, 117781. [CrossRef]
30. *ASTM C618*; Standard Specification for Coal Fly Ash and Raw or Calcined Natural Pozzolan for Use in Concrete. ASTM: West Conshohocken, PA, USA, 2003.
31. FHWA (Federal Highway Administration). *User Guidelines for Waste and by Product Materials in Pavement Construction*; FHWA Report 97148; Federal Highway Administration: Washington, DC, USA, 2016.
32. Ameli, A.; Babagoli, R.; Norouzi, N.; Jalali, F.; Mamaghani, F.P. Laboratory evaluation of the effect of coal waste ash (CWA) and rice husk ash (RHA) on performance of asphalt mastics and Stone matrix asphalt (SMA) mixture. *Constr. Build. Mater.* **2020**, *236*, 117557. [CrossRef]
33. Kütük-Sert, T.; Kütük, S. Physical and Marshall properties of borogypsum used as filler aggregate in asphalt concrete. *J. Mater. Civ. Eng.* **2013**, *25*, 266–273. [CrossRef]
34. Read, J.; Whiteoak, D. *Shell Bitumen Handbook*, 5th ed.; Thomas Telford: Telford, UK, 2003.
35. Srivastava, V.C.; Mall, I.D.; Mishra, I.M. Characterization of mesoporous rice husk ash (RHA) and adsorption kinetics of metal ions from aqueous solution onto RHA. *J. Hazard. Mater. B* **2006**, *134*, 257–267. [CrossRef] [PubMed]
36. Kandhal, P.S.; Chakraborty, S. Effect of Asphalt Film Thickness on Short- and Long-Term Aging of Asphalt Paving Mixtures. *Transp. Res. Rec.* **1996**, *1535*, 83–90. [CrossRef]
37. Tapkin, S. Mechanical Evolution asphalt-aggregate mixtures prepared with fly ash as a filler replacement. *Can. J. Civ. Eng.* **2008**, *35*, 27–40. [CrossRef]
38. Fini, E.H.; Hung, A.M.; Roy, A. Active mineral fillers arrest migrations of alkane acids to the interface of bitumen and siliceous surfaces. *ACS Sustain. Chem. Eng.* **2019**, *7*, 10340–10348. [CrossRef]
39. Oldham, D.; Mallick, R.; Fini, E.H. Reducing susceptibility to moisture damage in asphalt pavements using polyethylene terephthalate and sodium montmorillonite clay. *Constr. Build. Mater.* **2021**, *269*, 121302. [CrossRef]
40. Karnati, S.R.; Oldham, D.; Fini, E.H.; Zhang, L. Application of Surface-Modified Silica Nanoparticles with Dual Silane Coupling Agents in Bitumen for Performance Enhancement. *Constr. Build. Mater* **2020**, *244*, 118324. [CrossRef]
41. Mousavi, M.; Fini, E. Silanization Mechanism of Silica Nanoparticles in Bitumen Using 3-Aminopropyl Triethoxysilane (APTES) and 3-Glycidyloxypropyl Trimethoxysilane (GPTMS). *ACS Sustain. Chem. Eng.* **2020**, *8*, 3231–3240. [CrossRef]
42. El Badawy, A.; Rahim, A. Evaluation of Nanoclay Additives for Improving Resistance to Moisture Damage in Hot Mix. 2023. Available online: https://rosap.ntl.bts.gov/view/dot/66358/dot_66358_DS1.pdf (accessed on 11 January 2023).
43. WBPWD (West Bengal Public Works Department). *Schedule of Rate for Road and Bridge Works-Vol III.*; West Bengal Public Works Department: Kolkata, West Bengal, India, 2018.

Disclaimer/Publisher's Note: The statements, opinions and data contained in all publications are solely those of the individual author(s) and contributor(s) and not of MDPI and/or the editor(s). MDPI and/or the editor(s) disclaim responsibility for any injury to people or property resulting from any ideas, methods, instructions or products referred to in the content.

Article

Agro-Industrial Waste Biochar Abated Nitrogen Leaching from Tropical Sandy Soils and Boosted Dry Matter Accumulation in Maize

Michael Egyir [1,2], Innocent Yao Dotse Lawson [1,*], Daniel Etsey Dodor [1] and Deogratius Luyima [2,*]

[1] Department of Soil Science, College of Basic and Applied Sciences, University of Ghana, Legon P.O. Box LG 25, Ghana

[2] Department of Agricultural Chemistry, College of Agriculture and Life Sciences, Chungnam National University, Daejeon 34134, Republic of Korea

* Correspondence: idlawson@ug.edu.gh (I.Y.D.L.); deoluyima@gmail.com or deoluyima@o.cnu.ac.kr (D.L.)

Citation: Egyir, M.; Lawson, I.Y.D.; Dodor, D.E.; Luyima, D. Agro-Industrial Waste Biochar Abated Nitrogen Leaching from Tropical Sandy Soils and Boosted Dry Matter Accumulation in Maize. C 2023, 9, 34. https://doi.org/10.3390/c9010034

Academic Editors: Indra Pulidindi, Pankaj Sharma, Aharon Gedanken and Dimitrios Kalderis

Received: 19 January 2023
Revised: 22 February 2023
Accepted: 3 March 2023
Published: 14 March 2023

Copyright: © 2023 by the authors. Licensee MDPI, Basel, Switzerland. This article is an open access article distributed under the terms and conditions of the Creative Commons Attribution (CC BY) license (https:// creativecommons.org/licenses/by/ 4.0/).

Abstract: This study was conducted to assess the effects of amending tropical sandy soils with biochar derived from agro-industrial wastes on the leaching and utilization of nitrogen (N) by maize. The experiment was conducted in pots in a greenhouse with two sandy soil types and two different biochars. The biochars used in this experiment were preselected in a preliminary column experiment that assessed the N retention capacities of the different biochars and those that exhibited the best retention capacities chosen for experimentation. The biochars evaluated included saw dust, rice husk and corncob pyrolyzed at 500 °C and the results from the column leaching experiment showed that sawdust biochar had superior retention capacities for both NO_3^- and NH_4^+, followed by rice husk biochar. The pot experiment utilized sawdust and rice husk biochars applied at rates of 0, 20 and 40 t/ha to the soil treated with different N sources including cow dung and ammonium sulfate and growing maize on the amendments for two seasons with each season lasting for five weeks. The soils were leached on the 14th and 28th days after planting to determine the amount of leachable N. Biochar amendments reduced the leaching of NO_3^-N and NH_4^+N with no significant differences observed between biochar types, but between soil types. The abatement of leaching by biochar amendments consequently enhanced N uptake by maize and dry matter production and thus, agro-industrial waste biochar amendment is recommended for reducing leaching in tropical sandy soils.

Keywords: biochar; dry matter accumulation; leaching of nitrogen; nitrogen uptake; tropical sandy soils

1. Introduction

Despite the fact that nitrogen (N) is by far the largest component in the atmospheric air, it is the major limiting factor for the productivity of agricultural crops and is hence an indispensable input in crop production [1,2]. However, N applied to the soil is highly prone to leaching and volatilization, with nitrous oxide (N_2O) and ammonia (NH_3) being the gaseous emissions of great environmental concern, while NO_3^- ions are the main form of N leached [3]. Indeed, McAllister et al. [4] indicated that about 50% to 70% of N applied to the soil is lost through a combination of pathways, including leaching, erosion, denitrification, incorporation into microbial biomass and volatilization. N lost from the soil causes or exacerbates environmental pollution through triggering/heightening eutrophication, global warming, loss of biodiversity and depletion of ozone in the stratosphere [1,2]. The vast majority of West Africa is rich in low activity clay soils [5], on which leaching of nutrients, including N, is severe. The low nutrient retention capacities of these soils coupled with high infiltration rates, low organic matter (OM) content and high water conductivity culminate into low nutrient uptake by plants, fertilizer use efficiency and yield [6,7].

The use of biochar to check the leaching of applied N fertilizers has attracted a lot of interest in recent years and several studies have been conducted. An early study by Yao et al. [8] showed that only a few of the thirteen biochars used in their experiment could reduce leaching of nitrates in a batch experiment, but nine of the thirteen biochars attenuated ammonium leaching. Another study by Major et al. [9] found that biochar applied to a Columbian savanna oxisol at a rate of 20 tons per hectare increased nitrate leaching up to a soil depth of 0.6 m but the leaching was reduced by 8% at a 1.2 m soil depth. Sika and Hardie [10] showed that the pine wood sawmill waste biochar applied to a South African sandy soil at rates of 0.5, 2.5 and 10% reduced the leaching of ammonium nitrogen by 12, 50 and 86%, respectively, while the reduction rates of nitrate leaching stood at 26, 42 and 96%, respectively. Later on, Xu et al. [11] found that the application of maize straw biochar at 2, 4 and 8% to fluvo-aquic soil layered in columns reduced the leaching of urea by 18.8, 19.5 and 20.2%, respectively. A study by Sun et al. [12] showed that wheat straw biochar applied to saline coastal soils at rates of 0.5, 1.0, 2.0 and 4.0% reduced the leaching of ammonium by 11.64–27.68% and nitrate by 13.19–36.26%, with the leaching reduction power of the biochar increasing with increasing application rates.

However, its worth noting that the biochar-induced reductions in N leaching may not be agronomically beneficial, as has been demonstrated by numerous studies. For example, Sika and Hardie [10] noted that the leached biochar-amended soils contained infinitesimal amounts of exchangeable ammonium (0–7.3 mg kg^{-1}) and nitrate (5.8–8.0 mg kg^{-1}), which could negatively affect crop yield even though they did not grow any crops in their experiment. Indeed, in a four-year field experiment, Haider et al. [13] found that although biochar reduced nitrate leaching from the temperate soil, there was no positive effect on the yield of the grown maize. Contrary to this observation, Liu et al. [14] indicated that the concomitant reduction in N leaching caused by biochar applied to the sandy soil boosted N uptake and the dry weight of ryegrass in the first season of the experiment, although this positive effect was ephemeral and could not be reproduced in the second season. It is important to note that both experiments that involved the growing of crops were conducted in temperate soils and, to the best of our knowledge, there are no data that examined the effects of biochar on the leaching of N and subsequent effects on the uptake and yield of the crops grown in a tropical soil. Therefore, this experiment was conducted to assess the effects of biochar on N leaching from two different tropical sandy soils and to discern if the biochar's influence on N leaching affects N uptake and the growth of crops using maize as a test crop.

2. Materials and Methods

2.1. Descriptions of the Soils and Biochar Used in the Experient

The soils used belonged to the Keta (K) and Nyankpala (Ny) series whose samples were taken from Anloga in the Volta and Nyanpkala in Tolon-Kumbumgu district of the northern regions of Ghana, respectively. The Volta region is situated within the coastal savannah zone of Ghana with a mean temperature of 28 °C and an average annual rainfall of about 900 mm which is evenly spread over the year. On the other hand, Nyanpkala is situated within the guinea savannah zone of Ghana with a unimodal rainfall pattern of 1000–1300 mm per annum and a mean temperature of 32 °C. K series belongs to an Entisol order and Psamment suborder (*Quartzipsamment*) of the USDA soil taxonomy. Although the K series has little agricultural prospect due its low fertility status, with heavyfertilization the soil has been used for intensive maize and vegetable production over the years (Obeng, 2000). The Ny series soils are classified as Plinthic acrisols according to the FAO soil classification system. Both soil types were sampled at the depth of 0–20 cm, transported to the laboratory, air-dried, sieved through a 2 mm sieve, analyzed and used for the study. Biochar was produced at 500 °C in a kiln from three different feedstock biomasses and these were rice husk, saw dust and corn cob, following the pyrolysis method described by Lehmann et al. [15]. After pyrolysis, the biochar samples obtained were ground and the particles were homogenized by sieving through a 0.5 mm sieve.

2.2. Biochar and Soil Analysis

Biochar's pH measurement with a pH meter followed extraction of biochar with distilled water in ratios of 1:10, respectively. The total phosphorus and cations were extracted by wet ashing with concentrated nitric acid. The orthophosphate was then determined colorimetrically following a method espoused by Murphy and Rilley [16] after neutralization of the digest's pH with NaOH, while the cations were quantified with atomic absorption spectrometery (AAS). The total surface area was determined by following the Brunauer–Emmett–Teller (BET) method. The soil pH was measured both in water and calcium chloride. The mixing ratios adopted in the former method were 1:5, while 1:2.5 (soil: calcium chloride) was adopted for the latter. Particles size distributions were determined by using the Bouyoucos method [17], while the soil bulk densities were determined through strict adherence to the Blake and Hartge [18] method. The total organic carbon was determined by strictly adhering to the Walkley and Black [19] method, whilst the total nitrogen content was analyzed through the Kjeldahl method. The available phosphorus was determined by strictly adhering to the Bray P1 extraction procedure as outlined by Jones [20] and quantifying the orthophosphate colorimetrically at 880 nm by following the Murphy and Rilley [16] method. The basic cations were extracted from the soil samples with 1 M neutral ammonium acetate and analyzed by the AAS. The selected properties of the two soils used in the experiment are shown in Table 1, while those of the biochars are exhibited in Table S1 in the Supplementary File.

Table 1. Properties of the soils used in the experiment.

Parameters Assessed		Soil Type	
		K	Ny
Particle size (%)	Sand	90.04 ± 0.44	68.01 ± 0.65
	Silt	7.01 ± 0.07	24.01 ± 0.19
	Clay	2.95 ± 0.05	7.98 ± 0.09
Bulk density (Kg/m^3)		1.63 ± 0.04	1.58 ± 0.06
pH	H$_2$O (1:5)	6.60 ± 0.22	5.35 ± 0.34
	CaCl$_2$ (1:2.5)	6.31 ± 0.16	5.10 ± 0.25
TOC (g kg^{-1})		3.77 ± 0.30	9.95 ± 0.69
TN (g kg^{-1})		0.20 ± 0.01	0.71 ± 0.05
Avail. P (mg kg^{-1})		1.71 ± 0.13	2.23 ± 0.28
CEC (cmol$_c$ kg^{-1})		3.03 ± 0.11	8.14 ± 0.59
Exchangeable bases (cmol$_c$ kg^{-1})	Ca	1.02 ± 0.07	1.10 ± 0.15
	Mg	0.50 ± 0.00	0.54 ± 0.03
	K	0.30 ± 0.05	0.40 ± 0.01
	Na	0.50 ± 0.09	0.23 ± 0.06

K: Keta series, Ny: Nyankpala series.

2.3. Assessment of the NH_4^+ and NO_3^- Retention Capacities of the Biochars

Before the leaching experiment, the biochars were tested for their potentials to retain both the ammonium and nitrate ions. To execute this objective, biochar was packed into acrylic cylinders, the bottoms of which were covered with Whatman No 42 filter paper followed by a 25 μm pore size nylon mesh. The filter paper and nylon mesh were secured at the mouth with circular metal clips to prevent biochar particles from falling. Then, 150 g of each of the biochar sample was weighed into the acrylic cylinders and packed to 200 cm^3 by gently tapping the sides of the cylinders. The set up was replicated three times for each biochar type. Subsequently, 2.1 g of $(NH_4)_2SO_4$ and 3.42 g of KNO_3 as sources of NH_4^+ and NO_3^-, respectively, were each dissolved in 500 mL of deionized water and allowed to pass through the biochar sample in the column. A constant head of 50 cm was maintained and the leachate was collected. The concentrations of NH_4^+-N and NO_3^--N in every 50 mL of the leachate collected were determined colorimetrically. The former was determined through the indophenol blue method after extraction with 2 M KCl following a method adopted by Hood-Nowotny et al. [21], while the latter was extracted with 2M

KCl and determined following the chromotropic acid procedure espoused by West and Ramachandran [22]. The NH_4^+-N and NO_3^--N retention powers of the biochar were determined according to the equation below:

$$A = M_1 - M_2/W,$$

where A = amount of NH_4^+-N or NO_3^--N retained by the biochar, M_1 = mass of NH_4^+-N or NO_3^--N applied, M_2 = mass of NH_4^+-N or NO_3^--N in leachate, W = weight of biochar in the column. Apart from fertilizers that supplied single forms of ionic N, a retention experiment with a fertilizer source that supplied both NH_4^+ and NO_3^-, i.e., NH_4NO_3 was also conducted with NH_4NO_3 used in such quantities as to maintain the amounts of NH_4^+ and NO_3^- employed in the above-mentioned cases. Two biochars with superior retention capabilities of both NH_4^+ and NO_3^- were selected out of the three biochars tested for the leaching and used for the maize growing experiments in the greenhouse. The quantities of the different ionic forms of N retained by the different biochar types are shown in Table S2 in the supplementary file. The biochar impregnated with ammonium sulfate was dried and used as a nitrogen-enriched biochar-based fertilizer, denoted as ASS.

2.4. Greenhouse Experiment

The greenhouse experiment was conducted to assess the effects of biochar amendments on leaching of nitrogen from two distinctly different sandy soils and elucidating whether the biochar-induced effects on leaching has any bearing on N uptake and dry matter accumulation by maize. The two soils used for the experiment were the K series soil (Quartzipsamment) and the Ny series soil (Plinthic Acrisol) collected from the Volta and northern regions of Ghana, respectively. Under heavy fertilization regimes, both soils have been used for intensive maize and vegetable production over the years. A 2.3 kg sample of each of the soils was weighed into experimental pots measuring 15 cm in height and 8 cm in diameter. Four holes were created at the bottom of the pots which were then plugged with cotton wool to prevent soil particles from falling. The moisture content of the soil was maintained at 80% field capacity except on the days of collecting the leachate when the soils were saturated. Each pot was placed in a bowl to allow easy collection of leachate. Each of the biochars was applied at rates of 20 and 40 tons per hectare, while N in form of ammonium sulfate (ASP) was applied at the recommended rate of 265 kg per hectare and each of the treatments was replicated thrice. Additionally, the two biochars were impregnated with ASP and applied as nutrient-enriched biochar-based fertilizers. The ASP embedded into the biochar was denoted as ASS. A treatment with cow dung (CD) as a nitrogen source instead of the mineral fertilizers was included and it was applied at a rate that satisfied the nitrogen requirement of maize. Another treatment, CDASS was constituted by combining ASS and CD in equal proportions in terms of their nitrogen contents. The experiment was organized in a completely randomized design. The sawdust biochar applied at 20 tons and 40 tons per hectare was denoted as SD20 and SD40, respectively. The rice husk biochar applied to the soil at 20 tons and 40 tons per hectare was denoted as RH20 and RH40, respectively. All the soil amendments, including N, were applied once at the beginning of the first growing season. Four seeds of maize were sown per pot and thinned to two plants per pot after germination. Maize was grown for two seasons and each season lasted for a period of 5 weeks. Each of the treatments was replicated thrice. The leachate was collected on the 14th and 28th days after planting (DAP) in both seasons. The harvested maize plants were separated into shoots and roots, and dried in an oven at a temperature of 68 °C for 48 h to determine dry matter weight and N content of the maize which was determined through the Kjeldhal method. In order to investigate whether biochar amendments could lead to the preservation of soil N, maize was grown for a second season without any additional amendments to the soil.

2.5. Statistical Analysis

The data collected were subjected to analysis of variance (ANOVA) using Genstats (9th edition) and the means were separated at a least significance level of 5%.

3. Results and Discussion

3.1. Nitrogen Retentions by Biochar and Leaching from the Soil

As shown in Table S2, the saw dust biochar retained the highest amounts of NH_4^+ from both the $(NH_4)_2SO_4$ and NH_4NO_3 fertilizers with the quantity adsorbed from the former totaling to 2273.40 mg kg^{-1} while that from the latter amounted to 2475.1 mg kg^{-1}. This was followed by the rice husk biochar which retained 1809.57 mg kg^{-1} and 1703.88 mg kg^{-1} of NH_4^+ from $(NH_4)_2SO_4$ and NH_4NO_3 fertilizers, respectively. The corncob biochar on the other hand adsorbed 1756.70 mg kg^{-1} of NH_4^+ from the $(NH_4)_2SO_4$ fertilizer whilst the quantity retained from NH_4NO_3 was a meagre 1022.38 mg kg^{-1}. This same trend was observed with the adsorption of NO_3^- from NH_4NO_3 where the sawdust biochar retained 2283.93 mg kg^{-1}, whereas rice husk and corncob biochars adsorbed 1860.32 mg kg^{-1} and 1569.08 mg kg^{-1} of NO_3^-, respectively. With regard to the adsorption of NO_3^- from KNO_3, the saw dust biochar adsorbed the highest quantity which totaled up to 2283.93 mg kg^{-1} and was followed in a descending order by the corn cob biochar at 1881.31 mg kg^{-1} and the rice husk biochar at 1743.33 mg kg^{-1}. Therefore, in the presence of both NH_4^+-N and NO_3^--N (when NH_4NO_3 was used), the sawdust biochar exhibited a preference for NH_4^+ while both the rice husk and corncob biochars showed more affinity for NO_3^- than NH_4^+. The differences in the affinity for different forms of N by biochar have been documented by Yao et al. [8] through a batch experiment, where they employed one hydrochar as well as twelve biochars produced from four different biomass feedstocks at three different temperatures of 300 °C, 450 °C and 600 °C. They found that only three biochars pyrolyzed at the highest temperature of 600 °C could adsorb 0.12% to 3.7% of NO_3^--N from NH_4NO_3 while nine out of the twelve biochars adsorbed 1.8% to 15.7% of NH_4^+-N. Fidel et al. [23] later supported these deductions by demonstrating that the sorption of NO_3^--N increased with the pyrolysis temperature and that differences in sorption existed between red oak and corn stover biochars at most pyrolysis temperatures. The relatively high amount of NH_4^+-N and NO_3^--N retained by the sawdust biochar when the biochar samples were loaded with the fertilizer solutions could be due to the relatively high surface area of the sawdust biochar in comparison with other biochar types (Table S2), which provided more surfaces for N adsorption. Indeed, Zhou et al. [24] elucidated that adsorptions of nitrate and phosphates by biochars were influenced by their surface areas and porosity characteristics.

Leachates collected from the K series soil contained significantly higher amounts of both NH_4^+-N and NO_3^--N ($p < 0.05$) than the leachates obtained from the Ny series soil on both the leaching events and in both maize growing seasons as shown in Tables 2 and 3. The differences in the quantities of inorganic N leached from the two soils might have ensued from the differences in their clay contents. Additionally, control treatments without any biochar amendments leached significantly higher quantities of both NH_4^+-N and NO_3^-N ($p < 0.05$) than the biochar-amended soils, implying that biochar greatly reduced leaching of the aforementioned ions from the soil. In comparison with the control experiment, the worst performing biochar amendments reduced NH_4^+-N leaching from the K series soils fertilized with ASP, ASS, CD and CDASS by 229.8%, 329.6%, 121.4% and 180.3%, respectively, in the first leaching event of the first season (see Table 2). In the second leaching event of the first season, the worst performing biochar amendments reduced NH_4^+-N leaching by 269.0%, 160.6%, 177.2% and 87.5% from the K series soils fertilized with ASP, ASS, CD and CDASS, respectively.

Table 2. Amount of NH_4^+-N leached from the soil.

Season	Soil Type	Biochar Amendment	Nitrogen Fertilizers							
			ASP		ASS		CD		CDASS	
			L1	L2	L1	L2	L1	L2	L1	L2
1st Season	K	Control	23.58 ± 2.12a	16.20 ± 1.08a	25.99 ± 1.85a	15.27 ± 2.39a	17.93 ± 1.05b	16.44 ± 0.90a	20.60 ± 1.85a	10.97 ± 3.10b
		RH20	7.15 ± 1.24d	4.03 ± 2.10d	4.89 ± 1.33e	3.87 ± 0.89d	6.45 ± 1.52d	3.48 ± 1.66d	5.94 ± 2.00de	5.85 ± 1.72c
		RH40	4.51 ± 0.79e	3.91 ± 1.05d	6.05 ± 0.68d	3.60 ± 1.09d	8.10 ± 1.11d	5.93 ± 0.95c	7.35 ± 0.88d	4.10 ± 1.05d
		SD20	6.56 ± 1.88d	3.68 ± 0.93d	5.67 ± 1.27de	5.86 ± 0.78c	6.76 ± 2.16d	4.24 ± 1.92d	5.51 ± 2.21de	3.69 ± 0.69d
		SD40	6.17 ± 2.22d	4.39 ± 1.76d	5.08 ± 0.85e	3.75 ± 0.73d	8.01 ± 1.66d	5.64 ± 1.99d	4.03 ± 0.78e	3.41 ± 1.62d
	Ny	Control	10.44 ± 3.10c	9.70 ± 2.07b	11.72 ± 0.97c	8.08 ± 1.54b	10.59 ± 1.39c	10.91 ± 1.07b	9.34 ± 1.59c	9.98 ± 0.85b
		RH20	2.10 ± 0.67f	3.98 ± 0.92d	1.74 ± 1.05f	4.12 ± 0.82d	3.96 ± 0.63e	4.14 ± 0.93d	2.75 ± 0.69f	1.95 ± 0.55e
		RH40	1.66 ± 0.23f	3.68 ± 0.87d	2.14 ± 0.64f	4.09 ± 0.73d	1.83 ± 0.87f	5.01 ± 0.91cd	1.78 ± 0.56f	3.33 ± 0.39d
		SD20	2.17 ± 0.86f	2.63 ± 0.75d	3.68 ± 0.58def	1.86 ± 0.63e	2.90 ± 0.97f	3.53 ± 0.81d	1.89 ± 0.55f	2.77 ± 0.36d
		SD40	1.98 ± 0.85f	1.85 ± 0.36e	2.24 ± 0.94f	2.79 ± 0.81d	1.76 ± 0.46f	2.56 ± 0.65d	1.66 ± 0.59f	2.04 ± 0.88de
2nd Season	K	Control	8.78 ± 0.69ab	3.99 ± 0.56b	9.24 ± 0.39a	4.65 ± 0.25a	10.50 ± 0.48a	3.81 ± 0.67b	6.05 ± 2.10b	4.51 ± 0.92ab
		RH20	1.75 ± 0.96d	0.56 ± 0.44d	2.00 ± 0.72d	0.37 ± 0.13d	3.03 ± 0.45d	0.55 ± 0.33d	4.67 ± 0.96c	0.64 ± 0.17d
		RH40	2.25 ± 0.13d	0.46 ± 0.18d	3.13 ± 0.43d	0.51 ± 0.11d	1.96 ± 0.52d	0.63 ± 0.30d	2.40 ± 0.33d	0.50 ± 0.46d
		SD20	3.00 ± 0.91d	0.72 ± 0.12d	4.05 ± 1.22c	0.65 ± 0.26d	2.21 ± 0.54d	0.51 ± 0.42d	2.47 ± 0.81d	0.48 ± 0.21d
		SD40	2.53 ± 0.54d	0.67 ± 0.31d	2.14 ± 0.72d	0.54 ± 0.39d	2.44 ± 0.42d	0.66 ± 0.11d	1.07 ± 0.65d	0.42 ± 0.36d
	Ny	Control	5.70 ± 1.06b	1.98 ± 0.73c	6.01 ± 0.98b	2.09 ± 0.56c	7.09 ± 2.10b	3.07 ± 0.91bc	10.02 ± 2.17a	1.52 ± 0.82c
		RH20	1.91 ± 0.33d	0.53 ± 0.23d	2.11 ± 0.68d	0.68 ± 0.16d	1.84 ± 0.18d	0.49 ± 0.29d	3.06 ± 0.76d	0.52 ± 0.13d
		RH40	2.31 ± 0.22d	0.48 ± 0.16d	1.87 ± 0.30d	0.53 ± 0.17d	2.11 ± 0.73d	0.62 ± 0.35d	2.01 ± 0.52d	0.40 ± 0.29d
		SD20	2.07 ± 0.28d	0.54 ± 0.16d	1.91 ± 0.26d	0.60 ± 0.24d	1.59 ± 0.22d	0.50 ± 0.18d	2.78 ± 0.36d	0.38 ± 0.34d
		SD40	1.85 ± 0.42d	0.66 ± 0.41d	1.64 ± 0.18d	0.46 ± 0.13d	2.88 ± 0.53d	0.39 ± 0.10d	1.77 ± 0.17d	0.47 ± 0.20d

K: Keta series, Ny: Nyankpala series, L1: leachate collected on the 14th day after planting, L2: leachate collected on the 28th day after planting, ASP: Ammonium sulfate fertilizer, ASS: Ammonium sulfate embedded into the biochar, CD: Cow dung, CDASS: Cow dung + Ammonium sulfate embedded into the biochar, RH20: 20 t ha^{-1} Rice husk biochar, SD20: 20 t ha^{-1} sawdust biochar, RH40: 40 t ha^{-1} rice husk biochar, SD40: 40 t ha^{-1} sawdust biochar. Values for leachate collected on the same day with the same letter are not significantly different $p = 0.05$.

Table 3. Amount of NO_3^--N leached from the soil.

Season	Soil Type	Biochar Amendment	Nitrogen Fertilizers									
			ASP		ASS		CD		CDASS	
			L1	L2	L1	L2	L1	L2	L1	L2
1st Season	K	Control	8.86 ± 0.75a	6.88 ± 1.09a	10.05 ± 1.73a	7.53 ± 0.66a	9.13 ± 0.87a	7.01 ± 1.39a	7.49 ± 0.84b	8.03 ± 0.55a
		RH20	2.01 ± 0.68d	0.79 ± 0.07c	2.35 ± 0.39d	0.83 ± 0.07c	3.02 ± 0.51d	0.71 ± 0.09c	2.64 ± 0.75d	0.66 ± 0.08c
		RH40	3.09 ± 0.79d	0.89 ± 0.09c	2.70 ± 0.51d	0.65 ± 0.03c	1.69 ± 0.91d	0.72 ± 0.06c	2.51 ± 0.94d	0.79 ± 0.07c
		SD20	2.17 ± 0.87d	0.67 ± 0.09c	2.69 ± 0.73d	0.75 ± 0.05c	2.87 ± 0.51d	0.69 ± 0.08c	1.97 ± 0.91d	0.80 ± 0.07c
		SD40	1.99 ± 0.90d	0.82 ± 0.06c	2.26 ± 0.33d	0.69 ± 0.08c	2.66 ± 0.72d	0.59 ± 0.08c	2.99 ± 0.51d	0.77 ± 0.03c
	Ny	Control	6.67 ± 0.62bc	3.13 ± 0.57b	5.10 ± 0.81c	4.57 ± 0.76b	5.02 ± 1.00c	3.33 ± 0.32b	7.46 ± 2.54b	2.98 ± 0.54b
		RH20	1.88 ± 0.74d	0.30 ± 0.08c	0.77 ± 0.19e	0.29 ± 0.06c	0.93 ± 0.04e	0.41 ± 0.09c	0.57 ± 0.08e	0.31 ± 0.09c
		RH40	1.05 ± 0.71d	0.27 ± 0.09c	3.23 ± 0.80d	0.32 ± 0.06c	1.56 ± 0.79d	0.35 ± 0.05c	2.23 ± 1.03d	0.49 ± 0.06c
		SD20	2.22 ± 0.43d	0.51 ± 0.04c	1.55 ± 0.61d	0.27 ± 0.05c	1.89 ± 0.18d	0.29 ± 0.08c	1.58 ± 0.35d	0.33 ± 0.07c
		SD40	1.54 ± 0.69d	0.39 ± 0.07c	1.91 ± 0.22d	0.39 ± 0.07c	3.89 ± 0.07d	0.27 ± 0.03c	2.22 ± 1.09d	0.32 ± 0.06c
2nd Season	K	Control	4.29 ± 0.81a	1.97 ± 0.90b	3.93 ± 1.07a	5.46 ± 1.13a	4.09 ± 1.67a	3.16 ± 1.89b	1.99 ± 0.98b	2.39 ± 1.07b
		RH20	1.04 ± 0.82c	0.50 ± 0.06c	0.87 ± 0.08cd	0.63 ± 0.07c	0.60 ± 0.09d	1.77 ± 0.06b	0.28 ± 0.04d	0.79 ± 0.07c
		RH40	0.73 ± 0.08d	0.80 ± 0.05c	0.97 ± 0.06cd	1.81 ± 0.06b	0.87 ± 0.10cd	0.63 ± 0.04c	1.19 ± 0.30c	0.75 ± 0.09c
		SD20	0.69 ± 0.11d	0.45 ± 0.03c	0.75 ± 0.08d	0.63 ± 0.07c	0.57 ± 0.05d	0.61 ± 0.04c	0.86 ± 0.04cd	0.50 ± 0.06c
		SD40	0.47 ± 0.17d	0.66 ± 0.05c	0.55 ± 0.05d	0.60 ± 0.04c	0.85 ± 0.03d	0.48 ± 0.07c	0.93 ± 0.05d	0.57 ± 0.05c
	Ny	Control	2.16 ± 0.90b	1.89 ± 0.69c	3.78 ± 1.07a	0.90 ± 0.07c	1.95 ± 0.66b	2.03 ± 0.11b	4.10 ± 2.10a	1.86 ± 0.71b
		RH20	0.66 ± 0.09d	0.57 ± 0.05c	0.52 ± 0.06d	0.47 ± 0.06c	0.53 ± 0.09d	0.71 ± 0.08c	0.65 ± 0.07d	0.48 ± 0.03c
		RH40	0.70 ± 0.05d	0.69 ± 0.09c	0.75 ± 0.09d	0.82 ± 0.04c	0.47 ± 0.07d	0.66 ± 0.09d	0.71 ± 0.08d	0.52 ± 0.06c
		SD20	0.54 ± 0.08d	0.46 ± 0.06d	0.69 ± 0.07d	0.40 ± 0.07c	0.73 ± 0.08d	0.54 ± 0.07d	0.55 ± 0.05d	1.03 ± 0.08c
		SD40	0.99 ± 0.16cd	0.55 ± 0.04c	0.44 ± 0.05d	0.56 ± 0.09c	0.60 ± 0.07d	0.39 ± 0.05d	0.73 ± 0.03d	0.77 ± 0.09c

K: Keta series, Ny: Nyankpala series, L1: leachate collected on the 14th day after planting, L2: leachate collected on the 28th day after planting, ASP: Ammonium sulfate fertilizer, ASS: Ammonium sulfate embedded into the biochar, CD: Cow dung, CDASS: Cow dung + Ammonium sulfate embedded into the biochar, RH20: 20 t ha^{-1} Rice husk biochar, SD20: 20 t ha^{-1} sawdust biochar, RH40: 40 t ha^{-1} rice husk biochar, SD40: 40 t ha^{-1} sawdust biochar. Values for leachate collected on the same day with same letter are not significantly different $p = 0.05$.

In the Ny series soils, the worst performing biochar amendments abated NH_4^+-N leaching by 381.1%, 218.5%, 167.4% and 239.6% from pots fertilized with ASP, ASS, CD and CDASS, respectively, in the first leaching event of the first season (see Table 2). In the second leaching event, NH_4^+-N leaching reduced by 143.7%, 96.1%, 117.8% and 199.7% in the ASP, ASS, CD and CDASS fertilized pots, respectively. In the first leaching event of the second season, the worst performing biochar amendments reduced NH_4^+-N leaching from the ASP, ASS, CD and CDASS fertilized K series soil by 192.7%, 128.1%, 246.5% and 29.5%, respectively, while in the second leaching event, the reductions stood at 454.2%, 615.4%, 477.3% and 604.7%, respectively. The worst performing biochar amendments reduced NH_4^+-N leaching from the Ny series soil fertilized with ASP, ASS, CD and CDASS by 146.8%, 184.8%, 146.2% and 227.5%, respectively, in the first leaching event, whereas in the second leaching event, the reductions amounted to 200.0%, 207.4%, 395.2% and 192.3%, respectively. This observation accorded with the results obtained by Yao et al. [8], Sika and Hardie [10] and Haider et al. [13], who reported decrements in the amounts of nitrates and ammonium leached from biochar-amended soils.

The leachates generally contained more NO_3^--N than NH_4^+-N in all the leaching events in both soils, which was in agreement with the observation made by Liu et al. [14], who found that NO_3^--N accounted for more than 90% of the total amount of inorganic N leached from the soil. This can be attributed to the presence of negatively charged carboxylic and phenolic compounds on the biochar surface with limited ability to retain NO_3^- [25] and accelerated nitrification brought about by biochar which resulted in greater amounts of leachable NO_3^--N [14]. However, the amount of leached NO_3^--N were far less in biochar-amended soils than in the control. In comparison with the control, the worst performing biochar amendments on ASP, ASS, CD and CDASS fertilized K series soil reduced NO_3^--N leaching by 186.7%, 272.2%, 202.3% and 150.5%, respectively, in the first leaching event, and by 673.0%, 807.2%, 873.6% and 903.8%, respectively, in the second leaching event of the first season (see Table 3). In the Ny series soil, the worst performing biochar amendments waned NO_3^--N leaching by 200.5%, 57.9%, 29.0% and 234.5% in the first leaching event, and by 513.7%, 1071.8%, 7122.2% and 508.2%, respectively, in the second leaching event.

In the second season, the worst performing biochar amendments on K series soil fertilized with ASP, ASS, CD and CDASS lessened NO_3^--N leaching by 312.5%, 305.2%, 370.1% and 67.2%, respectively, in the first leaching event, and by 146.3%, 201.7%, 78.5% and 202.5%, respectively, in the second leaching event. In the N series, the worst performing biochar amendments lowered NO_3^--N leaching by 118.2%, 40.4%, 167.1% and 461.6% in the first leaching event, and by 173.9%, 9.8%, 185.9% and 8.1%, respectively, in the second leaching event. This is because biochar also has positive surface charges which aid in its attraction of the negatively charged ions including NO_3^- and phosphate ions, as Chintala et al. [26] noted. Great statistical differences existed both between the biochar types and application rates in the first leaching event as far as the amounts of NO_3^--N leached were concerned. In this regard, the sawdust biochar exhibited the highest leaching attenuation power and the leached quantities of NO_3^--N decreased with increasing application rates of either biochars. These statistical differences, however, diminished with subsequent leaching events to the extent that there were hardly any statistical differences in the amounts of NO_3^--N leached from the biochar-amended soils by the last leaching event, as seen in Table 3. On the other hand, the amount of NH_4^+-N leached from both soils in similar leaching events did not exhibit statistical variations. The amount of N retained in the soil at the end of the experiment was at least three times more in the biochar-amended soils than in the control, as shown in Table S3 which confirms the great N retention power of the biochar.

3.2. Dry Matter Accumulation and Nitrogen Uptake by Maize

The shoot and root dry matter (DM) accumulations in maize grown in the various amended soils are shown in Figures 1 and 2, respectively. The accumulated shoot dry matter of maize grown in the first and second seasons are shown in Figure 1a,b, respectively, while Figure 2a,b exhibits the root dry matter accumulated in the first and second seasons, respectively. The shoot and root dry weights produced by the control treatments in both soils were not significantly ($p > 0.05$) different from each other. This could be attributed to the inherent inability of the soils used to retain applied plant nutrients especially N as a result of their low CEC (Table 1). Amending the two soils with biochar significantly ($p < 0.05$) increased the dry matter yield. In comparison with the control experiment, the worst performing biochar amendments on ASP, ASS, CD and CDASS fertilized K series soil boosted shoot dry matter accumulation by 127.8%, 240.9%, 119.4% and 110.5%, respectively, in the first season. In the Ny series, the dry matter enhancement amounted to 101.8%, 128.5%, 220.0% and 162.7%, respectively. In the second season, the worst performing biochar amendments on ASP, ASS, CD and CDASS fertilized pots increased shoot dry matter accumulation by 928.9%, 534.7%, 488.9% and 378.8%, respectively, in the K series soil, and by 1629.6%, 1209.7%, 661.5% and 741.6%, respectively, in the Ny series soil.

The root dry matter increased by a minimum of 495.8%, 752.8%, 452.8% and 415.8% in the ASP, ASS, CD and CDASS fertilized K series soils, respectively, in the first season, and by 819.2%, 529.4%, 445.0% and 375.6%, respectively, in the second season in comparison to the control. On the other hand, the worst performing biochar amendments on the Ny series increased root dry matter accumulation by 498.0%, 555.3%, 700.0% and 590.7% in the first season, and by 1025.8%, 1050.0%, 311.0% and 691.5% in the second season when applied together with ASP, ASS, CD and CDASS nitrogen sources, respectively. The observations made as far as dry matter accumulations are concerned contradict the one made by Haider et al. [13] who indicated that while biochar was able to reduce leaching from temperate soils in a four-year field trial, the effects on dry matter accumulation and yield were null. Additionally, Liu et al. [14] showed that biochar increased the dry matter accumulation in ryegrass grown on a sandy soil only in the first season of a two-season experiment, while the effects in the second season were null in comparison to the control.

The observed increment in the dry matter accumulation observed in the current study could be ascribed to the increased availability of the applied N brought about by biochar's high adsorption power for the ionic forms of the N fertilizer. According to Taghizadeh-Toosi et al. [27], recent evidence has indicated that N adsorbed by biochar is eventually made available for plant uptake. Ma and Matsunaka [28] reported that when biochar is applied as a sole amendment or together with N fertilizer, it significantly improved the dry matter of shoots and roots of lettuce. A similar trend was also observed during the second planting, as shown in Figure 2a,b. In both seasons, maize grown on the Ny series accumulated higher shoot and root dry matter than the one grown on the K series. It is worth noting, however, that the differences in the shoot dry matter of the maize grown on the amended soils were not statistically significantly different in the first season while the differences in the root dry matter in both seasons were not significantly different. The differences in the dry matter accumulated in the maize grown on the K and Ny series can be attributed to the differences in the clay contents of the two soils, whereby the Ny series soil contained almost thrice as much clay as did the K series (see Table 1). This observation concurs with the one made by Chintala et al. [26] who observed differences in dry matter accumulations in maize grown on two different soils and with different biochar types. However, they attributed the observed differences to the differing biochar properties rather than the differences in soil properties. Therefore, the reasons behind the observed differences require further investigations.

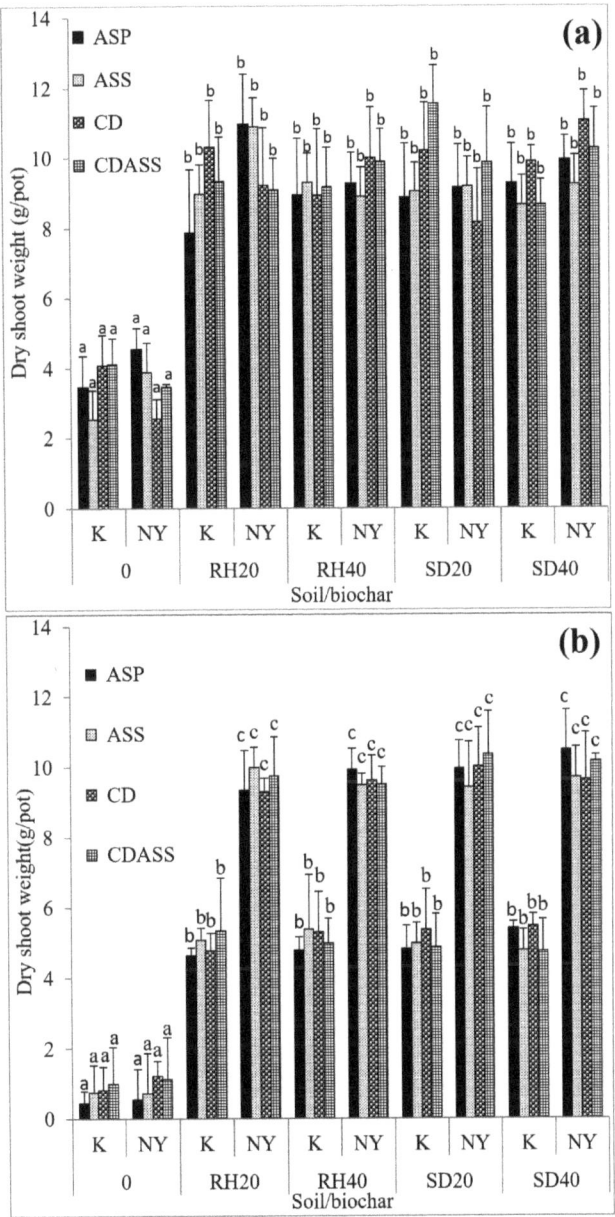

Figure 1. Dry shoot weight of the maize grown with the different amendments in the (**a**) first and (**b**) second growing seasons. K: Keta series, Ny: Nyankpala series, ASP: Ammonium sulfate fertilizer, ASS: Ammonium sulfate embedded into the biochar, CD: Cow dung, CDASS: Cow dung + Ammonium sulfate embedded into the biochar, RH20: 20 t ha^{-1} Rice husk biochar, SD20: 20 t ha^{-1} sawdust biochar, RH40: 40 t ha^{-1} rice husk biochar, SD40: 40 t ha^{-1} sawdust biochar. Values for dry shoot weight with same letter(s) under the same planting season were not significantly different at $p = 0.05$.

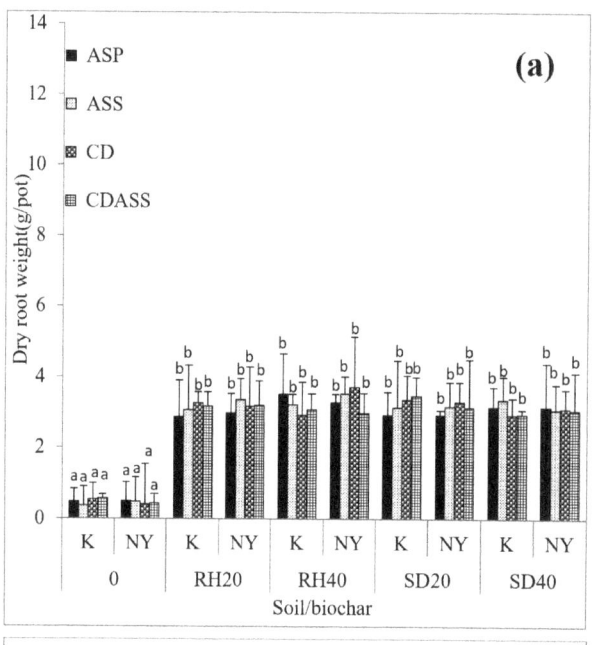

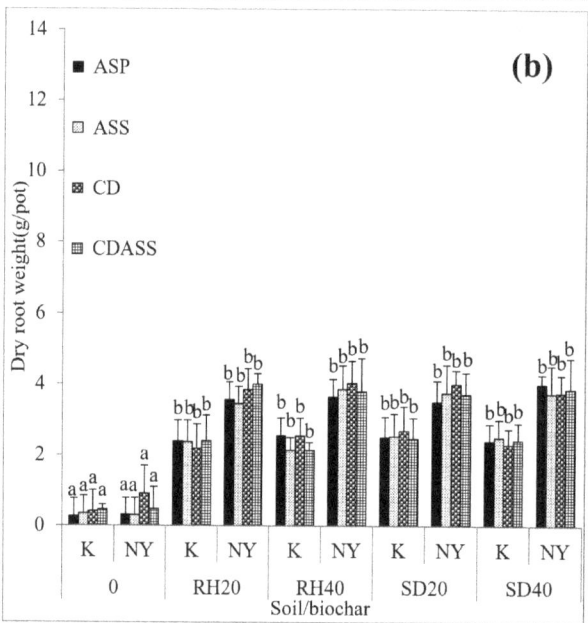

Figure 2. Dry root weights of the maize grown with the different amendments in the (**a**) first and (**b**) second growing seasons. K: Keta series, Ny: Nyankpala series, ASP: Ammonium sulfate fertilizer, ASS: Ammonium sulfate embedded into the biochar, CD: Cow dung, CDASS: Cow dung + Ammonium sulfate embedded into the biochar, RH20: 20 t ha^{-1} Rice husk biochar, SD20: 20 t ha^{-1} sawdust biochar, RH40: 40 t ha^{-1} rice husk biochar, SD40: 40 t ha^{-1} sawdust biochar. Values for dry root weight with same letter(s) under the same planting season were not significantly different at $p = 0.05$.

Enriching biochar with nitrogen before application did not have any impact on the dry matter accumulation in maize as the shoot and root dry weights obtained with ASS were not significantly different from the rest of the treatments. This observation is in contrast to the one made by Dietrich et al. [29] and Luyima et al. [30]. The former found that biochar enriched with biogas digestates produced heavier dry shoot and root weights of maize than the normal biochar without any enrichment, while the latter indicated that sorbing urea into co-pyrolyzed cow dung and bone meal produced leaf lettuce with heavier shoots and roots than the control. Unlike the dry shoot weight, which exhibited significant statistical differences in the second season across the amended soils, there were no significant statistical differences in the dry root weights obtained from the biochar-amended soils in both seasons, even though the roots obtained from the K series were lighter than those from the Ny series.

Concerning N uptake by maize, the uptake was higher in the maize grown in the Ny series than in the K series in both seasons, as shown in Table 4. This may still be attributed to the fact that the clay content in the K series was lower than that in the Ny series, as shown in Table 1. Indeed, differences in N uptake ensuing from variations in soil properties have been well documented by Liu et al. [14], who found an increased N uptake by ryegrass grown on a sandy soil amended with biochar but failed to observe any improvement in N uptake when biochar was applied to a loam soil. Amending the soils with biochar significantly ($p < 0.05$) enhanced N uptake, as shown in Table 4. The addition of biochar to the two soils enhanced N retention in the soils, which might have helped in making the retained N available for possible uptake by maize. In contrast to the observation made by Liu et al. [14], who indicated that the ability of biochar to increase N uptake by ryegrass diminished in the second season, the capacity of biochar amendments to increase N uptake by maize remained strong even in the second season in the present study. This observation indicates that biochar is capable of enhancing soil retention of N for extended periods. There are several pathways through which biochar improves N retention in the soil including reductions in gaseous emissions, adsorptions of both NO_3^- and NH_4^+ ions, abatement of denitrification of NO_2, etc., as Rashid et al. [31], Luyima et al. [2,32,33] and others have elaborated. Taghizadeh-Toosi et al. [27] indicated that N adsorbed by biochar is bioavailable and is easily released for plant uptake.

Table 4. Nitrogen uptake in maize.

Soil Type	Biochar Amendment	Nitrogen Fertilizers							
		ASP		ASS		CD		CDASS	
		N1	N2	N1	N2	N1	N2	N1	N2
K	Control	8.97 ± 0.69a	9.64 ± 0.95a	9.54 ± 0.52a	7.94 ± 0.72a	8.89 ± 0.44a	6.56 ± 0.67a	9.02 ± 0.29a	7.46 ± 0.18a
	RH20	51.47 ± 4.17b	43.80 ± 3.10b	59.37 ± 4.98b	51.38 ± 2.10b	52.01 ± 3.21b	40.64 ± 5.01b	60.10 ± 3.97b	48.05 ± 2.87b
	RH40	51.60 ± 3.96b	50.49 ± 3.89b	47.87 ± 2.67b	52.71 ± 4.17b	56.37 ± 3.88b	43.36 ± 2.98b	54.12 ± 5.11b	50.66 ± 3.07b
	SD20	51.29 ± 2.86b	47.37 ± 2.77b	54.53 ± 2.85b	54.64 ± 3.62b	50.52 ± 2.99b	45.02 ± 5.40b	47.85 ± 2.84b	50.31 ± 3.52b
	SD40	52.36 ± 3.18b	44.19 ± 1.98b	55.77 ± 3.17b	51.71 ± 2.73b	53.62 ± 3.77b	42.87 ± 1.98b	52.70 ± 5.06b	50.42 ± 4.37b
Ny	Control	9.47 ± 0.71a	8.04 ± 0.45a	8.29 ± 0.32a	6.69 ± 0.51a	11.60 ± 0.62a	6.68 ± 0.39a	13.58 ± 0.83a	8.38 ± 0.28a
	RH20	84.13 ± 5.78c	77.09 ± 4.89c	78.70 ± 5.67c	66.36 ± 6.19c	81.77 ± 5.89c	69.89 ± 4.90c	73.34 ± 5.19c	74.11 ± 3.49c
	RH40	78.43 ± 4.13c	76.21 ± 6.67c	86.91 ± 5.81c	67.55 ± 4.96c	87.88 ± 5.33c	72.57 ± 6.13c	86.37 ± 5.76c	68.93 ± 4.88c
	SD20	81.09 ± 6.89c	65.46 ± 5.93c	75.99 ± 4.88c	73.68 ± 5.73c	84.78 ± 7.24c	68.93 ± 6.85c	83.29 ± 5.12c	71.48 ± 5.33c
	SD40	79.79 ± 5.88c	74.31 ± 4.89c	64.66 ± 6.02c	67.26 ± 3.94c	81.41 ± 6.88c	66.50 ± 3.99c	71.65 ± 6.03c	73.09 ± 6.42c

K: Keta series, Ny: Nyankpala series, N1: Nitrogen uptake by maize in the first season, N2: Nitrogen uptake by maize in the second season, ASP: Ammonium sulfate fertilizer, ASS: Ammonium sulfate embedded into the biochar, CD: Cow dung, CDASS: Cow dung + Ammonium sulfate embedded into the biochar, RH20: 20 t ha^{-1} Rice husk biochar, SD20: 20 t ha^{-1} sawdust biochar, RH40: 40 t ha^{-1} rice husk biochar, SD40: 40 t ha^{-1} sawdust biochar. Values for nitrogen uptake by maize obtained in the same season with same letter are not significantly different $p = 0.05$.

4. Conclusions

Although biochar exhibited significant differences in the sorption of ionic forms of N from solution during the column experiment, no such differences were observed in their capacities to reduce leaching of N from the soil. This means that even biochars that have been found to be inferior in sorbing ionic forms of N in batch experiments can be helpful in attenuating N leaching from the soil. Secondly, the ability of biochar to reduce N leaching from the tropical sandy soils is long-lived and, hence, biochar can result in a prolonged conservation of N in the soil. The abatement in the amounts of N leached out of the soil in biochar-amended sandy soils consequently resulted in more N being available to the growing maize crop which increased both N uptake and dry matter accumulations. From the observations made in this study, therefore, biochar amendments are recommended for sustainable maize production on tropical sandy soils since they result in the conservation of N and higher maize yields than the conventional fertilizer amendments.

Supplementary Materials: The following supporting information can be downloaded at: https://www.mdpi.com/article/10.3390/c9010034/s1, Table S1: Some physico-chemical properties of the biochar used; Table S2: Amount of NH_4^+–N and NO_3^-–N retained by biochar; Table S3: Residual soil available N.

Author Contributions: M.E.: Conceptualization, Methodology, Formal analysis, Investigation, Writing—original draft. I.Y.D.L.: Conceptualization, Software, Resources, Writing—review & editing, Supervision. D.E.D.: Conceptualization, Validation, Data curation, Writing—original draft. D.L.: Conceptualization, Methodology, Writing—review & editing, Supervision. All authors have read and agreed to the published version of the manuscript.

Funding: No funding was received for this study but we received enormous material support from the department of soil science of the University of Ghana including greenhouse space.

Data Availability Statement: The data to support the conclusions made in the study are included in the manuscript while small amounts of the produced biochars can be provided to anyone upon request.

Conflicts of Interest: The authors declare that they have no conflict of interest.

References

1. Rütting, T.; Aronsson, H.; Delin, S. Efficient use of nitrogen in agriculture. *Nutr. Cycl. Agroecosystems* **2018**, *110*, 1–5. [CrossRef]
2. Luyima, D.; Egyir, M.; Lee, J.H.; Yoo, J.H.; Oh, T.K. A review of the potentiality of biochar technology to abate emissions of particulate matter originating from agriculture. *Int. J. Environ. Sci. Technol.* **2021**, *19*, 3411–3428. [CrossRef]
3. Sutton, M.A.; Howard, C.M.; Erisman, J.W.; Bleeker, A.; Billen, G.; Grennfelt, P.; Van Grinsven, H.; Grizzetti, B. (Eds.) *The European Nitrogen Assessment Sources, Effects and Policy Perspectives*; Cambridge University Press: Cambridge, UK, 2011.
4. McAllister, C.H.; Beatty, P.H.; Good, A.G. Engineering nitrogen use efficient crop plants; the current status. *J. Plant Biotechnol.* **2012**, *10*, 1467–7652. [CrossRef]
5. Buol, S.W. Mineralogy Classes in Soil Families with Low Activity Clays. In *Mineral Classification of Soils*; Kittrick, J.A., Ed.; SSSA Special Publication: Madison, WI, USA, 1985; Volume 16, pp. 169–178.
6. Zotarelli, L.; Scholberg, J.M.; Dukes, M.D.; Carpena, R.M. Monitoring of nitrate leaching in sandy soils: Comparison of three methods. *J. Environ. Qual.* **2007**, *36*, 953–962. [CrossRef]
7. Sitthaphanit, S.; Limpinuntana, V.; Toomsan, B.; Panchaban, S.; Bell, R.W. Fertiliser strategies for improved nutrient use efficiency on sandy soils in high rainfall regimes. *Nutr. Cycl. Agroecosystems* **2009**, *85*, 123–139. [CrossRef]
8. Yao, Y.; Gao, B.; Zhang, M.; Inyang, M.; Zimmerman, A.R. Effect of biochar amendment on sorption and leaching of nitrate, ammonium, and phosphate in a sandy soil. *Chemosphere* **2012**, *89*, 1467–1471. [CrossRef]
9. Major, J.; Rondon, M.; Molina, D.; Riha, S.J.; Lehmann, J. Nutrient Leaching in a Colombian Savanna Oxisol Amended with Biochar. *J. Environ. Qual.* **2012**, *41*, 1076. [CrossRef]
10. Sika, M.P.; Hardie, A.G. Effect of pine wood biochar on ammonium nitrate leaching and availability in a South African sandy soil. *Eur. J. Soil Sci.* **2013**, *65*, 113–119. [CrossRef]
11. Xu, N.; Tan, G.; Wang, H.; Gai, X. Effect of biochar additions to soil on nitrogen leaching, microbial biomass and bacterial community structure. *Eur. J. Soil Biol.* **2016**, *74*, 1–8. [CrossRef]
12. Sun, H.; Lu, H.; Chu, L.; Shao, H.; Shi, W. Biochar applied with appropriate rates can reduce N leaching, keep N retention and not increase NH3 volatilization in a coastal saline soil. *Sci. Total Environ.* **2017**, *575*, 820–825. [CrossRef]
13. Haider, G.; Steffens, D.; Moser, G.; Müller, C.; Kammann, C.I. Biochar reduced nitrate leaching and improved soil moisture content without yield improvements in a four-year field study. *Agric. Ecosyst. Environ.* **2016**, *237*, 80–94. [CrossRef]
14. Liu, Z.; He, T.; Cao, T.; Yang, T.; Meng, J.; Chen, W. Effects of biochar application on nitrogen leaching, ammonia volatilization and nitrogen use efficiency in two distinct soils. *J. Soil Sci. Plant Nutr.* **2017**, *17*, 515–528. [CrossRef]
15. Lehmann, J.; Pereira da Silva, J., Jr.; Steiner, C.; Nehls, T.; Zech, W.; Glaser, B. Nutrient availability and leaching in an archaeological Anthrosol and a Ferralsol of the Central Amazon basin: Fertilizer, manure and charcoal amendments. *Plant Soil* **2003**, *249*, 343–357. [CrossRef]

16. Murphy, J.; Riley, J.P. A modified single solution method for the determination of phosphate in natural waters. *Anal. Chim. Acta* **1962**, *27*, 31–36. [CrossRef]
17. Gee, G.W.; Bauder, J.W. Particle-size Analysis. In *Methods of Soil Analysis: Part 1—Physical and Mineralogical Methods*, 2nd ed.; Klute, A., Ed.; SSSA Book Series: Madison, WI, USA, 1986; pp. 383–411.
18. Blake, G.R.; Hartge, K.H. Bulk Density. In *Methods of Soil Analysis: Part 1—Physical and Mineralogical Methods*, 2nd ed.; Klute, A., Ed.; SSSA Book Series: Madison, WI, USA, 1986; pp. 363–375.
19. Walkley, A.; Black, I.A. An examination of the Degtjareff method for determining soil organic matter, and a proposed modification of the chromic acid titration method. *Soil Sci.* **1934**, *37*, 29–38. [CrossRef]
20. Jones, J.B., Jr. *Laboratory Guide for Conducting Soil Tests and Plant Analysis*; CRC Press: Boca Raton, FL, USA, 2001.
21. Hood-Nowotny, R.; Umana NH, N.; Inselbacher, E.; Oswald-Lachouani, P.; Wanek, W. Alternative Methods for Measuring Inorganic, Organic, and Total Dissolved Nitrogen in Soil. *Soil Sci. Soc. Am. J.* **2010**, *74*, 1018. [CrossRef]
22. West, P.W.; Ramachandran, T.P. Spectrophotometric determination of nitrate using chromotropic acid. *Anal. Chim. Acta* **1966**, *35*, 317–324. [CrossRef]
23. Fidel, R.B.; Laird, D.A.; Spokas, K.A. Sorption of ammonium and nitrate to biochars is electrostatic and pH-dependent. *Sci. Rep.* **2018**, *8*, 17627. [CrossRef]
24. Zhou, L.; Xu, D.; Li, Y.; Pan, Q.; Wang, J.; Xue, L.; Howard, H. Phosphorus and Nitrogen Adsorption Capacities of Biochars Derived from Feedstocks at Different Pyrolysis Temperatures. *Water* **2019**, *11*, 1559. [CrossRef]
25. Jin, Z.; Chen, X.; Chen, C.; Tao, P.; Han, Z.; Zhang, X. Biochar impact on nitrate leaching in upland red soil, China. *Environ. Earth Sci.* **2016**, *75*, 1–10. [CrossRef]
26. Chintala, R.; Gelderman, R.H.; Schumacher, T.E.; Malo, D.D. Vegetative Corn Growth and Nutrient Uptake in Biochar Amended Soils from an Eroded Landscape. In Proceedings of the Joint Annual Meeting of the Association for the Advancement of Industrial Crops and the USDA National Institute of Food and Agriculture, Washington, DC, USA, 12–16 October 2013; pp. 200–216.
27. Taghizadeh-Toosi, A.; Clough, T.J.; Sherlock, R.R.; Condron, L.M. Biochar adsorbed ammonia is bioavailable. *Plant Soil* **2011**, *350*, 57–69. [CrossRef]
28. Ma, Y.L.; Matsunaka, T. Biochar derived from dairy cattle carcasses as an alternative source of phosphorus and amendment for soil acidity. *Soil Sci. Plant Nutr.* **2013**, *59*, 628–641. [CrossRef]
29. Dietrich, C.C.; Rahaman, M.A.; Robles-Aguilar, A.A.; Latif, S.; Intani, K.; Müller, J.; Jablonowski, N.D. Nutrient Loaded Biochar Doubled Biomass Production in Juvenile Maize Plants (*Zea mays* L.). *Agronomy* **2020**, *10*, 567. [CrossRef]
30. Luyima, D.; Sung, J.; Lee, J.H.; Woo, S.A.; Park, S.J.; Oh, T.K. Sorption of urea hydrogen peroxide by co-pyrolysed bone meal and cow dung slowed-down phosphorus and nitrogen releases but boosted agronomic efficiency. *Appl. Biol. Chem.* **2020**, *63*, 52. [CrossRef]
31. Rashid, M.; Hussain, Q.; Khan, K.S.; Alwabel, M.I.; Hayat, R.; Akmal, M.; Ijaz, S.S.; Alvi, S.; Obaid-ur-Rehma. Carbon-Based Slow-Release Fertilizers for Efficient Nutrient Management: Synthesis, Applications, and Future Research Needs. *J. Soil Sci. Plant Nutr.* **2021**, *21*, 1144–1169. [CrossRef]
32. Luyima, D.; Lee, J.H.; Sung, J.K.; Oh, T.K. Co-pyrolysed animal manure and bone meal-based urea hydrogen per-615 oxide (UHP) fertilisers are an effective technique of combating ammonia emissions. *J. Mater. Cycles Waste Manag.* **2020**, *22*, 1887–1898. [CrossRef]
33. Luyima, D.; Egyir, M.; Yun, Y.U.; Park, S.J.; Oh, T.K. Nutrient Dynamics in a Sandy Soil and Leaf Lettuce following the Application of Urea and Urea-Hydrogen Peroxide Impregnated Co-Pyrolysed Animal Manure and Bone Meal. *Agronomy* **2021**, *11*, 1664. [CrossRef]

Disclaimer/Publisher's Note: The statements, opinions and data contained in all publications are solely those of the individual author(s) and contributor(s) and not of MDPI and/or the editor(s). MDPI and/or the editor(s) disclaim responsibility for any injury to people or property resulting from any ideas, methods, instructions or products referred to in the content.

Article

Efficient Removal of Tannic Acid from Olive Mill Wastewater Using Carbon Steel Slag

Otmane Sarti [1], Fouad El Mansouri [2,*], El Habib Yahia [1], Emilia Otal [3], José Morillo [3] and Mohamed Saidi [1]

[1] Laboratory of LAMSE, Faculty of Sciences and Techniques of Tangier, P.O. Box 416, Tangier 90000, Morocco
[2] Research Team: Materials, Environment and Sustainable Development (MEDD), Faculty of Sciences and Techniques of Tangier, Abdelmalek Essaâdi University, P.O. Box 416, Tangier 93000, Morocco
[3] Department of Chemical and Environmental Engineering, University of Seville, Camino de Los Descubrimientos, s/n, 41092 Seville, Spain
* Correspondence: fouad.elmansouri@etu.uae.ac.ma; Tel.: +212-662-102-847

Citation: Sarti, O.; El Mansouri, F.; Yahia, E.H.; Otal, E.; Morillo, J.; Saidi, M. Efficient Removal of Tannic Acid from Olive Mill Wastewater Using Carbon Steel Slag. C 2023, 9, 32. https://doi.org/10.3390/c9010032

Academic Editors: Indra Pulidindi, Pankaj Sharma and Aharon Gedanken

Received: 8 February 2023
Revised: 4 March 2023
Accepted: 8 March 2023
Published: 12 March 2023

Copyright: © 2023 by the authors. Licensee MDPI, Basel, Switzerland. This article is an open access article distributed under the terms and conditions of the Creative Commons Attribution (CC BY) license (https://creativecommons.org/licenses/by/4.0/).

Abstract: Mediterranean countries experience a large production of olive oil, thus generating huge quantities of non-biodegradable vegetation waters. The discharge of these effluents into aquatic environments seriously affects the quality of surface waters. This study investigated the potential use of carbon steel slag (SS) as an adsorbent and improver for reducing olive mill wastewater (OMWW) toxicity. The elemental and structural characterization of SS was carried out using inductively coupled plasma-optical emission spectrometry (ICP/EOS), X-ray fluorescence (XRF), X-ray powder diffraction (XRD), scanning electron microscopy (SEM), and Brunauer–Emmett–Teller (BET) analysis. OMWW characterization indicated that the effluent was acidic in nature, with a pH of 4.8, a higher conductivity reaching 14.92 mS/cm, higher COD of 157.31 g/L, rich in organic matter 112.33 g/L, and total phenolic compounds of 11.13 g/L. The neutralization capacity of SS was demonstrated by reducing the OMWW's acidic character. Afterward, the adsorption of tannic acid (TA) was investigated using SS. Parameters such as contact time, initial TA concentration, adsorbent dosage, pH, and temperature were investigated. The kinetic study indicated that the adsorption of TA onto SS fitted well with the second pseudo-order (r = 0.99) and Elovich (r = 0.98) models, indicating that the adsorption of TA was mainly chemical and depends on the reactions of oxide hydrolysis and hydroxides dissolution. Moreover, Langmuir isotherm has greatly described the adsorption of TA on SS (R = 0.997), suggesting that the surface of SS is homogenous, and the adsorption occurs mainly in monolayer. The maximum adsorption capacity reached 714.28 mg/g, indicating the higher capacity of SS to reduce the polyphenolic compounds in OMWW. This study demonstrated that SS residue from the steelmaking industry could present a highly interesting material for OMWW remediation.

Keywords: OMWW; adsorption; carbon steel slag; tannic acid

1. Introduction

In the Mediterranean region, olive oil (*Olea europaea* L.) is increasingly consumed for its bioactive content, organoleptic properties, medicinal value, and protection against certain diseases [1–3]. In addition, the Mediterranean countries are responsible for 98% of olive oil production worldwide [4]. However, regardless of its economic importance, the extraction of olive oil is associated with some environmental impacts, such as water pollution, soil deterioration, and air emissions [5]. Olive oil extraction generates solid wastes composed of wet pomace formed from the pulp and pits of olives and a liquid effluent called olive mill wastewater (OMWW) or sometimes vegetable waters. Generally, the solid by-product does not constitute disposal problems as long as specific industrial oil mills use it to produce a so-called pomace oil by (chemical) solvent extraction. After the extraction, the pomace serves as fuel in industrial boilers, ovens, and public baths. In contrast, the liquid effluent is rarely treated and dumped directly into natural waters. OMWW is composed mainly of vegetation water from olive, process water (washing and treatment), a portion of the pulp, and residual oil [6]. The olive oil extraction generates quantities of OMWWs ranging from 0.3 to 1.1 m^3 per ton of olives processed, depending on the olive oil extraction technique [7–9]. This liquid effluent is characterized by a cloudy appearance, a strong odor of olive oil, an intense brown-red to black color, and is heavily loaded with organic matter and suspended solids [10]. Eroğlu et al. [11] reported that the black-brownish color characteristic of this effluent is due to the low biodegradability of phenolic compounds.

Recently, great efforts have been made to reduce the toxicity of OMWWs, and various authors discussed the possibility of OMWW treatment through physical, chemical, and biological processes. Among these methods, combined coagulation–flocculation with hydrogen peroxide oxidation [12], biological treatments [13], photocatalysis [14], solvent extraction followed by photo-Fenton oxidation [15], electrocoagulation [16], aerobic biological treatments [17], and anaerobic digestion treatment [18] were investigated. The removal of phenolic compounds from OMWWs has recently gained more attention. Among the existing methods, adsorption is regarded as the most appropriate treatment for removing polyphenols since it is a cost-effective operation that can be handled without needing higher temperatures, specific methodology, or large energy input [19,20].

Oil mill waste contains a high concentration of phenolic compounds such as polyphenols, including condensed and hydrolysable tannins, which require proper disposal to avoid environmental hazard [21]. Besides the olive mills wastewater, tannic acid may also be found in effluent from the coir and cork processing industries, as well as the plant medicine, paper, and leather industries [22,23]. Tannic acid (TA) is a natural organic matter (NOM) component present in surface and groundwater generated from the degradation of biomass [23,24]. Because it can generate carcinogenic disinfection by-products (DBPs) during the chlorination process, TA might contaminate drinking water [23,24]. Since it is a water-soluble polyphenolic molecule, TA is also hazardous to aquatic organisms such as algae, phytoplankton, fish, and invertebrates [23]. Consequently, eliminating TA from water and wastewater is critical for both human health and the ecosystem [23]. Several attempts were made to remove TA from aqueous solution, e.g., polyaniline (PANI) prepared by chemical oxidation [25], electrochemical processes [26], coagulation-adsorption [27], membranes ultrafiltration [28], and biological processes [29]. Among these methods, adsorption has been widely used to remove TA from water and wastewater using different adsorbents such as carbon nanotubes [30], amino-functionalized magnetic mesoporous silica [23], silk fiber [31], polystyrene microplastics [32], attapulgite/CoFe$_2$O$_4$ [33], chitosan-montmorillonite composites [34], chitosan and activated clay [35], and functionalized zeolites [36].

Steel slag (SS) is a solid by-product of the steel-making industry generated in huge quantities (thousands of tons) in different regions of the world. The production of 1 ton of steel is associated with the production of 0.13–0.2 tons of slags [37,38]. The steel slag (SS) is an alkaline residue representing a higher surface area and porosity and mainly consisting of oxides (CaO, MgO, SiO$_2$, Al$_2$O$_3$, and Fe$_2$O$_3$). Investigations on the use of steel slag in water treatment have been carried out to achieve effective and comprehensive reuse of this solid waste, thereby presenting a low-cost solution for wastewater treatment [39]. Some researchers have already demonstrated satisfactory results when applying SS as an adsorbent to remove ammonia, phosphate, hydrogen sulfide, cadmium Cd (II), and arsenic As (V) [40–44]. Furthermore, due to the greater amount of metal oxides such as CaO and MgO, the SS residues displayed effectiveness in terms of CO$_2$ capture [45–48].

This research presents a new low-cost and simple approach for treating olive mill wastewaters (OMWWs) utilizing steel slag (SS) waste from industrial steelmaking. The physicochemical properties of OMWW were analyzed. Afterward, various techniques were employed to characterize the steel slag used in this work (ICP/OES, XRF, XRD, SEM, and BET). Before Bach experiments, OMWW was treated with different ratios of raw SS to investigate the direct effect of SS addition on the OMWWs' pH neutralization. The adsorption behavior of tannic Acid (TA) under different conditions (contact time, initial

TA concentration, adsorbent mass, pH, and temperature) was investigated. Adsorption isotherms, kinetics, and adsorption mechanisms were discussed. The contribution of this study is expected to reduce the acute acidity of vegetation waters and the high polyphenolic content of OMWWs.

2. Materials and Methods

2.1. Olive Mill Waste Water (OMWW)

The olive mill effluent investigated in this study derived from a traditional oil-producing plant located in Chefchaouen city in the Northwest of Morocco. Sampling was carried out in an oil mill during the pressing phase of oil production. The newly collected olive mill wastewater was transported to the laboratory in polyethylene bottles (5 L), carefully filled, and tightly sealed to maintain an oxygen-free environment. To avoid fermentation, OMWW samples were stored in the dark at 4 °C in a non-oxygenated environment.

2.2. Olive Mill Wastewater Characterization

The pH and Electrical conductivity (EC) measurements were made for a sample of undiluted raw olive vegetable water using a pH and EC meter brand HANNA HI2550 (Hanna Instruments, Woonsocket, RI, USA). After calibrating the pH meter, the measuring electrode was immersed in a flask containing 50 mL of effluent. The OMWW water content was determined by the difference between the fresh weight of the effluent sample and its dry weight (after drying in an oven at 105 °C for 24 h) and expressed as a percentage of humidity, while the solid residue was considered as dry matter (DM) expressed in (g/L). The organic matter (OM) was determined by the loss on ignition (LOI) method at 525 °C for 2 h and expressed as g/L. The COD was determined using JP SELECTA BLOC DIGEST apparatus (JP Selecta, Barcelona, Spain). The oxidation was carried out by the potassium dichromate method. This method is based on boiling oxidation (150 °C for 2 h) of reducing materials with an excess of ($K_2Cr_2O_7$) in an acidic medium (H_2SO_4), and in the presence of silver sulfate as a catalyst and sulfate mercury as a chloride complexing agent. The COD is evaluated at the end of the reaction by taking a suitably diluted sample before oxidation 100 times. This measurement was made for raw OMWW samples according to the NFT 90–101 standardized method. Determining the biological oxygen demand (BOD) is an essential parameter for treating liquid effluents such as OMWWs. The 5-day biochemical oxygen demand BOD5 was determined by a respirometry method using an OxiTop system (WTW, Xylem, Weilheim, Germany). The total phenolic content was determined following the Folin–Ciocaltau method [49] using tannic acid (tannic acid powder, puriss; Sigma Aldrich, St. Louis, MO, USA) as a standard and equivalent of the total polyphenolic concentration contained in OMWW. The concentration of tannic acid was quantitatively determined at 760 nm using the JASCO 630 Double Beam UV/Visible spectrophotometer (JASCO, Tokyo, Japan).

2.3. Steel Slag (SS) Characterization

The steel manufacturing process in electric arc furnaces consists of two stages: the first, called primary metallurgy or fusion, in which raw materials are fused in an electric arc furnace (EAF), and the second, called secondary metallurgy or molten bath refining, which begins in the electric furnace and ends in the ladle furnace. The steel slag used in this study was produced in Sevilla, Spain. The plant produces two types of SS. Black slags generated during the fusion step are very hard materials, unlike the white slags produced during the refining phase, which is a relatively soft and finer material and has high basicity due to the percentage of free lime. The white steel slag used in this study was collected from the ladle furnace. The material was characterized by a crumbly structure and a softer texture, the sample was sieved using a 200-micron mesh, and no modification was performed on the raw SS (Figure 1). The pH of the SS sample was measured following the 9045D method using a pH meter (Thermo Fisher Scientific, Waltham, MA, USA). The determination of major elements and trace elements in the SS was carried out using inductively coupled plasma–optical emission spectroscopy (ICP/OES, Agilent 5100, Tokyo, Japan) after acid digestion (HNO_3-HCL 1:3 (v/v)) using a DigiPREP heating and digestion blocks (SCP Science, Montreal, QC, Canada). The Brunauer–Emmett–Teller (BET) analysis was employed to investigate the pore size distribution, specific surface area, and nitrogen adsorption–desorption curve of SS. The analysis was performed using BET Surface Area Analyzer Micromeritics Tristar II 3020 (Micromeritics Instr. Corps., Norcross, GA, USA). The structural and morphological surface of the steel slag sample was determined using scanning electron microscopy (SEM) (SEM, Hitachi, Tokyo, Japan, S-4800). Simultaneously, the elemental composition of the SS was determined using an X-ray fluorescence spectrometer (XRF, PANalytical Axios FAST simultaneous WDXRF, Malvern PANalytical Ltd., Almelo, The Netherlands). The crystalline composition of SS was analyzed by X-ray diffraction (XRD) using a PANalytical X'Pert Pro X-ray diffractometer (Malvern PANalytical Ltd., Almelo, The Netherlands). All the experiments were carried out using ultrapure Mllii-Q water.

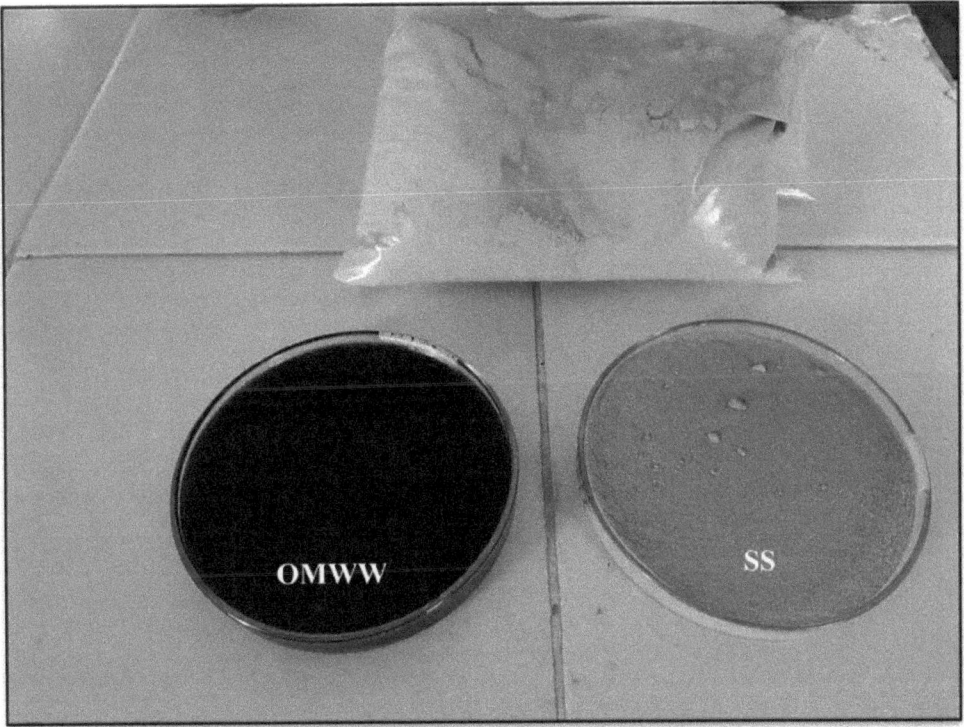

Figure 1. Olive mill wastewater (OMWW) and steel slag (SS) samples.

2.4. Effect of SS Addition on Raw OMWW

In order to investigate the effect of SS addition on the OMWW pH variation, The OMWW was treated with different ratios of SS. Different doses of SS (0.5-1-2-3-4-5-10-20-30-40 g/L of OMWW) were dispersed in the raw OMWW samples and stirred. Afterward, the pH variation of OMWW was measured at different contact times (30-60-90-120-150-180 min) and after 24 h.

2.5. Adsorption Study

For the adsorption study, the influence of different parameters on the TA-SS adsorption mechanism, namely the time of equilibration, the pH, the initial concentration of the solute, and the influence of the temperature, were investigated.

The batch experiments were performed to assess the TA adsorption isotherms onto SS. Accordingly, 100 mg of steel slag SS was added to a 150 mL flask containing 100 mL of TA solution with initial concentrations ranging from 50 to 1000 mg/L and a pH of 6 (pH solution). The flasks were placed inside an incubator and stirred at 25 °C for 24 h. After achieving adsorption equilibrium, the adsorbent particles were centrifuged from the aqueous solution, filtered, and the residual TA concentrations in the aliquot were evaluated using a UV-vis spectrometer with a detection wavelength of 760 nm. The equilibrium of TA adsorption capacity was calculated using Equation (1):

$$qe = \frac{(C0 - Ce)V}{M} \quad (1)$$

Noting that qe is the adsorbed amount of tannic acid in mg/g, V the volume of solution in mL, Ce concentration at the equilibrium in mg/L, and $C0$ initial concentration of TA in mg/L, and M is the SS mass in (g).

In order to study the time effect, the experiments were carried out by varying the contact time from 30 to 240 min. The pH solution, temperature, and SS mass were kept constant. The studied pH values range from 2 to 11 (the acidic and basic pH values obtained by adjusting the initial pH of the adsorbate solutions using NaOH and HCl solutions). The suspensions were stirred to the equilibrium time determined by the kinetic study of adsorption. The other parameters were held constant. The optimal temperature of adsorption was carried out by Bach experiments at 15, 30, 45, and 60 °C. The remaining parameters were kept constant. Adsorption kinetics were investigated using the pseudo-first-order, pseudo-second-order, Elovich, and intraparticle diffusion models. The modeling of isotherms was performed by plotting the linear form of Langmuir and Freundlich isotherms.

3. Results and Discussion

3.1. The Results of OMWW Characterization

The OMWWs are generally characterized by an acid pH of 3–6, rich in organic matter, and low in biodegradable polyphenols (0.5–24 g/L) [19,50–53]. In addition, these effluents displayed higher electrical conductivity (EC), chemical oxygen demand (COD), and biological oxygen demand (BOD) ranging from 40–220 g/L and 35–110 g/L, respectively [54,55]. The pH of the sampled OMWW was 4.8. As a result, the nature of this effluent is acidic. The existence of organic acids (phenolic acids, fatty acids, etc.) capable of releasing positively charged protons H^+ is primarily responsible for this acidity. Table 1 represents the main results of OMWW characterization. The measured electrical conductivity (EC) of OMWW was 14.92 mS/cm, resulting mainly from the higher concentration of dissolved minerals and salts. Olives acquire a significant amount of water during the pressing process. The OMWW contained 83.04% water content and a relative dry matter (DM) of 177.04 g/L. Total suspended matter (TSM) concentration was high in OMWW, showing a value of 98.85 g/L, theses solid materials are essentially organic in nature and generate increased turbidity in the receiving environment. The OM loading in the OMWW sample was higher, reaching 112.33 g/L. Furthermore, the COD and BOD_5 of OMWW were 157.31 and 35.40 g/L, respectively, indicating severe organic pollution of this effluent and its limited epuration capability. Similarly, OMWW had a higher polyphenolic concentration of 11.13 g/L, indicating an abundance of non-biodegradable phytotoxic compounds. According to Bouknana et al. and El Gnaoui et al. the poor biodegradability of OMWW is mainly related to their high polyphenol and COD concentrations, limiting thus any biological decomposition [56,57].

Table 1. Physicochemical characterization of OMWW.

Parameter	Unit	Value
pH	–	4.80
EC	ms/S	14.92
Water content	%	83.04
Dry matter (DM)	g/L	177.04
Total Suspended matter (TSM)	g/L	98.85
Organic Matter (OM)	g/L	112.33
DOC	g/L	157.31
BOD_5	g/L	35.40
Total phenols (as tannic acid)	g/L	11.13

3.2. Steel Slag (SS) Characterization

3.2.1. Physicochemical Characterization

The main physicochemical characterization results are shown in Table 2. The white slag sampled from the ladle furnace has a higher alkalinity with a pH of 12.58; This higher value is due to the dissociation of calcium and magnesium hydroxides, which allows the formation of OH^- ions, and thus a more elevated alkalinity. Indeed, the calcium content was higher and represented 15.05% of the total SS residue. This higher calcium content in SS is attributable to the introduction of CaO (free lime) and dolomite as fluxing agents to remove impurities from the molten steel contained in the Ladle furnace [58]. Evaluating SS toxicity is a necessary step to ensure the safe utilization of SS residue and avoid the environmental hazard. The SS leaching experiments are conducted to gather more information about the possible release of contaminants from the slag samples into the nearby environment as a result of weathering mechanism. Additionally, trace elements concentrations in SS leachates could be compared to water and soil quality standards to determine the possibly harmful effects. There are generally two methodologies regarding trace elements leaching from solid wastes. The first being the total concentration test of trace elements which is intended to estimate the total trace elements content through heating and aggressive acid digestion following (U.S. EPA. "Method 3050B"), methodology followed during this study. While a second methodology based on (EPA method 1311-Toxicology Characteristic Leaching Test) is designed to determine the mobility of trace element in the liquid phase. As a result, the SS toxicity depends on their leaching capacity, and not directly on the totally available trace elements. In this study, the total concentration of trace elements in SS has decreased in the following sequence Zn > Cu > Cr > Pb > Ni > As > Mo > Cd > Co > Hg. Among the analyzed trace elements, only Zinc concentration exceeded 390 mg/Kg. According to De Martino et al. [59] Zinc-Aluminum hydroxide could decontaminate OMWW by reducing the phenolic compounds up to 90%. The concentrations of Cu, Cr, Pb, and Ni displayed natural background levels and did not exceed the permissible level for waste disposal. At the same time, trace elements such as As, Cd, Co, and Hg were relatively low in the SS. A study of SS toxicity of heavy metals showed a relatively higher total concentrations than the present study, e.g., Cd = 89.65; As= 23.13; Cu = 70.38; and Pb = 978.41 mg/Kg [60]. However, according to this study, the Toxicology Characteristic Leaching Test (method 1315) displayed limited mobility of heavy metals indicating a secure disposal condition. In addition, the SS used in this study is intended to remediate a highly loaded organic wastewater and not directly discharged into the environment. Generally, the white SS from the Ladle furnace is usually less toxic because most trace elements are accumulated in the black slags produced during the fusion process. Furthermore, due to the limited solubility of its mineral phases, Ladle furnace slag has negligible environmental consequences [61]. The trace elements analysis of SS generally demonstrated low concentrations with a total weight of less than 0.1% in volume. These results confirm that white SS from the refining phase of the steelmaking industry could be used for OMWWs treatment.

Table 2. Physicochemical characterization of SS.

Parameter	Unit	Value
pH	–	12.58
Ca	%	15.05
As	mg/kg	8.86
Cd	mg/kg	0.91
Co	mg/kg	0.53
Cu	mg/kg	71.23
Hg	mg/kg	ND
Mo	mg/kg	4.65
Ni	mg/kg	23.19
Pb	mg/kg	24.56
Zn	mg/kg	390.86
Cr (total)	mg/kg	44.58

3.2.2. Elemental Composition by XRF

Table 3 shows the elemental composition of SS. CaO, MgO, SiO_2, and Al_2O_3 were the main oxides present in SS, representing a percentage higher than 80%, with a predominance of Ca followed by Mg oxides. Other oxides such as FeO, MnO, NaO, K_2O, TiO, P_2O_5, and SO_3 were present in a lower percentage of 3.55%. In contact with water, oxides such as CaO, MgO, and other alkaline metals dissociate and form hydroxide groups. The mixed hydrolysis of these hydroxides liberates hydroxyl (OH^-) groups and cationic metals such as Ca^{2+} and Mg^{2+} in water [39]. The free (OH^-) contained in an aqueous solution could be responsible for the adsorption, ion exchange, and precipitation processes of contaminants in wastewater [39].

Table 3. Chemical composition of steel slag by XRF spectrometry.

Element	SiO_2	Al_2O_3	Fe_2O_3	MnO	MgO	CaO	Na_2O	K_2O	TiO_2	P_2O_5	SO_3	L.O.I
SS	11.64	5.92	1.09	0.64	22.6	40.2	0.26	0.06	0.27	0.03	1.2	15.11

3.2.3. XRD Characterization of Steel Slag

The XRD patterns show the main identified crystalline phases contained in the raw SS (Figure 2). The profile displays a heterogeneous composition with various mineralogical phases composed mainly of oxides and hydroxides. The following mineralogical phases were the major identified patterns in the SS sample: (1) katoïte $Ca_3Al_{12}(SiO_4)(OH)_8$; (2) Magnesium hydroxide base ($Mg(OH)_2$); (3) Calcium Silicate (Ca_2SiO_4), (4) portlandite ($Ca(OH)_2$); and (5) calcium silicate hydrate (3 CaO, $2SiO_2$, $3H_2O$). These results confirm that the dominant oxides CaO, MgO, SiO_2, and Al_2O_3 previously determined by XRF were the main constituent of SS mineralogical phases, demonstrating heterogeneity in the distribution of SS oxyhydroxides. The obtained XRD pattern phases were much similar to those provided by Navarro et al. [62]. Furthermore, Rađenović et al. [61] stated that silicate calcium was the main constituent of Ladle furnace slags under its different mineralogical forms. The same authors have reported that the presence of CaO, MgO, and SiO_2, in addition to aluminum silicates, confers to SS the adsorption properties.

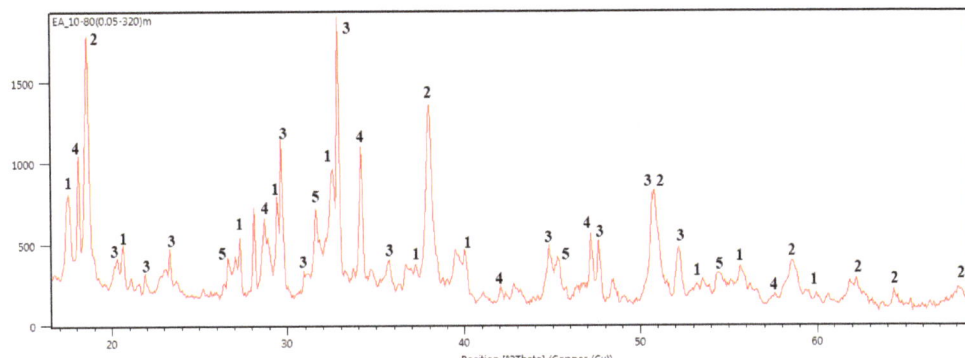

Figure 2. Main identified phases detected in steel slag sample: (1) Calcium Aluminum Silicate Hydroxide; (2) Magnesium Hydroxide; (3) Calcium Silicate; (4) Calcium Hydroxide; (5) Calcium Silicate Hydroxide Hydrate.

3.2.4. BET Characterization of the Steel Slag Adsorbent

Table 4 represents the characteristics of the SS surface by the BET method. The studied steel slag was characterized by a surface area of 10.90 m^2/g, representing a great surface area compared with other SS studies. Furthermore, the SS demonstrated a larger Langmuir surface area and pore diameter distribution of 15.61 m^2/g and 12.55 nm, respectively. The total pore volume was 0.035 cm^3/g. It is also noted that the micropore surface area was low in SS, representing only 0.86 (m^2/g). Wang et al. [23] studied the adsorption of TA on amino-functional magnetic mesoporous silica. They reported that efficient and rapid adsorption of TA is achieved due to the mesoporous nature of the adsorbent. The N$_2$ adsorption–desorption curve shows that SS's maximum N2 adsorption capacity reached 31.16 (cm^3/g STP) (Figure 3). The SS adsorption–desorption isotherm belongs to type IV, which is characteristic of mesoporous material. Effectively, the pore size distribution confirms that steel slag contains meso- and macropores structure mainly distributed between 10 and 100 nm.

Table 4. Steel slag (SS) BET surface area characteristics.

Sample	BET Surface Area (m^2/g)	Langmuir Surface Area (m^2/g)	Micropore Area (m^2/g)	Total Pore Volume (cm^3/g)	Pore Diameter (nm)
SS	10.90	15.61	0.86	0.035	12.55

3.2.5. Morphological Characterization of Steel Slag by SEM

The SEM micrographs of SS are shown in Figure 4. Firstly, the particle size distribution shows that SS exclusively has a low grain size distribution of <5 μm according to the SEM scale. While some minerals, probably Quartz (SiO$_2$) represent a relatively higher diameter. The morphology of SS was quite heterogeneous, displaying a rich mineralogical surface. In addition, SEM micrographs show pores of various sizes distributed over the surface of SS and oxides, confirming that the SS contains macro and mesopores as reported by BET analysis. These pores were predominantly large in diameter and volume, with low microporosity observed. Furthermore, it was clear that the surface of SS was quite rich in oxides, which display heterogeneous shapes and forms.

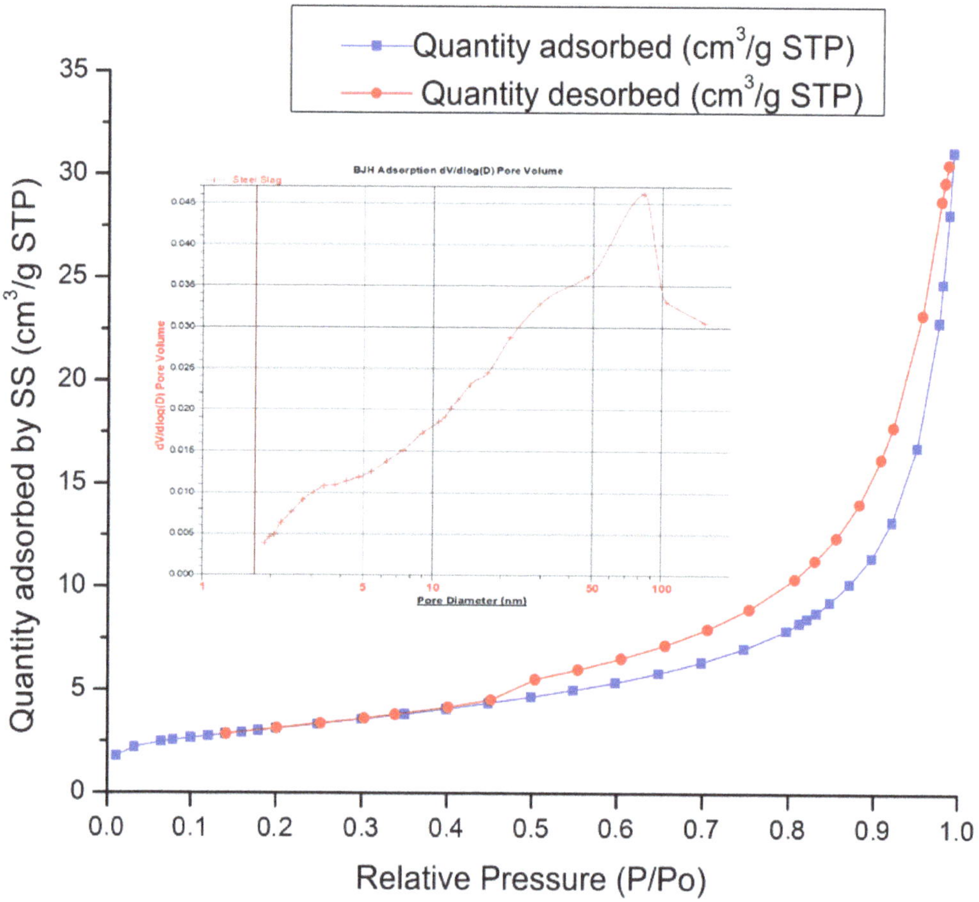

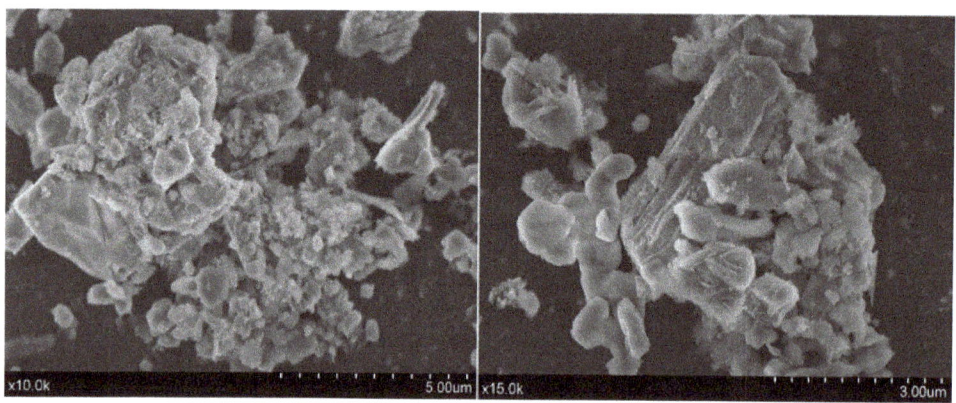

Figure 3. N2 adsorption desorption curve and pore diameter distribution of steel slag.

Figure 4. SEM micrographs of steel slag adsorbent at different magnitude.

3.3. Study of OMWW Neutralization by Steel Slag (SS)

Olive mill wastewater's acidity is one of the most preoccupations that affect the quality of natural waters. This acidity is mainly due to the organic acids and polyphenols-rich wastewater.

Depending on disposal specifications, this experiment could determine the amount of SS needed to raise the OMWW to a given pH. For this reason, OMWW was treated with different dosages of SS. The variation of OMWW pH as a function of SS dosage is highlighted in Figure 5. It was obvious that the pH of OMWW increases parallelly with increasing time and SS mass. By considering neutral condition (pH = 7) as the appropriate pH value for OMWW disposal/treatment. The addition of 0.5, 1, 2, 4, 6, and 8 $g_{(ss)}/L_{(OMWW)}$ increased the pH of OMWW gradually from 4.8 to 6.54 after 24 h of agitation. While a neutral pH of 7.18 was achieved only after 24 h of agitation by adding 10 $g_{(ss)}/L_{(OMWW)}$ as the minimum required dose. For the remaining doses, 20 g/L and 40 g/L, neutral pH of 7.33 and 7.3 was reached rapidly after 120 and 60 min, respectively. It is to note that the maximum recorded pH was 9.02 after 24 h of agitation and 40 $g_{(ss)}/L_{(OMWW)}$. The process of pH augmentation in OMWW could be explained by the dissociation of $(Ca(OH)_2)$ and $(Mg(OH)_2)$ hydroxides resulting from the hydrolysis of metal oxides such as CaO, MgO, and Al_2O_3. This dissociation results in the release of free cations and hydroxyl groups (OH-) in the solution, which raises the pH by neutralizing the OMWW's organic acids. According to XRD and XRF characterization results, the studied SS was composed mainly of Hydrated lime $(Ca(OH)_2)$, dolomitic lime $(Mg(OH)_2)$, free lime CaO, and other alkaline metals. Lime is widely reported for its efficiency in OMWW treatment [63–66], e.g., Aktas et al. found that applying 40 g/L of lime for OMWW treatment raised the pH to 12 and removed 73% and 40% of polyphenols and COD, respectively [67]. Accordingly, SS could be considered a liming agent given its high content of dolomite $(CaMg(CO_3)_2)$ and portlandite $(Ca(OH)_2)$. Similarly, Sarti et al. demonstrated the potential use of olive pomace biomass slag, which contains a similar composition as SS, to neutralize acidic agricultural sandy soils [68]. Therefore, regardless of the potential unexplored effects of OMWW treatment by SS. It could be deduced that the application of SS could at least neutralize OMWW and probably reduce COD and other polyphenolic compounds.

3.4. Adsorption of Tannic Acid (TA) on Steel Slag (SS)

3.4.1. Adsorption Kinetics

The kinetic of TA adsorption on SS is highlighted in Figure 6. Results show that TA is adsorbed on the SS matrix with the same kinetics for the three initial concentrations; a maximum adsorption of TA was noticed (0–45 min), during which more than 92, 80, and 75% of the adsorption capacity was reached for an initial TA concentration of 100, 200, and 300 mg/L, respectively. While a relatively slow to constant adsorption was observed between 45 and 225 min, after 24 h (1440 min), no effective change was highlighted in the adsorption rate allowing us to define 225 min as the optimum equilibrium time since more than 90% of TA removal is achieved during this contact time. These results suggest that the adsorption equilibrium of TA depends either on the initial concentration of TA or the equilibrium time. Tannic acid is a weak acid, and therefore its reaction with a strong base such $Ca(OH)_2$ and/or $Mg(OH)_2$ could be responsible for the formation of insoluble precipitates. In this study, the primary preproposal process of TA adsorption on SS was regarded as chemical because of the rich accessibility of oxides and hydroxides on the SS surface. Therefore, the following reactions are considered in the process of adsorption.

$$CaO + H_2O \rightarrow Ca(OH)_2 \qquad (2)$$

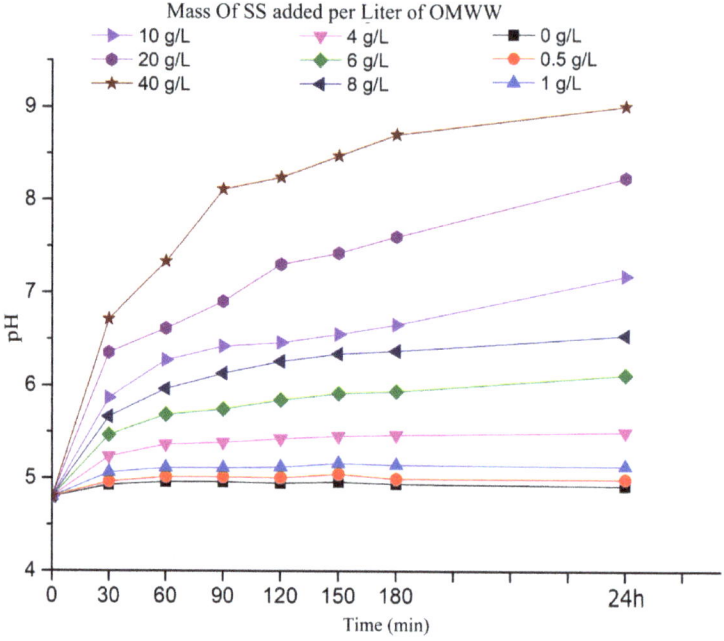

Figure 5. pH variation of OMWW as a function of time and steel slag dosage.

$$Ca(OH)_2 \rightarrow Ca^{2+} + 2OH^- \quad (3)$$

$$MgO + H_2O \rightarrow Mg(OH)_2 \rightarrow Mg^{2+} + 2OH^- \quad (4)$$

$$Ca_2SiO_4 + H_2O \rightarrow 2CaO \cdot 2SiO_2 \cdot 3H_2O \rightarrow 2Ca^{2+} + H_2SiO_4^{2-} + 2OH^- \quad (5)$$

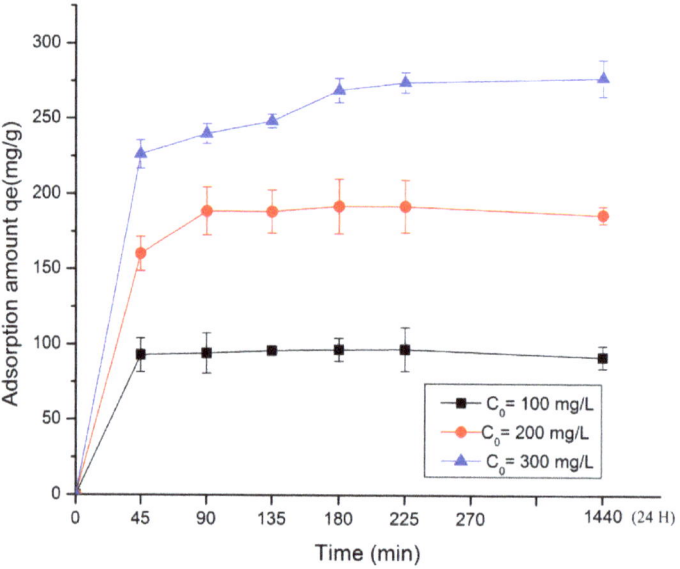

Figure 6. Effect of time on the adsorption of TA by SS (m = 100 mg; V = 100 mL; concentration = 100, 200 and 300 mg/L; pH = 5.8 ± 0.1; T = 293 ± 1 K; Agitation = 500 rpm).

Murdiati et al. reported that the reaction between TA and Ca(OH)$_2$ generates binded insoluble complexes [69]. Other authors added that complexation through chelation is possible between calcium and TA [70]. In comparison, other authors confirmed that the presence of Ca^{2+} in the solution greatly enhanced the adsorption of TA on polystyrene microplastics. In another study, Murdiati et al. studied the capacity of polyvinylpyrolidone (PVP), Ca(OH)$_2$, and activated charcoal to precipitate TA in solution and found that Ca(OH)$_2$ was more effective in the precipitation process of tannin and tannic acid [71]. This explains the fast adsorption of low TA concentration by SS, and may suggest that the formation of SS–TA complex depends on the initial TA concentration and the accessible free cations and hydroxyl groups in the solution. However, the adsorption of TA onto accessible mesopores should not be ignored.

In order to investigate the main mechanisms intervening in the adsorption of TA on SS as function of time. The following kinetic models were used in this study: Pseudo-first-order Equation (6), Pseudo-second-order Equation (7), Elovich Equation (8), and Intra-particle models Equation (9).

$$Ln(q_e - q_t) = Lnq_e - K_1 t \tag{6}$$

where, q_e (mg·g^{-1}) represents adsorption amount of TA at equilibrium; q_t (mg·g^{-1}) adsorption amount at time t; K_1 (1·min^{-1}) pseudo-first-order rate constant.

$$\frac{t}{q_t} = \frac{1}{K_2 \cdot q_e^2} + \frac{t}{q_e} \tag{7}$$

where q_e (mg·g^{-1}) is the adsorption amount of TA at equilibrium; q_t (mg·g^{-1}) refers to TA absorbed onto SS at time t; K_1 (1·min^{-1}) represents the constant of pseudo-first-order rate, K_2 (g·m^{-1}·min^{-1}) represents the constant of pseudo-second-order rate

$$qt = \frac{1}{\beta} \ln(\alpha\beta) + \frac{1}{\beta} ln\, t \tag{8}$$

where α refers to the initial adsorption coefficient (mg·g^{-1}·min^{-1}); β is the desorption coefficient (g·mg^{-1});

$$qt = K_{int} \sqrt{t} + C \tag{9}$$

K_{int} the constant of the rate (g·mg^{-1}·min$^{-1/2}$) and C is a constant of intra-particle diffusion model.

Figure 7 displays the fitting results of the four studied models. According to the fitting parameters in Table 5, the pseudo-second-order model provides the most significant correlation coefficients (R^2 = 0.997). According to these characteristics, the pseudo-second-order kinetic model was more accurate and helpful in simulating TA adsorption on steel slag. Considering its mathematical characteristics, the pseudo-second order fitted well the experimental adsorption data in this study and the chemical kinetics provide the best correlation for the adsorption of TA on SS. The adsorption process, known as chemisorption, chemically controls the rate-determining step of the sorption process and kinetically provides the highest correlation due to the mathematical characteristics of the pseudo-second-order model. Furthermore, a study on the adsorption of OMWW phenolic compounds onto local natural clay showed that chemisorption was mainly the governing process of adsorption and followed the pseudo-second-order [72]. Among the remaining models, the Elovich kinetic model showed in turn, a good correlation coefficient of 0.9841, confirming thus the chemical nature of TA sorption on SS.

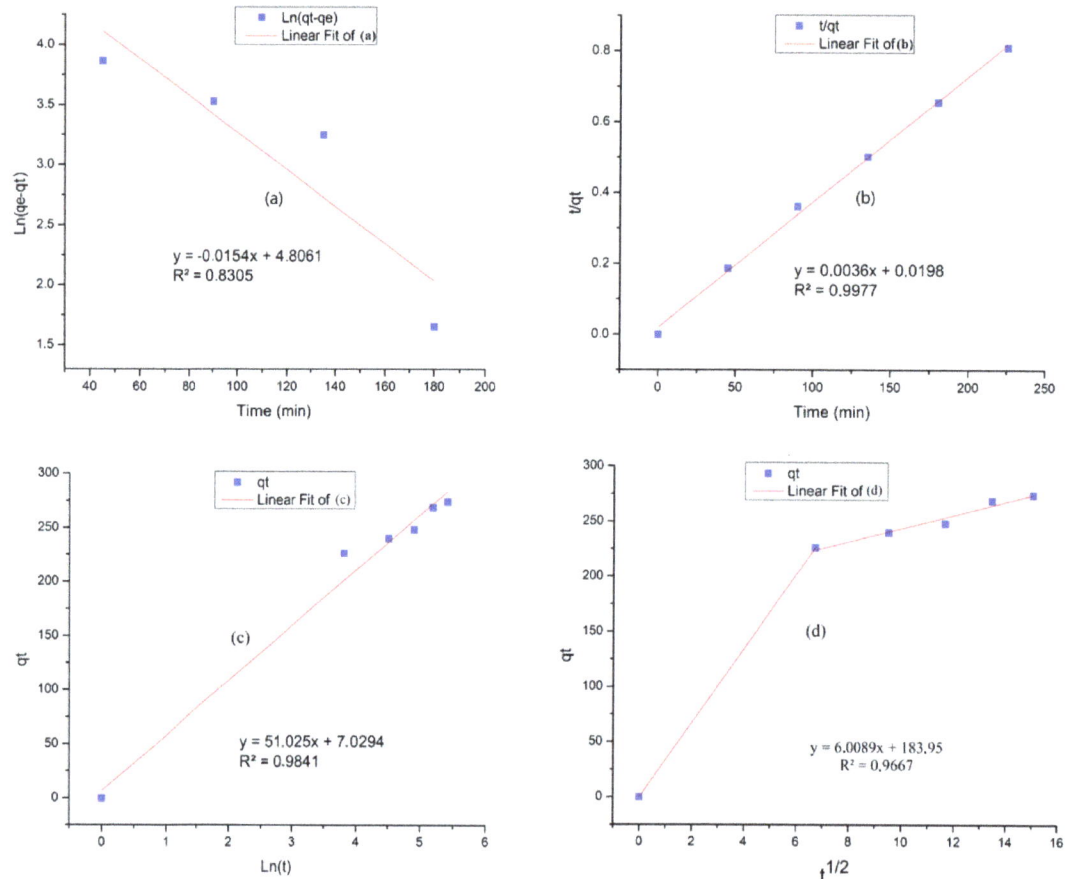

Figure 7. Kinetics curves for the TA adsorption by SS: (**a–c**) pseudo-first-order, pseudo-second-order and Elovich kinetics fitting, respectively, and (**d**) intraparticle diffusion model. Adsorption conditions: pH = 6.6; T = 293 K; time = 225 min.

Table 5. Fitting parameters of TA adsorption to SS following pseudo-first-order, pseudo-second-order, Elovich kinetics, and Intraparticle diffusion model.

Models	Pseudo-First-Order			Pseudo-Second-Order			Elovich			Intraparticle Diffusion		
Parameters	q_e	K_1	R^2	q_e	K_2	R^2	α	β	R^2	K_i	C	R^2
Value	241.048	0.0222	0.8305	294.117	$1.92\ 10^{-4}$	0.9977	58.545	0.0196	0.9841	17.649	44.122	0.8496

3.4.2. Effect of SS Dosage on Adsorption of TA

In order to investigate the adsorption amount % and adsorption capacity qe (mg/g) of SS, the effect of adsorbent dosage on TA adsorption within a range of 0.5 g/L to 10 g/L of SS was investigated. In this experiment, the initial TA concentration was 300 mg/L. Figure 8 shows the removal of TA according to SS adsorbed amount (a) and adsorption capacity (b). It was observed that increasing the dosage of SS from 50 mg to 1000 mg resulted in an increased TA removal rate from 80.50% to 98.91 %, respectively. This result indicates that a small dose of adsorbent might not achieve the equilibrium adsorption of TA on SS. However, in terms of efficiency, the adsorption capacity (qe) of SS decreased parallelly with increased SS dosage at fixed TA concentration, indicating that higher SS dosage affects the adsorption capacity of TA. According to these findings, 1 g/L of SS, which demonstrates a higher removal rate of 92 %, was selected to investigate the effect of pH and initial concentrations on TA adsorption.

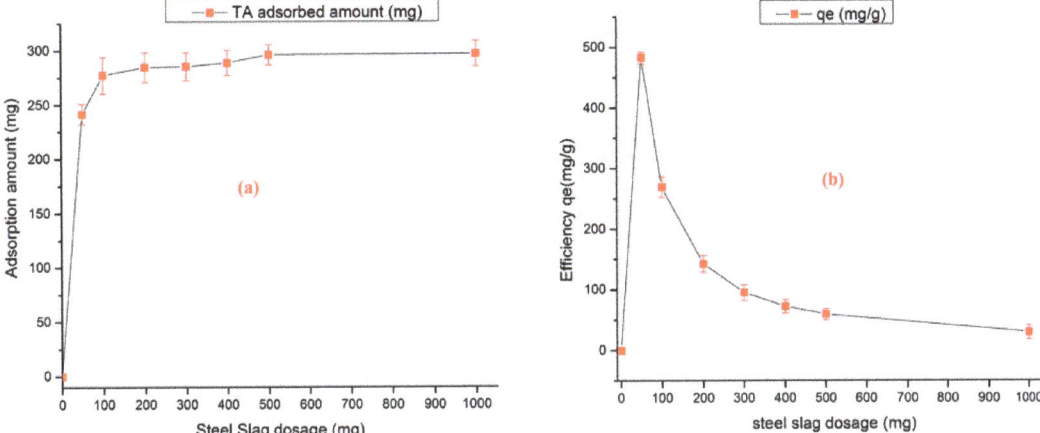

Figure 8. Effect of SS dosage on TA adsorbed amount (**a**) and efficiency of removal (**b**). Adsorption conditions: pH = 6.6; T = 293 K; time = 225 min; initial concentration = 300 mg/L, volume = 100 mL.

3.4.3. Effect of pH

The initial pH is one of the main factors influencing the adsorption behavior. As mentioned before, steel slag has high alkalinity due to its high concentration of basic oxides, which allows it to neutralize strong acids. The adsorption of TA on SS at different pH is shown in Figure 9. The results indicate that the adsorption of TA increased slightly with increasing pH values. Whereas a maximum adsorption capacity was achieved at pH 11. Generally, the transition between acidic and alkaline pH has no significant effect on TA adsorption, indicating only a slight removal difference of 5.43%. This low difference could be explained by the overdominance of the SS hydroxyl group (OH^-) in the solution, allowing the maintenance of an alkaline pH medium. It is assumed that the free cations in the solution react with the negatively charged TA molecules to create a precipitate, therefore removing TA from the aqueous solution. According to research, steel slag was able to raise the pH of acidic wastewater, thus reaching the solubility product (K_{sp}) of the contaminant and resulting in the formation of a precipitate easily separated from wastewater. A similar study reported that removing total phenolic compounds by olive pomace-activated carbon was greatly achieved at pH 10 [73]. Similarly, the adsorption of organic acids onto clay was greatly affected by higher pH and/or $Ca(OH)_2$ concentration [74]. Furthermore, polyphenol dissociation can occur only when the pH of the solution is greater than the Pka of the polyphenols [75]. Given that TA Pka is approximately 6, it was evident that SS had a significant influence on TA solubility, paving the path for the adsorption process. In accordance with the findings of this investigation regarding the influence of SS on raw OMWW, and because the adsorption study is intended to evaluate the SS capacity to reduce polyphenolic substances, it is possible to infer that the amount of steel slag could be the main pH regulating agent when treating OMWW.

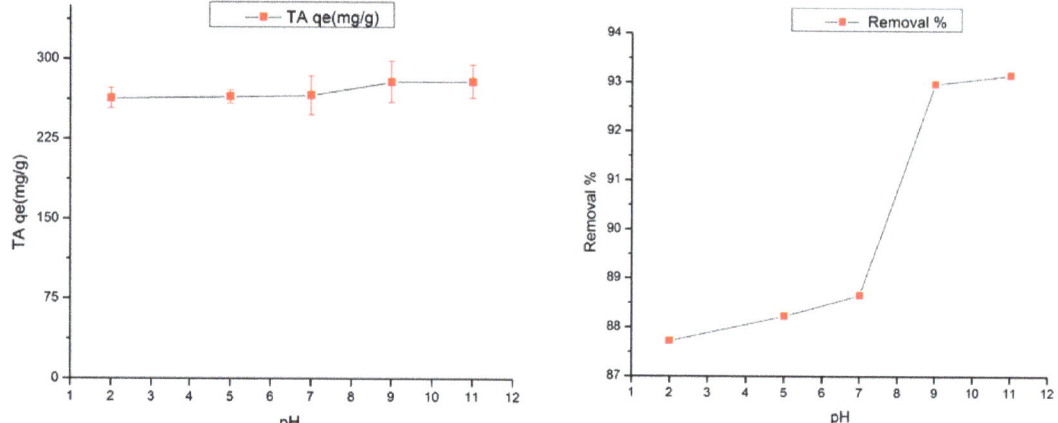

Figure 9. Effect of pH on TA adsorption. Adsorption conditions: SS dosage = 100 mg; T = 293 K; time = 225 min; initial concentration = 300 mg/L, volume = 100 mL.

3.4.4. Effect of TA Initial Concentration and Isothermal Study

The adsorption of TA on SS at different initial concentrations (50–1000 mg/L) was investigated in this study. TA adsorption according to different equilibrium concentrations (Ce) is presented in Figure 10. An apparent linearity was observed regarding the removal of tannic acid, indicating a substantial removal of more than 90 % for each investigated initial TA concentration. These results demonstrate that the adsorption capacity increases with increasing TA initial concentration. The higher initial concentrations of TA could increase the driving force required to overcome the mass transfer barrier of phenols between the liquid and solid phases, resulting in a more significant possibility of contact between TA and sorbent [75,76].

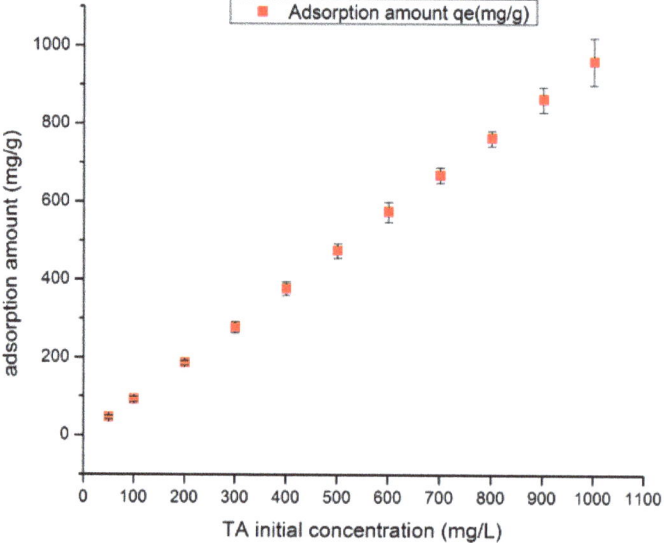

Figure 10. Adsorption isotherm of TA on SS as function of initial concentrations. Adsorption conditions: pH = 6.6; T = 293 K; time = 225 min; SS dosage= 100 mg, volume = 100 mL.

Currently, several theoretical models have been developed to describe adsorption isotherms, gathering information on surface adsorption, and types of interaction between Sorbent and Sorbate. Langmuir and Freundlich models listed below were used to model the experimental results in this study. The experimental data obtained from initial concentration were used to fit linear equation of Langmuir [77] (Equation (10)) and Freundlich [78] (Equation (11)) isotherms:

$$Ln(q_e - q_t) = Ln q_e - K_1 t \qquad (10)$$

$$\frac{t}{q_t} = \frac{1}{K_2 \cdot q_e^2} + \frac{t}{q_e} \qquad (11)$$

where q_{max} (mg·g^{-1}) is the Langmuir maximum adsorption capacity; C_e (mg·L^{-1}) is the equilibrium concentration of the TA solution; q_e (mg·g^{-1}) is the amount of TA absorbed at equilibrium; K_L (L·mg^{-1}) is the Langmuir bonding term related to the interaction energies, K_F (L·mg^{-1}) is the Freundlich affinity coefficient; n is Freundlich linearity constant.

The results presented in Table 6 summarizes the different constants obtained for each model, according to the Langmuir and Freundlich plotting models Figure 11.

Table 6. Parameters of the Freundlich and Langmuir models for TA adsorption on steel slag.

Types	Langmuir Isotherm				Freundlich Isotherm		
Parameter	q_{max} (mg·g^{-1})	K_L (L·mg^{-1})	R_L	R^2	K_F (L·mg^{-1})	1/n	R^2
Value	714.28	0.01380	0.59	0.9969	9.76	0.6906	0.9782

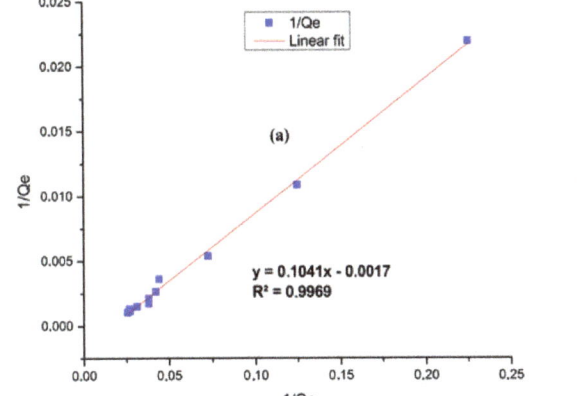
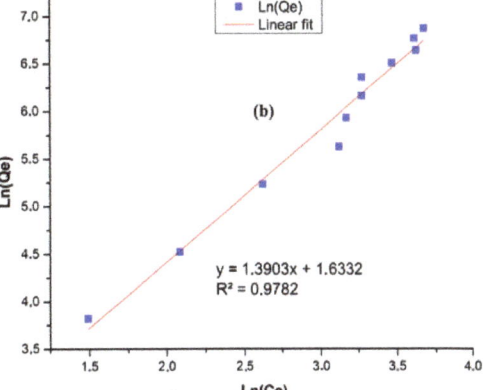

Figure 11. Adsorption isotherm for the TA adsorption by Steel Slag: Langmuir (a), Freundlich (b).

According to the fitting results, the Langmuir isotherm provides higher correlation coefficient to experimental data R = 0.9969, while a significative correlation was also observed by Freundlich model displaying a coefficient of R = 0.9782. These findings indicate that the adsorption of TA on SS conforms suitably to both Langmuir and Freundlich models. The assumption regarding Langmuir's model is that TA adsorption occurs on a homogeneous monolayer surface of SS with no interaction between adsorbed molecules. Previous studies have shown that TA adsorption onto polyaniline [25], commercial resins XAD-7, XAD-4, and NDA-7 [24], surfactant-modified zeolite [79], chitosan-coated attapulgite [33], chitosan/NaOH/fly ash [80], soy protein isolate–alginate hybrid spheres (SPIAHS) [81], and green synthesized nano-hydroxyapatite [82], has similar affinity for the Langmuir model reporting a q(max) values of 1023, 286, 111, 95.2, 286, 243.90, 1076.5, and 94.8 mg/g, respectively. According to Shi et al., the Langmuir model may adequately represent the adsorption process of numerous pollutants onto SS [39]. In this study, the maximum adsorption capacity (qmax) of SS following the Langmuir model was 714.28 (mg/g), indicating a promising adsorption capacity when compared to the aforementioned synthesized adsorbents.

The separation factor (RL) shows whether the form of the isotherms is unfavorable (RL > 1), linear (RL = 1), favorable (0 < RL < 1), or irreversible (RL = 0). The calculated RL parameter was 0.59, implying that the adsorption of TA on SS was favorable.

3.4.5. Adsorption Thermodynamics

The thermodynamic parameters of free energy change (ΔG_0), enthalpy change (ΔH_0), and entropy change (ΔS_0), might be valuable in gathering further information on the thermodynamic adsorption behavior of TA on steel slag. The parameters were calculated using Equations (12)–(14):

$$\Delta G° = -RTLn(Kd) \text{ and } \Delta G° = \Delta H° - T\Delta S° \tag{12}$$

$$Ln(Kd) = \frac{\Delta S°}{R} - \frac{\Delta H°}{RT} \tag{13}$$

$$Kd = \frac{Q_e}{C_e} \tag{14}$$

where: T is the temperature in kelvin (K); Kd is the distribution constant in ($l \cdot g^{-1}$); R is the gas constant (8.314 $J \cdot mol \cdot K^{-1}$).

The thermodynamic equilibrium constant was calculated by plotting ln(qe/Ce) vs qe (Figure 12). $\Delta H°$ and $\Delta S°$ were determined from the slope and intercept of ln(qe/Ce) vs. 1/T plotting. The results reported in Table 7 show a positive value of $\Delta H°$, suggesting an endothermic mechanism for TA adsorption on SS. This is supported by the fact that increasing the temperature decreases TA concentration in the solution. In addition, the absolute value of $\Delta H°$ was 5.5 kJ/mol, demonstrating that hydrogen bonding intervenes in the adsorption process of TA in steel slag [81]. The negative value $\Delta G°$ suggests that the adsorption occurred spontaneously. The decrease in $\Delta G0$ as temperature increases suggests that higher removal rates could be achieved in higher temperatures. The high positive value of $\Delta S°$ (33.38 $J \cdot mol^{-1} \cdot K^{-1}$) indicates that TA adsorption on SS is an entropy-increasing mechanism that is traduced by the higher degree of freedom of TA and its affinity of SS surface.

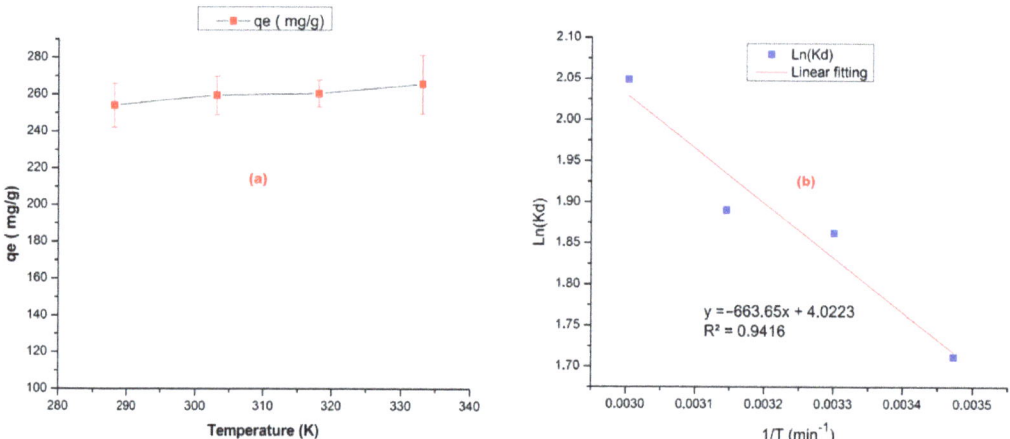

Figure 12. Effect of temperature on the adsorption of TA by steel slag (C_0 = 300 mg/L; m = 100 mg; V = 100 mL; pH = 6.6 ± 0.1; time = 225 min; Agitation = 500 rpm) (**a**), adsorption thermodynamic for the TA adsorption by steel slag (**b**).

Table 7. Thermodynamics parameters for the adsorption of acid tannic on steel slag.

Thermodynamic Parameters	$\Delta G°$ (kJ·mol^{-1})				$\Delta H°$ (kJ·mol^{-1})	$\Delta S°$ (J·mol^{-1}·K^{-1})
Temperature	288	303	318	333		
Value	−4.107	−4.608	−5.108	−5.776	5.506	33.3798

3.5. Characterization of SS–TA before Adsorption

In order to explore the SS-loaded TA characteristics after adsorption, XRD patterns of raw SS and SS-loaded TA were compared in Figure 13. By comparing the peaks characteristics, it was observed that calcite CaCO$_3$ phase was appeared in the SS–TA composition, noting that no calcium carbonate phase was detected in the raw SS. On the other hand, calcium hydroxide Ca(OH)$_2$ peaks were reduced after the adsorption of TA onto SS. The precipitation of TA in water solution and subsequent formation of CaCO$_3$ may be attributed to the interaction of the TA carbon molecules (C$_{76}$H$_{52}$O$_{46}$) with calcium hydroxide Ca(OH)$_2$. It was also noticed that SS composition after TA adsorption was approximately the same structure as before adsorption. Zhao et al. [83] compared the steel slag XRD characteristics before and after adsorption of p-Nitrophenol and reported no significative changes in the crystalline phase of SS. Similarly, the same authors indicated the reduction in Ca(OH)$_2$ and Ca$_2$SiO$_4$ peaks after adsorption. The significant difference in peaks intensities before and after adsorption suggests that TA could affect the crystalline size of SS resulting in higher and sharper peaks. In addition, since the material maintains approximately the same structure as before adsorption the potential of its reuse still possible. Among the techniques suggested is pyrolysis of SS-loaded TA to remove TA organic carbon. This regeneration procedure could result in a great substance with a larger surface area that can be reused in other applications.

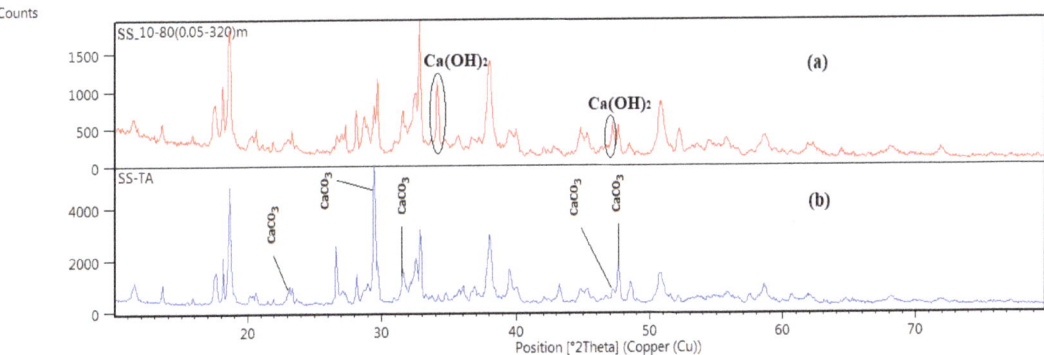

Figure 13. XRD patterns of raw steel slag after adsorption (**a**), and SS–TA before adsorption (**b**).

Significant differences between the raw SS and SS-loaded TA were observed by scanning electron microscopy (Figure 14). It can be seen that there was in fact a reaction between the TA and the SS surface functions (micrograph f). The raw SS showed sharp edges composed mainly from different shapes (micrographs a–c). The surface morphology of TA-adsorbed SS, on the other hand, differed from that of SS, as evidenced by a decrease in SS cracks and mesopores. After the adsorption, the SS-loaded TA showed a relatively rounded structure on which a thin carbon coating layer has been developed. This thin layer was characterized by a woolly texture covering the surface of SS (micrographs d,e). These findings demonstrated that TA molecules were well adsorbed onto the surface of SS.

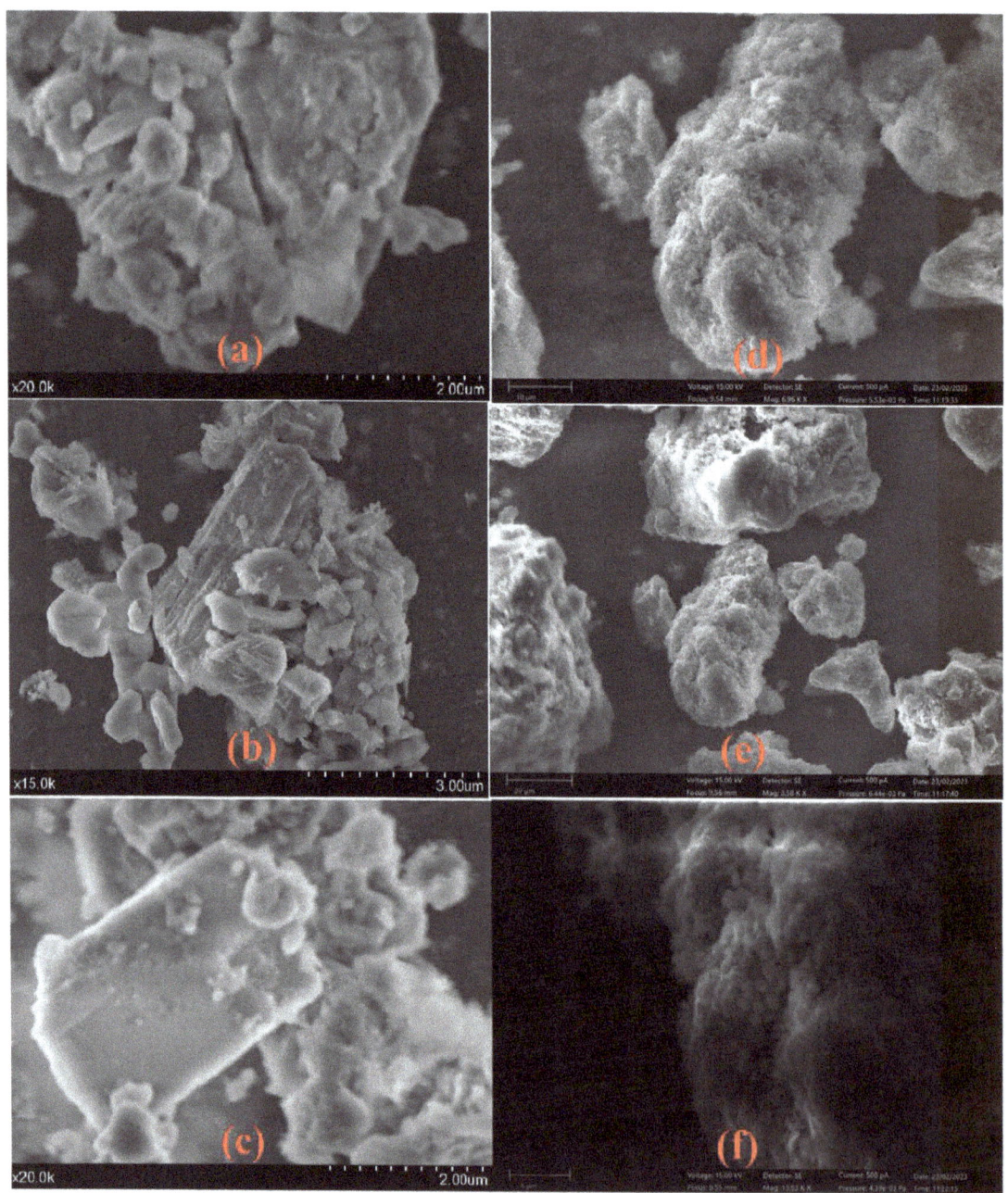

Figure 14. Scanning electron microscope (SEM) of SS before adsorption (**a–c**), and SS-loaded TA after adsorption (**d–f**).

4. Conclusions

In this study the potential treatment of olive mill wastewater (OMWW) by steel slag (SS) as an improver and adsorbent was investigated. The OMWW characterization demonstrated that this effluent is acidic, rich in organic matter and polyphenols, and displayed higher chemical and biological oxygen demands, whereas SS characterization showed that this residue is non-toxic, rich in alkaline metals oxides and hydroxides, and possesses a meso-macroporous structure, which provides an excellent material to be valorized in wastewater treatment.

The buffer capacity of SS was investigated in order to promote neutral disposal conditions of OMWW. The results confirm that SS could be used as a liming material, given its high content in alkaline oxyhydroxides. Moreover, the neutralization process could be achieved by controlling the optimal SS dosage in the function of time. In addition, the OMWW neutralization by SS might be responsible for reducing COD and phytotoxic polyphenols.

The study of tannic Acid (TA) adsorption onto steel slag (SS) was carried out to assess the SS capacity in reducing the total polyphenolic compounds in OMWWs. The following points were the main conclusions:

Batch experiments revealed that an alkaline medium could be a favorable sorption condition for TA on SS.

The maximum adsorption efficiency can reach 98.91% with an adsorbent dosage of 10 g/L. However, a decrease in adsorption efficiency was related to higher SS dosages.

The TA adsorption increased with the contact time until reaching the equilibrium which was higher than 90 % of initial TA concentrations, indicating that the adsorption capacity increased while increasing the initial TA concentration.

The adsorption capacity increased with increasing the temperature.

Adsorption kinetic and mechanism investigations revealed that TA adsorption fitted the pseudo-second-order kinetic model and Elovich. While Langmuir, isotherm described the TA adsorption on SS suitably, providing a maximum adsorption capacity (q_{max}) of 714.28 mg/g.

The chemical alterations in SS resulting from the interactions between SS and TA chemical bonds were emphasized by the characterization of SS after TA adsorption. The affinity towards Langmuir and pseudo-second-order models observed during the TA adsorption on SS indicate that chemisorption was the mechanism underlying TA adsorption onto SS.

This study shows that steel slag, an industrial waste, is a low-cost and readily accessible adsorbent for treating OMWWs. Further research should be carried out on the recovery of valuable chemicals from SS.

Author Contributions: Conceptualization, O.S. and F.E.M.; methodology, O.S.; software, O.S.; validation, O.S., F.E.M. and M.S.; formal analysis, J.M.; investigation, O.S. and E.H.Y.; resources, F.E.M., E.O. and M.S.; writing—original draft preparation, O.S. and F.E.M.; writing—review and editing, E.O., J.M. and F.E.M.; visualization, J.M.; supervision, M.S.; project administration, M.S. All authors have read and agreed to the published version of the manuscript.

Funding: This research received no external funding.

Institutional Review Board Statement: Not applicable.

Informed Consent Statement: Not applicable.

Data Availability Statement: Not applicable.

Acknowledgments: The authors are thankful to Mustapha EL HADRI, for his assistance in characterizing steel slag using XRD and SEM techniques after adsorption. The analysis was carried out in the Faculty of Sciences, University Abdelmalek Essaâdi, Tetouan.

Conflicts of Interest: The authors declare no conflict of interest.

References

1. Lechhab, T.; Lechhab, W.; Cacciola, F.; Salmoun, F. Sets of Internal and External Factors Influencing Olive Oil (*Olea europaea* L.) Composition: A Review. *Eur. Food Res. Technol.* **2022**, *248*, 1069–1088. [CrossRef]
2. Lechhab, T.; Salmoun, F.; Lechhab, W.; El Majdoub, Y.O.; Russo, M.; Camillo, M.R.T.; Trovato, E.; Dugo, P.; Mondello, L.; Cacciola, F. Determination of Bioactive Compounds in Extra Virgin Olive Oils from 19 Moroccan Areas Using Liquid Chromatography Coupled to Mass Spectrometry: A Study over Two Successive Years. *Eur. Food Res. Technol.* **2021**, *247*, 2993–3012. [CrossRef]
3. Lechhab, W.; Cincotta, F.; Lechhab, T.; Condurso, C.; Salmoun, F.; Cacciola, F.; Verzera, A. Preliminary Assessment of Occurrence, Potential Origin, and Human Health Risk of Volatile Organic Compounds in Uncontrolled Springs, North Morocco. *Metabolites* **2022**, *12*, 1213. [CrossRef]
4. Alaoui, S.; Aissam, H.; Merzouki, M.; Benlemlih, M. Characterization of the Harmful Effect of Olive Mill Wastewater on Spearmint. *J. Hazard Mater.* **2009**, *170*, 779–785.
5. Banias, G.; Achillas, C.; Vlachokostas, C.; Moussiopoulos, N.; Stefanou, M. Environmental Impacts in the Life Cycle of Olive Oil: A Literature Review. *J. Sci. Food Agric.* **2017**, *97*, 1686–1697. [CrossRef]
6. Scioli, C.; Vollaro, L. The Use of *Yarrowia lipolytica* to Reduce Pollution in Olive Mill Wastewaters. *Water Res.* **1997**, *31*, 2520–2524. [CrossRef]
7. Tsioulpas, A.; Dimou, D.; Iconomou, D.; Aggelis, G. Phenolic Removal in Olive Oil Mill Wastewater by Strains of *Pleurotus* spp. in Respect to Their Phenol Oxidase (Laccase) Activity. *Bioresour. Technol.* **2002**, *84*, 251–257. [CrossRef]
8. Khdair, A.I.; Abu-Rumman, G.; Khdair, S.I. Pollution Estimation from Olive Mills Wastewater in Jordan. *Heliyon* **2019**, *5*, e02386. [CrossRef]
9. El Hassani, F.Z.; Fadile, A.; Faouzi, M.; Zinedine, A.; Merzouki, M.; Benlemlih, M. The Long Term Effect of Olive Mill Wastewater (OMW) on Organic Matter Humification in a Semi-Arid Soil. *Heliyon* **2020**, *6*, e03181. [CrossRef]
10. Elabdouni, A.; Haboubi, K.; Merimi, I.; El Youbi, M.S.M. Olive Mill Wastewater (OMW) Production in the Province of Al-Hoceima (Morocco) and Their Physico-Chemical Characterization by Mill Types. *Mater. Today Proc.* **2020**, *27*, 3145–3150. [CrossRef]
11. Eroğlu, E.; Eroğlu, İ.; Gündüz, U.; Yücel, M. Treatment of Olive Mill Wastewater by Different Physicochemical Methods and Utilization of Their Liquid Effluents for Biological Hydrogen Production. *Biomass Bioenergy* **2009**, *33*, 701–705. [CrossRef]
12. Ginos, A.; Manios, T.; Mantzavinos, D. Treatment of Olive Mill Effluents by Coagulation–Flocculation–Hydrogen Peroxide Oxidation and Effect on Phytotoxicity. *J. Hazard. Mater.* **2006**, *133*, 135–142. [CrossRef]
13. Jaouani, A.; Vanthournhout, M.; Penninckx, M.J. Olive Oil Mill Wastewater Purification by Combination of Coagulation-Flocculation and Biological Treatments. *Environ. Technol.* **2005**, *26*, 633–642. [CrossRef] [PubMed]
14. Michael, I.; Panagi, A.; Ioannou, L.A.; Frontistis, Z.; Fatta-Kassinos, D. Utilizing Solar Energy for the Purification of Olive Mill Wastewater Using a Pilot-Scale Photocatalytic Reactor after Coagulation-Flocculation. *Water Res.* **2014**, *60*, 28–40. [CrossRef] [PubMed]
15. Papaphilippou, P.C.; Yiannapas, C.; Politi, M.; Daskalaki, V.M.; Michael, C.; Kalogerakis, N.; Mantzavinos, D.; Fatta-Kassinos, D. Sequential Coagulation–Flocculation, Solvent Extraction and Photo-Fenton Oxidation for the Valorization and Treatment of Olive Mill Effluent. *Chem. Eng. J.* **2013**, *224*, 82–88. [CrossRef]
16. Inan, H.; Dimoglo, A.; Şimşek, H.; Karpuzcu, M. Olive Oil Mill Wastewater Treatment by Means of Electro-Coagulation. *Sep. Purif. Technol.* **2004**, *36*, 23–31. [CrossRef]
17. Pelendridou, K.; Michailides, M.K.; Zagklis, D.P.; Tekerlekopoulou, A.G.; Paraskeva, C.A.; Vayenas, D. V Treatment of Olive Mill Wastewater Using a Coagulation–Flocculation Process Either as a Single Step or as Post-treatment after Aerobic Biological Treatment. *J. Chem. Technol. Biotechnol.* **2014**, *89*, 1866–1874. [CrossRef]
18. Marques, I.P. Anaerobic Digestion Treatment of Olive Mill Wastewater for Effluent Re-Use in Irrigation. *Desalination* **2001**, *137*, 233–239. [CrossRef]
19. Yangui, A.; Abderrabba, M. Towards a High Yield Recovery of Polyphenols from Olive Mill Wastewater on Activated Carbon Coated with Milk Proteins: Experimental Design and Antioxidant Activity. *Food Chem.* **2018**, *262*, 102–109. [CrossRef]
20. Abu-Dalo, M.; Abdelnabi, J.; Bawab, A. Al Preparation of Activated Carbon Derived from Jordanian Olive Cake and Functionalized with Cu/Cu2 o/Cuo for Adsorption of Phenolic Compounds from Olive Mill Wastewater. *Materials* **2021**, *14*, 6636. [CrossRef]
21. Pepi, M.; Lampariello, L.R.; Altieri, R.; Esposito, A.; Perra, G.; Renzi, M.; Lobianco, A.; Feola, A.; Gasperini, S.; Focardi, S.E. Tannic Acid Degradation by Bacterial Strains Serratia Spp. and Pantoea Sp. Isolated from Olive Mill Waste Mixtures. *Int. Biodeterior. Biodegrad.* **2010**, *64*, 73–80. [CrossRef]
22. Wang, J.; Zheng, C.; Ding, S.; Ma, H.; Ji, Y. Behaviors and Mechanisms of Tannic Acid Adsorption on an Amino-Functionalized Magnetic Nanoadsorbent. *Desalination* **2011**, *273*, 285–291. [CrossRef]
23. Wang, J.; Zheng, S.; Liu, J.; Xu, Z. Tannic Acid Adsorption on Amino-Functionalized Magnetic Mesoporous Silica. *Chem. Eng. J.* **2010**, *165*, 10–16. [CrossRef]
24. Wang, J.; Li, A.; Xu, L.; Zhou, Y. Adsorption of Tannic and Gallic Acids on a New Polymeric Adsorbent and the Effect of Cu (II) on Their Removal. *J. Hazard. Mater.* **2009**, *169*, 794–800. [CrossRef] [PubMed]
25. Sun, Q.; Xiong, B.; Pan, Y.; Cui, H. Adsorption Removal of Tannic Acid from Aqueous Solution by Polyaniline: Analysis of Operating Parameters and Mechanism. *J. Colloid Interface Sci.* **2017**, *487*, 175–181. [CrossRef]
26. Mansouri, K.; Elsaid, K.; Bedoui, A.; Bensalah, N.; Abdel-Wahab, A. Application of Electrochemically Dissolved Iron in the Removal of Tannic Acid from Water. *Chem. Eng. J.* **2011**, *172*, 970–976. [CrossRef]
27. Jung, C.; Phal, N.; Oh, J.; Chu, K.H.; Jang, M.; Yoon, Y. Removal of Humic and Tannic Acids by Adsorption–Coagulation Combined Systems with Activated Biochar. *J. Hazard. Mater.* **2015**, *300*, 808–814. [CrossRef] [PubMed]

28. Ulbricht, M.; Ansorge, W.; Danielzik, I.; König, M.; Schuster, O. Fouling in Microfiltration of Wine: The Influence of the Membrane Polymer on Adsorption of Polyphenols and Polysaccharides. *Sep. Purif. Technol.* **2009**, *68*, 335–342. [CrossRef]
29. Afify, A.S.; Mahmoud, M.A.; Emara, H.A.; Abdelkreem, K.I. Phenolic Compounds and COD Removal from Olive Mill Wastewater by Chemical and Biological Procedures. *Aust. J. Basic Appl. Sci.* **2009**, *3*, 1087–1095.
30. Lin, D.; Xing, B. Tannic Acid Adsorption and Its Role for Stabilizing Carbon Nanotube Suspensions. *Environ. Sci. Technol.* **2008**, *42*, 5917–5923. [CrossRef]
31. Zhang, W.; Yang, Z.-Y.; Cheng, X.-W.; Tang, R.-C.; Qiao, Y.-F. Adsorption, Antibacterial and Antioxidant Properties of Tannic Acid on Silk Fiber. *Polymers* **2019**, *11*, 970. [CrossRef] [PubMed]
32. Li, J.; Ma, S.; Li, X.; Wei, W. Adsorption of Tannic Acid and Macromolecular Humic/Fulvic Acid onto Polystyrene Microplastics: A Comparison Study. *Water* **2022**, *14*, 2201. [CrossRef]
33. Teng, Y.; Liu, Z.; Yao, K.; Song, W.; Sun, Y.; Wang, H.; Xu, Y. Preparation of Attapulgite/CoFe2O4 Magnetic Composites for Efficient Adsorption of Tannic Acid from Aqueous Solution. *Int. J. Environ. Res. Public Health* **2019**, *16*, 2187. [CrossRef] [PubMed]
34. An, J.-H.; Dultz, S. Adsorption of Tannic Acid on Chitosan-Montmorillonite as a Function of PH and Surface Charge Properties. *Appl. Clay Sci.* **2007**, *36*, 256–264. [CrossRef]
35. Chang, M.-Y.; Juang, R.-S. Adsorption of Tannic Acid, Humic Acid, and Dyes from Water Using the Composite of Chitosan and Activated Clay. *J. Colloid Interface Sci.* **2004**, *278*, 18–25. [CrossRef]
36. Jevremović, A.; Božinović, N.; Arsenijević, D.; Marmakov, S.; Vasiljević, B.N.; Usković-Marković, S.; Bajuk-Bogdanović, D.; Milojević-Rakić, M. Modulation of Cytotoxicity by Consecutive Adsorption of Tannic Acid and Pesticides on Surfactant Functionalized Zeolites. *Environ. Sci. Process. Impacts* **2020**, *22*, 2199–2211. [CrossRef] [PubMed]
37. Yüksel, İ. A Review of Steel Slag Usage in Construction Industry for Sustainable Development. *Environ. Dev. Sustain.* **2017**, *19*, 369–384. [CrossRef]
38. Jiang, Y.; Ahmad, M.R.; Chen, B. Properties of Magnesium Phosphate Cement Containing Steel Slag Powder. *Constr. Build. Mater.* **2019**, *195*, 140–147. [CrossRef]
39. Shi, C.; Wang, X.; Zhou, S.; Zuo, X.; Wang, C. Mechanism, Application, Influencing Factors and Environmental Benefit Assessment of Steel Slag in Removing Pollutants from Water: A Review. *J. Water Process. Eng.* **2022**, *47*, 102666. [CrossRef]
40. Miyata, Y.; Hayashi, A.; Kuwayama, M.; Yamamoto, T.; Urabe, N. Reduction Test of Hydrogen Sulfide in Silty Sediment of Fukuyama Inner Harbor Using Steelmaking Slag. *ISIJ Int.* **2015**, *55*, 2686–2693. [CrossRef]
41. Scott, I.S.P.C.; Penn, C.J. Estimating the Variability of Steel Slag Properties and Their Influence in Phosphorus Removal Ability. *Chemosphere* **2021**, *276*, 130205. [CrossRef]
42. Liem-Nguyen, V.; Sjöberg, V.; Dinh, N.P.; Huy, D.H.; Karlsson, S. Removal Mechanism of Arsenic (V) by Stainless Steel Slags Obtained from Scrap Metal Recycling. *J. Environ. Chem. Eng.* **2020**, *8*, 103833. [CrossRef]
43. Wang, M.; Wang, X.; Zhang, M.; Han, W.; Yuan, Z.; Zhong, X.; Yu, L.; Ji, H. Treatment of Cd (II) and As (V) Co-Contamination in Aqueous Environment by Steel Slag-Biochar Composites and Its Mechanism. *J. Hazard. Mater.* **2023**, *447*, 130784. [CrossRef] [PubMed]
44. Jiang, Z.; Chen, X.; Zhao, D. Quantitative Removal of Ammonia Nitrogen and Phosphorus with Compound of Zeolite and Steel Slag. *Environ. Sci. Technol. (China)* **2016**, *39*, 133–138.
45. Wang, J.; Zhong, M.; Wu, P.; Wen, S.; Huang, L.; Ning, P. A Review of the Application of Steel Slag in CO_2 Fixation. *ChemBioEng Rev.* **2021**, *8*, 189–199. [CrossRef]
46. Tian, S.; Jiang, J.; Li, K.; Yan, F.; Chen, X. Performance of Steel Slag in Carbonation–Calcination Looping for CO_2 Capture from Industrial Flue Gas. *Rsc Adv.* **2014**, *4*, 6858–6862. [CrossRef]
47. Miranda-Pizarro, J.; Perejón, A.; Valverde, J.M.; Sánchez-Jiménez, P.E.; Pérez-Maqueda, L.A. Use of Steel Slag for CO_2 Capture under Realistic Calcium-Looping Conditions. *RSC Adv.* **2016**, *6*, 37656–37663. [CrossRef]
48. Li, H.; Tang, Z.; Li, N.; Cui, L.; Mao, X. Mechanism and Process Study on Steel Slag Enhancement for CO_2 Capture by Seawater. *Appl. Energy* **2020**, *276*, 115515. [CrossRef]
49. Singleton, V.L.; Orthofer, R.; Lamuela-Raventós, R.M. [14] Analysis of Total Phenols and Other Oxidation Substrates and Antioxidants by Means of Folin-Ciocalteu Reagent. In *Methods in Enzymology*; Elsevier: Amsterdam, The Netherlands, 1999; Volume 299, pp. 152–178. ISBN 0076-6879.
50. Romeo, R.; De Bruno, A.; Imeneo, V.; Piscopo, A.; Poiana, M. Impact of Stability of Enriched Oil with Phenolic Extract from Olive Mill Wastewaters. *Foods* **2020**, *9*, 856. [CrossRef] [PubMed]
51. Niaounakis, M.; Halvadakis, C.P. *Olive Processing Waste Management: Literature Review and Patent Survey*; Elsevier: Amsterdam, The Netherlands, 2006.
52. Daâssi, D.; Lozano-Sánchez, J.; Borrás-Linares, I.; Belbahri, L.; Woodward, S.; Zouari-Mechichi, H.; Mechichi, T.; Nasri, M.; Segura-Carretero, A. Olive Oil Mill Wastewaters: Phenolic Content Characterization during Degradation by *Coriolopsis gallica*. *Chemosphere* **2014**, *113*, 62–70. [CrossRef]
53. Dutournié, P.; Jeguirim, M.; Khiari, B.; Goddard, M.-L.; Jellali, S. Olive Mill Wastewater: From a Pollutant to Green Fuels, Agricultural Water Source, and Bio-Fertilizer. Part 2: Water Recovery. *Water* **2019**, *11*, 768. [CrossRef]
54. El Moudden, H.; El Idrissi, Y.; Belmaghraoui, W.; Belhoussaine, O.; El Guezzane, C.; Bouayoun, T.; Harhar, H.; Tabyaoui, M. Olive Mill Wastewater Polyphenol-based Extract as a Vegetable Oil Shelf Life Extending Additive. *J. Food Process. Preserv.* **2020**, *44*, e14990. [CrossRef]

55. Shabir, S.; Ilyas, N.; Saeed, M.; Bibi, F.; Sayyed, R.Z.; Almalki, W.H. Treatment Technologies for Olive Mill Wastewater with Impacts on Plants. *Environ. Res.* **2023**, *216*, 114399. [CrossRef] [PubMed]
56. El Gnaoui, Y.; Sounni, F.; Bakraoui, M.; Karouach, F.; Benlemlih, M.; Barz, M.; El Bari, H. Anaerobic Co-Digestion Assessment of Olive Mill Wastewater and Food Waste: Effect of Mixture Ratio on Methane Production and Process Stability. *J. Environ. Chem. Eng.* **2020**, *8*, 103874. [CrossRef]
57. Bouknana, D.; Hammouti, B.; Salghi, R.; Jodeh, S.; Zarrouk, A.; Warad, I.; Aouniti, A.; Sbaa, M. Physicochemical Characterization of Olive Oil Mill Wastewaters in the Eastern Region of Morocco. *J. Mater. Environ. Sci* **2014**, *5*, 1039–1058.
58. Manocha, S.; Ponchon, F. Management of Lime in Steel. *Metals* **2018**, *8*, 686. [CrossRef]
59. De Martino, A.; Arienzo, M.; Iorio, M.; Vinale, F.; Lorito, M.; Prenzler, P.D.; Ryan, D.; Obied, H.K. Detoxification of Olive Mill Wastewaters by Zinc–Aluminium Layered Double Hydroxides. *Appl. Clay Sci.* **2011**, *53*, 737–744. [CrossRef]
60. Hu, R.; Xie, J.; Wu, S.; Yang, C.; Yang, D. Study of Toxicity Assessment of Heavy Metals from Steel Slag and Its Asphalt Mixture. *Materials* **2020**, *13*, 2768. [CrossRef] [PubMed]
61. Rađenović, A.; Malina, J.; Sofilić, T. Characterization of Ladle Furnace Slag from Carbon Steel Production as a Potential Adsorbent. *Adv. Mater. Sci. Eng.* **2013**, *2013*, 198240. [CrossRef]
62. Navarro, C.; Díaz, M.; Villa-García, M.A. Physico-Chemical Characterization of Steel Slag. Study of Its Behavior under Simulated Environmental Conditions. *Environ. Sci. Technol.* **2010**, *44*, 5383–5388. [CrossRef]
63. Boukhoubza, F.; Jail, A.; Korchi, F.; Idrissi, L.L.; Hannache, H.; Duarte, J.C.; Hassani, L.; Nejmeddine, A. Application of Lime and Calcium Hypochlorite in the Dephenolisation and Discolouration of Olive Mill Wastewater. *J. Environ. Manag.* **2009**, *91*, 124–132. [CrossRef]
64. Uğurlu, M.; Kula, İ. Decolourization and Removal of Some Organic Compounds from Olive Mill Wastewater by Advanced Oxidation Processes and Lime Treatment. *Environ. Sci. Pollut. Res.-Int.* **2007**, *14*, 319–325. [CrossRef]
65. Achak, M.; Ouazzani, N.; Yaacoubi, A.; Mandi, L. Modern Olive Mill Effluent Characterization and Their Treatment by Coagulation–Flocculation Using Lime and Aluminium Sulphate. *Caractérisation Des Margines Issues D'une Huilerie Mod-Erne Essais Leur Trait. Par Coagul.-Floculation Chaux Sulfate D'aluminium* **2008**, *21*, 53–67.
66. Sağlık, S.; Ersoy, L.; İmre, S. Oil Recovery from Lime-treated Wastewater of Olive Mills. *Eur. J. Lipid Sci. Technol.* **2002**, *104*, 212–215. [CrossRef]
67. Aktas, E.S.; Imre, S.; Ersoy, L. Characterization and Lime Treatment of Olive Mill Wastewater. *Water Res.* **2001**, *35*, 2336–2340. [CrossRef]
68. Sarti, O.; El Mansouri, F.; Otal, E.; Morillo, J.; Ouassini, A.; Brigui, J.; Saidi, M. Assessing the Effect of Intensive Agriculture and Sandy Soil Properties on Groundwater Contamination by Nitrate and Potential Improvement Using Olive Pomace Biomass Slag (OPBS). *C* **2023**, *9*, 1. [CrossRef]
69. Murdiati, T.B.; McSweeney, C.S.; Campbell, R.S.F.; Stoltz, D.S. Prevention of Hydrolysable Tannin Toxicity in Goats Fed Clidemia Hirta by Calcium Hydroxide Supplementation. *J. Appl. Toxicol.* **1990**, *10*, 325–331. [CrossRef] [PubMed]
70. Fang, Y.; Wang, J.; Wang, L.; Qian, X.; Wang, X.; Liao, W.; Chen, P.; Ma, H. Densifying Hydration Products of Alite by a Bio-Inspired Admixture. *Mater. Des.* **2023**, *225*, 111490. [CrossRef]
71. Murdiati, T.B.; McSweeney, C.S.; Lowry, J.B. Complexing of Toxic Hydrolysable Tannins of Yellow-wood (*Terminalia oblongata*) and Harendong (*Clidemia hirta*) with Reactive Substances: An Approach to Preventing Toxicity. *J. Appl. Toxicol.* **1991**, *11*, 333–338. [CrossRef]
72. Allaoui, S.; Naciri Bennani, M.; Ziyat, H.; Qabaqous, O.; Tijani, N.; Ittobane, N. Kinetic Study of the Adsorption of Polyphenols from Olive Mill Wastewater onto Natural Clay: Ghassoul. *J. Chem.* **2020**, *2020*, 7293189. [CrossRef]
73. Galiatsatou, P.; Metaxas, M.; Arapoglou, D.; Kasselouri-Rigopoulou, V. Treatment of Olive Mill Waste Water with Activated Carbons from Agricultural By-Products. *Waste Manag.* **2002**, *22*, 803–812. [CrossRef] [PubMed]
74. Faschan, A.; Cartledge, F.; Tittlebaum, M. Effect of Calcium Hydroxide and PH on Organoclay Adsorption of Organic Compounds. *J. Environ. Sci. Health Part A* **1993**, *28*, 585–597. [CrossRef]
75. Stasinakis, A.S.; Elia, I.; Petalas, A.V.; Halvadakis, C.P. Removal of Total Phenols from Olive-Mill Wastewater Using an Agricultural by-Product, Olive Pomace. *J. Hazard. Mater.* **2008**, *160*, 408–413. [CrossRef] [PubMed]
76. Malkoc, E.; Nuhoglu, Y.; Dundar, M. Adsorption of Chromium (VI) on Pomace—An Olive Oil Industry Waste: Batch and Column Studies. *J. Hazard. Mater.* **2006**, *138*, 142–151. [CrossRef]
77. Langmuir, I. The Adsorption of Gases on Plane Surfaces of Glass, Mica and Platinum. *J. Am. Chem. Soc.* **1918**, *40*, 1361–1403. [CrossRef]
78. Freundlich, H. Über Die Adsorption in Lösungen. *Z. Für Phys. Chem.* **1907**, *57*, 385–470. [CrossRef]
79. Lin, J.; Zhan, Y.; Zhu, Z.; Xing, Y. Adsorption of Tannic Acid from Aqueous Solution onto Surfactant-Modified Zeolite. *J. Hazard. Mater.* **2011**, *193*, 102–111. [CrossRef]
80. Agarwal, S.; Rajoria, P.; Rani, A. Adsorption of Tannic Acid from Aqueous Solution onto Chitosan/NaOH/Fly Ash Composites: Equilibrium, Kinetics, Thermodynamics and Modeling. *J. Environ. Chem. Eng.* **2018**, *6*, 1486–1499. [CrossRef]
81. Liu, F.; Long, Q.; Gao, N.; Peng, Q.; Huo, Y.; Chen, Y.; Tang, Q.; Huang, Q.; Liu, M.; Chen, L. Effective Adsorption of Tannic Acid by Porous Dual Crosslinked Soy Protein Isolate-Alginate Hybrid Spheres from Aqueous Solution. *Chem. Eng. Res. Des.* **2023**, *189*, 250–261. [CrossRef]

82. Wei, W.; Li, J.; Han, X.; Yao, Y.; Zhao, W.; Han, R.; Li, S.; Zhang, Y.; Zheng, C. Insights into the Adsorption Mechanism of Tannic Acid by a Green Synthesized Nano-Hydroxyapatite and Its Effect on Aqueous Cu (II) Removal. *Sci. Total Environ.* **2021**, *778*, 146189. [CrossRef]
83. Zhao, Y.; Wang, L.; Zhu, L.; Gao, F.; Xu, X.; Yang, J. Removal of P-Nitrophenol from Simulated Sewage Using Steel Slag: Capability and Mechanism. *Environ. Res.* **2022**, *212*, 113450. [CrossRef] [PubMed]

Disclaimer/Publisher's Note: The statements, opinions and data contained in all publications are solely those of the individual author(s) and contributor(s) and not of MDPI and/or the editor(s). MDPI and/or the editor(s) disclaim responsibility for any injury to people or property resulting from any ideas, methods, instructions or products referred to in the content.

Article

Changes of C, H, and N Elements of Corn Straw during the Microwave Heating Process

Zhihong Liu [1], Weitao Cao [1], Man Zhang [2,*], Wenke Zhao [1] and Yaning Zhang [1,*]

[1] School of Energy Science and Engineering, Harbin Institute of Technology, Harbin 150001, China; lzh18702504053@163.com (Z.L.); cwt20010228@gmail.com (W.C.); zhaowenke@hit.edu.cn (W.Z.)
[2] Department of Energy and Power Engineering, Tsinghua University, Beijing 100084, China
* Correspondence: zhangman@tsinghua.edu.cn (M.Z.); ynzhang@hit.edu.cn (Y.Z.)

Abstract: Due to the rapid growth of the global economy, energy consumption has been steadily increasing, leading to increasing issues such as energy shortages and environmental concerns. Biomass energy, a critical renewable energy source, plays a vital role in advancing low-carbon energy development and resource sustainability. In this study, experiments were conducted to study the migration of C, H, and N elements of corn straw during the microwave heating process, and the effects of residence time, heating temperature, and microwave power were also investigated. The results showed that when the temperature rose, both the proportion of C and H elements fluctuated slightly. Specifically, when the temperature rose from 75 °C to 275 °C, there was a 1.02% increase in the proportion of the C element and a 0.25% decrease in the proportion of the H element. Residence time appeared to be a significant factor influencing the changes in C, H, and N elements. For a 40 min residence time, the proportion of the C element increased from 31.77% to 35.36%, while the proportion of the H element decreased from 4.50% to 3.83%. When there was an increase in the microwave power between 160 W and 200 W, higher temperatures were reached in the samples, leading to the carbonization process of corn straw being more complete. Consequently, the proportion of the C element rose with extended residence time, whereas the proportion of the H element decreased as the residence time increased.

Keywords: C; H; N; corn straw; microwave heating

Citation: Liu, Z.; Cao, W.; Zhang, M.; Zhao, W.; Zhang, Y. Changes of C, H, and N Elements of Corn Straw during the Microwave Heating Process. C 2023, 9, 117. https://doi.org/10.3390/c9040117

Academic Editors: Indra Neel Pulidindi, Pankaj Sharma, Aharon Gedanken and Dimitrios Kalderis

Received: 1 October 2023
Revised: 14 November 2023
Accepted: 15 November 2023
Published: 5 December 2023

Copyright: © 2023 by the authors. Licensee MDPI, Basel, Switzerland. This article is an open access article distributed under the terms and conditions of the Creative Commons Attribution (CC BY) license (https://creativecommons.org/licenses/by/4.0/).

1. Introduction

With the rapid advancement of the global economy and the steady rise in population, energy consumption is increasing year by year [1,2]. The excessive exploitation and consumption of fossil fuels have led to a growing list of drawbacks. Energy shortages and environmental crises have gradually emerged as the primary challenges confronting the international community. The issue of climate change resulting from excessive carbon dioxide emissions is widely recognized as the most severe environmental problem of our time, and it represents a crucial strategic concern for global sustainable development. Notably, more than 90% of carbon dioxide emissions linked to energy production originate from the burning of fossil fuels. Consequently, reducing the consumption of fossil energy is a necessary step to mitigate carbon emissions [3,4]. Renewable energy, characterized by abundant reserves and minimal environmental impact, is viewed as an effective alternative to fossil fuels and has garnered significant global attention [5].

Biomass energy, a key component of renewable energy, boasts remarkable qualities such as renewability, widespread availability, and low carbon emissions. It holds substantial potential for future development [6]. China, being a major agricultural nation with a dense population, possesses abundant biomass resources. In 2022, China's biomass production reached 3.7 billion tons, equivalent to 460 million tons of standard coal. However, the country's energy utilization rate of biomass is merely 12%, and its contribution to the overall energy consumption structure stands at only 4.6%, considerably lower than the global average [7]. Therefore, actively utilizing and exploiting biomass energy resources holds immense significance in promoting low-carbon energy development and fostering resource recycling.

Presently, China's utilization of biomass energy exhibits a diversified trend, encompassing various approaches such as direct combustion, biogas fermentation, biomass gasification, and biomass oil production [8,9]. Notably, the products derived from biomass gasification and biomass oil production can effectively serve as substitutes for fossil fuels like oil and natural gas and have thus experienced substantial growth in recent years.

Microwave heating represents a novel approach to heating. In contrast to traditional heating methods, microwave heating offers distinct advantages, including even heating, rapid heating rates, high efficiency, and minimal environmental impact. These attributes have gathered significant attention from researchers [10–12]. Moreover, microwave-assisted chemical conversion technology has emerged as a central pathway for the utilization of biomass energy.

Qiu et al. [13] conducted a study on the impact of electric pyrolysis versus microwave pyrolysis of corn straw on biochar yield. The findings indicated that when subjected to microwave pyrolysis at 800 °C for 180 min, the biochar yield increased by 21.24% compared to electric pyrolysis. Shvets et al. [14] investigated the influence of microwave power on the product composition during the microwave pyrolysis of woody biomass. Their research revealed that as microwave power increased from 840 W to 1760 W, the average yields of CO and CO_2 remained relatively stable. However, with a further increase in power to 2200 W, there was a notable increase in the average yields of CO and CO_2. Additionally, high microwave power was found to enhance the production of CH_4 and H_2. Zhang et al. [15] developed a microwave-assisted biomass air gasification system. They subsequently investigated the effects of the airflow rate, gasification temperature, and residence time on the yield of syngas and carbon conversion rate. The outcomes demonstrated that increasing the residence time significantly elevated both syngas yield and carbon conversion rate. Specifically, under conditions of an airflow rate of 600 mL/min and a temperature of 700 °C, the residence time played a critical role in the process: at 30 min, the syngas yield can reach 76.14%. Fu et al. [16] carried out a study on microwave-assisted biomass air gasification in a fluidized bed reactor and obtained the optimal reaction conditions for syngas production: a gasification temperature of 900 °C, equivalent ratio of 0.35, and syngas yield of 78.2%. Mao et al. [17] carried out microwave-assisted pyrolysis experiments of furfural residue in a continuous spiral reactor and explored the effects of kaolin, CaO, and K_2CO_3 additives on the yield of pyrolysis gas. The results showed that all three additives can increase the yield of H_2 and CH_4, and CaO can also reduce the yield of CO_2.

The microwave absorption capacity of biomass is weak. In order to meet the temperature conditions of biomass gasification and pyrolysis, it is often necessary to add microwave absorbers, such as SiC, activated carbon, graphite, etc. Shi et al. [18] used bamboo, cork, pine, and mahogany as raw materials to study the effect of activated carbon additives on the production of syngas under microwave irradiation and found that the coupling of activated carbon reforming and microwave-assisted pyrolysis not only increased the yield of gas products and reduced the yield of bio-oil, but also further improved the selectivity of H_2 generation and increased the overall low heat value of products. Zhou et al. [19] studied and developed a continuous microwave-assisted pyrolysis system with a large processing capacity. Using SiC as the microwave absorber, it could process 10 kg of biomass per hour. Huang et al. [20] used magnetite as a microwave absorber to pyrolyze chlorella and spirulina to produce bio-oil. The results showed that Fe_3O_4 accelerated the decarboxylation reaction in the decomposition process of microalgae and promoted the formation of nitrogen-containing fatty compounds.

The bio-oil obtained by microwave pyrolysis of biomass has high oxygen content, low calorific value, complex composition, and instability and often needs to be upgraded by catalysis. Commonly used catalysts include metal salts, metal oxides, and ZSM-5 catalysts.

Li et al. [21] modified the ZSM-5 catalyst with NaOH and TPAOH alkaline solutions and applied it to microwave catalytic pyrolysis of rice husks. The results showed that the catalyst treated with 2.0 mol/L TPAOH solution showed better catalytic performance and the relative content of hydrocarbons in the pyrolysis products was high. The relative contents of benzene, toluene, and xylene were up to 58.6%. Wan et al. [22] applied $MgCl_2$ to the microwave catalytic pyrolysis of poplar wood, and the results showed that the furfural content in the pyrolysis products increased with the increase in MgCl.

Previous studies focused on the pyrolysis and gasification characteristics of biomass with the addition of microwave absorbents and catalysts but ignored the effect of microwaves on the heating characteristics of biomass. In this study, the effects of residence time, heating temperature, and microwave power on the migration of C, H, and N elements of corn straw were investigated by experiments without adding microwave absorbent, which is of great significance for the efficient utilization of biomass.

2. Materials and Methods

2.1. Materials

China is rich in straw resources, accounting for about 20% of the total biomass resources. In this study, samples of corn straw were taken from Lianyungang City, China, in 2022. Ultimate analysis was performed using an Organic Elemental Analyzer (Thermo Fisher Scientific Inc, Waltham, MA, USA), and the results are shown in Table 1.

Table 1. Proximate analysis and ultimate analysis results of corn straw.

Proximate Analysis (wt.%)				Ultimate Analysis (wt.%)				
M_{ar}	V_{daf}	FC_{ar}	A_{ar}	C_{ar}	H_{ar}	O_{ar} [a]	N_{ar}	S_{ar}
5.11	37.55	12.14	75.45	31.77	4.50	1.24	0.66	0.90

[a] Calculated by difference. M, V, FC, and A indicate moisture, volatile, fixed carbon and ash, respectively. C, H, O, N, and S indicate elements C, H, O, N, and S, respectively. Subscript ar means "as received basis". Subscript daf means "dry ash-free basis".

The corn straw was crushed in a Baihaojia HY-608 pulverizer (Hongtaiyang Electromechanical Inc., Yongkang, China), and then the crushed material was screened with an 80-mesh sieve; the corn straw below the sieve was taken as the experimental sample, as shown in Figure 1. In order to ensure that the composition of the sample does not change, contact with air should be avoided as much as possible, and it should be stored in a dry and airtight container before use.

Figure 1. Corn straw used.

2.2. Experimental System

The microwave heating device used in this study is shown in Figure 2 and mainly consists of two systems: (a) a microwave heating system; and (b) a temperature measurement system. Microwave heating systems include quartz tubes and microwave ovens. The microwave oven is made by Beijing Yihuida Microwave Devices Co., LTD. (Beijing, China), with a length of 30 cm, a width of 30 cm, and a height of 35 cm. In the quartz tube setup, a small sieve plate is positioned in the middle section. This serves a dual purpose: it facilitates the loading of experimental samples and ensures that both the upper and lower layers of the sample are exposed to the surrounding air. To monitor temperature, a temperature measurement system composed of a digital thermometer and a K-type thermocouple is employed. The digital thermometer is responsible for gathering temperature data from the thermocouple probe and transmitting it to the computer through a USB connection. Once the arrangement of the quartz tube is finalized, the setup is insulated by adding thermal insulation cotton inside the furnace and wrapping the quartz tube. This precaution is taken to prevent any interference from radiation heat transfer, which could affect the accuracy of experimental results.

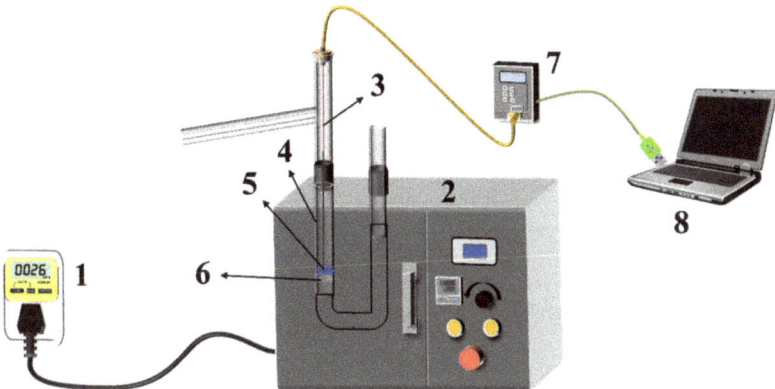

Figure 2. Microwave heating system. (1) power supply; (2) microwave; (3) K-type thermocouple; (4) reactor; (5) samples; (6) sieve plate; (7) digital thermometer; (8) computer.

2.3. Experimental Procedures

Connect the experimental device as shown in Figure 2 and open the temperature recorder to record the temperature data in real time. It should be emphasized that the diffusion rate of air in the sample was very slow without external force. Therefore, a small amount of sample was used in each experiment to ensure that the thermocouple temperature probe was just submerged by the sample.

Start heating by activating the microwave oven. Adjust the microwave power to reach the desired experimental temperature while maintaining a consistent temperature range or slight fluctuations within ±15 °C. This step aims to investigate the impact of temperature and residence time on the migration of C, H, and N elements. The effect of microwave power on the migration of C, H, and N elements was investigated by controlling the heating time. After the experiment was completed, the sample was cooled naturally and removed from the glass tube, and the change rule of C, H, and N elements was explored by ultimate analysis, which was performed using Thermo's Flash Smart Organic Element Analyzer.

To determine an appropriate experimental temperature, a preliminary experiment was conducted. Figure 3 shows the temperature of corn straw particles at 200 W. It was observed that the sample gradually heats up to 292 °C. Afterward, the temperature began to increase rapidly, reaching 1200 °C within 2 s. Simultaneously, a dense white smoke formed within the reactor. Evidently, at this point, a combustion reaction occurred within the sample rather than slow oxidation. To prevent a combustion reaction, the sample temperature must be maintained below 292 °C.

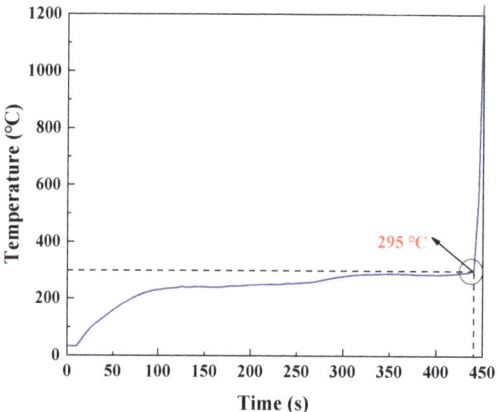

Figure 3. Temperatures of corn straw particles at 200 W.

3. Results and Discussion

3.1. Effect of Temperature on Migration of C, H, and N Elements

The reaction time was set at 10 min, and the proportions of C, H, and N elements changed at various temperatures were examined, as illustrated in Figure 4. When the temperature increased, the proportion of the N element remained nearly constant, with only a minimal 0.047% difference between the maximum and minimum values, which could be disregarded. This stability was due to the fact that N in biomass primarily existed in organic matter form, which was not easily vaporized at lower temperatures [23].

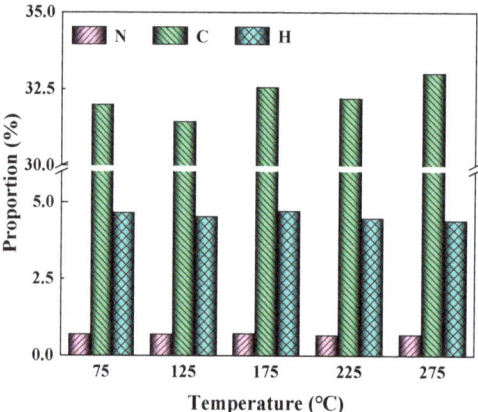

Figure 4. Proportions of C, H, and N elements at different temperatures (10 min).

Since the biomass contained a significant amount of water, it required lower temperatures for the water to vaporize. As the temperature rose, more water vaporized, causing the proportion of H element to steadily decrease. The proportion of the C element showed an overall upward trend. On the one hand, water and some volatiles were released, and the content of the H and O elements decreased, which in turn increased the proportion of the other elements. On the other hand, most C-containing components started to decompose at temperatures above 300 °C. Below 275 °C, the C element underwent only slow oxidation, resulting in minimal changes. For instance, at 275 °C, the proportion of the C element was 33.01%, while the proportion of the H element was 4.41%. In comparison to 75 °C, this represented a 1.02% increase in the C element proportion and a 0.25% decrease in the H element proportion, as shown in Figure 4.

3.2. Effect of Residence Time on Migration of C, H, and N Elements

Temperature plays a pivotal role in the release of volatile compounds and the oxidation of element C. At lower temperatures, biomass samples primarily underwent water release, with minimal involvement in other reactions. Based on pre-experiment findings, a temperature of 275 °C was selected for further investigation into the impact of residence time on the migration of C, H, and N elements, as depicted in Figure 5. Over the initial 0–20 min, an increase in time led to a continuous rise in the proportion of the C element and a concurrent decline in the proportion of the H element. This phenomenon could be attributed to the volatilization of water and certain biomass components containing H. By the 30 min mark, there was a notable shift compared to the state at 20 min. The proportion of the C element decreased from 34.05% to 32.24%, and the proportion of the H element continued to decrease. This was due to the continuous release of volatile fractions containing H, further reducing the proportion of the H element. Meanwhile, the C element began undergoing a noticeable oxidation reaction, with a portion converting into CO_2 and being released. Comparing the color of the samples after the experiment, it can be found that when the residence time was 30 min, some samples became black. This resulted in a significant decrease in the proportion of the C element, overtaking the release rate of the H element. As time progressed to 40 min, the proportion of the H element continued to decline, with an accelerated rate, decreasing from 4.32% at 30 min to 3.83%. Conversely, the proportion of the C element increased notably, from 32.24% at 30 min to 35.36%. At this stage, the slow oxidation reaction of the C element had essentially concluded. The reduction in the proportions of H elements became the primary factor driving the increase in the proportion of the C element. These findings demonstrated that residence time is a crucial factor influencing the slow oxidation reaction of the C element, with the 30 min mark serving as a crucial point. The period between 20 and 30 min represented the active reaction time, while after 30 min, the reaction nearly ceased.

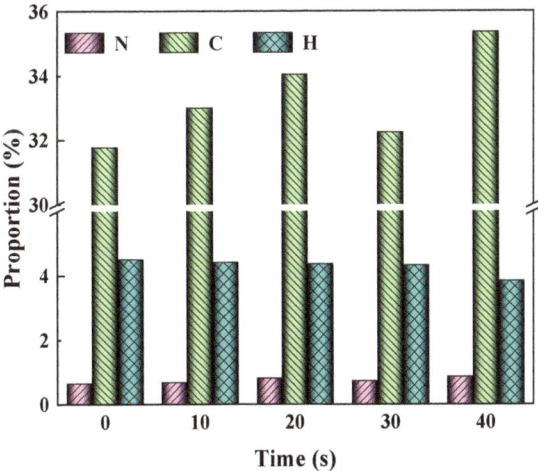

Figure 5. Proportion of C, H, and N elements at different times (275 °C).

3.3. Effect of Microwave Power on Migration of C, H, and N Elements

Based on the preliminary experiment, microwave power settings of 160 W, 180 W, and 200 W were selected. Figure 6 illustrates the temperature of corn straw particles during the initial 20 min at different microwave powers, with peak temperatures reaching 144 °C, 193 °C, and 233 °C, respectively. The temperature changes in corn straw followed a similar pattern at other time intervals. At the 40 min mark, the maximum attainable temperatures for the corn straw were 165 °C, 223 °C, and 257 °C, respectively.

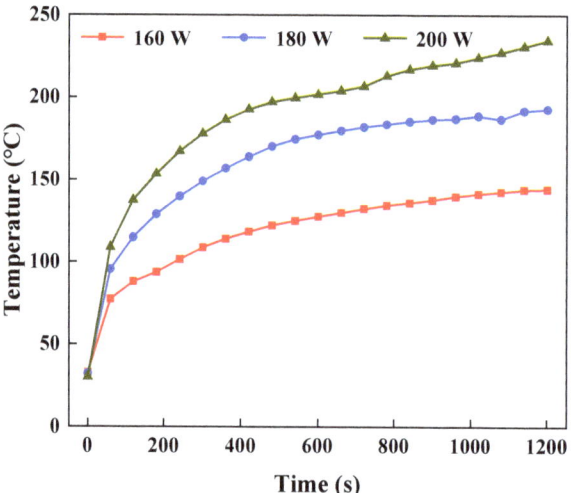

Figure 6. Temperatures of corn straw particles at different microwave powers.

Figure 7 displays the changes in the proportion of C, H, and N elements at various microwave powers. Across the three microwave power settings, the changes in C, H, and N elements followed a consistent pattern: the proportion of the C element increased with time, the proportion of the H element decreased with time, and the proportion of the N element remained relatively stable with time. At the 10 min mark, the proportions of the C element for 160 W, 180 W, and 200 W microwave power were 31.62%, 31.81%, and 32.28%, respectively. During this time, the sample heating rates corresponding to these microwave power levels were as follows: 16.5 °C/min, 17.8 °C/min, and 24.4 °C/min. According to the relevant literature, the heating rate significantly affects the release of volatile gases [24], and the change in the proportion of the C element confirmed this observation. With an increased heating rate, the release of volatile gases containing H was accelerated, resulting in a decrease in the H element content and an increase in the C element proportion.

In the case of 200 W, the proportion of the H element changed very little over time. At 10 min, the H element content was 4.48%, and at 40 min, it was 4.42%, which was negligible. This indicated that most of the volatiles containing H releasable under this power condition have been released within a certain time. Consequently, even as time progressed, the volatiles containing H were no longer released, and the proportion of the H element remained almost constant until a higher temperature condition for further volatile release was reached. Different from the previous section, the microwave power was not adjusted during the experiment in this section, so the migration rule of C and H elements was different.

When the microwave power was 160 W and the time exceeded 30 min, the proportion of the C element began to decline, which was because the oxidation rate of the C element exceeded the release rate of volatile fractions containing H. At the same time, when the microwave power was 180 W and 200 W, the proportion of the C element continued to maintain an upward trend. It was proved that the temperature required for the oxidation of the C element was significantly lower than the temperature containing H volatile release. With the increase in microwave power, the maximum temperature that core straw can reach increased, and the release rate of volatiles containing H was accelerated. Although the oxidation rate of element C was also accelerated, the former rate was faster, so the proportion of element C will continue to rise.

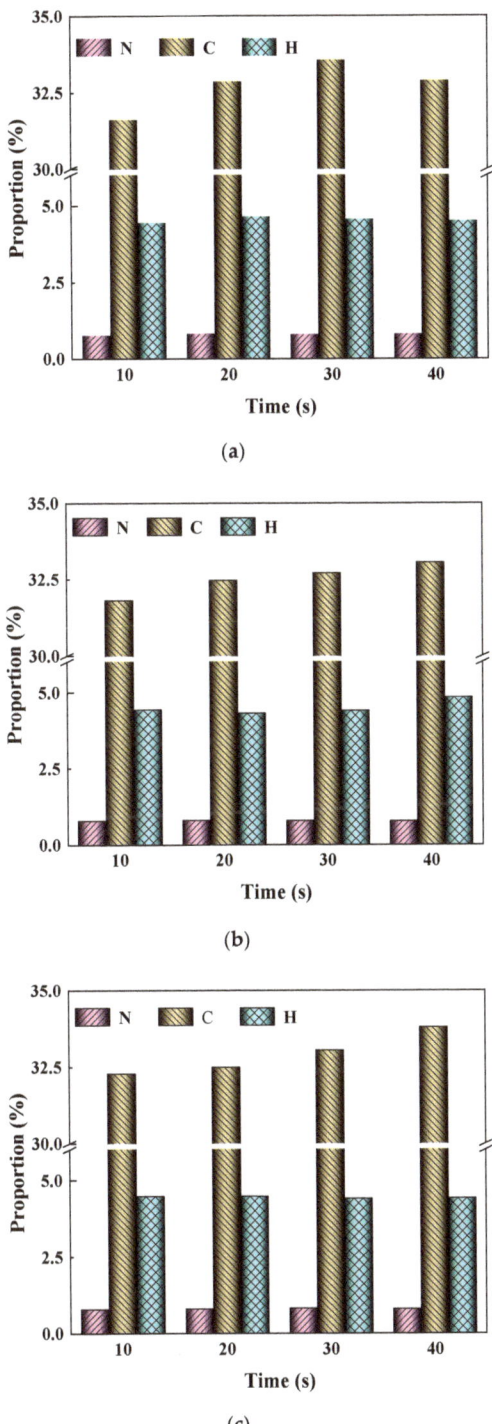

Figure 7. Proportions of C, H, and N elements at (**a**) 160 W, (**b**) 180 W, and (**c**) 200 W.

4. Conclusions

In this study, the effects of temperature, residence time, and microwave power on the migration of C, H, and N elements of corn straw during the microwave heating process were experimentally investigated. The following conclusions were obtained:

At temperatures below 275 °C, N elements of corn straw particles did not migrate much. As the temperature increased, the carbonization process of corn straw was more complete, and the proportion of the C element increased gradually. At 225 °C, the proportion of the C element decreased due to an oxidation reaction.

The residence time was critical for the migration of C, H, and N elements. The increase in residence time made the carbonization process of corn straw more complete, resulting in an increase in the C element proportion from 31.77% to 35.36%. A significant decrease in the proportion of the C element occurred around the 30 min mark, indicating a notable oxidation process.

Microwave power played a role in both the heating rate of the sample and the maximum temperature it could reach, which affected the changes in C, H, and N elements. At 160 W, the proportion of the C element increased initially and then decreased as residence time grew, with the turning point occurring at 30 min due to the oxidation of the C element. The proportion of the H element decreased as residence time increased due to the continuous release of H-containing volatiles. When the microwave power was 180 W and 200 W, the carbonization process of the C element was more complete with the increased residence time. Consequently, the proportion of the C element increased, while the proportion of the H element decreased.

Author Contributions: Z.L.: Experiment and writing original draft, W.C.: Experiment and writing original draft, M.Z.: Supervision, W.Z.: Validation, Y.Z.: Supervision. All authors have read and agreed to the published version of the manuscript.

Funding: National Natural Science Foundation of China (52076049), Heilongjiang Province "Double First-class" Discipline Collaborative Innovation Achievement Project (LJGXCG2023-080), and Heilongjiang Provincial Key R&D Program "Unveiling the Leader" Project (2023ZXJ02C04).

Data Availability Statement: Data are contained within the article.

Acknowledgments: Financial support was provided by the National Natural Science Foundation of China (52076049), Heilongjiang Province "Double First-class" Discipline Collaborative Innovation Achievement Project (LJGXCG2023-080), and Heilongjiang Provincial Key R&D Program "Unveiling the Leader" Project (2023ZXJ02C04).

Conflicts of Interest: The authors declare no conflict of interest.

References

1. Amalina, F.; Krishnan, S.; Zularisam, A.W.; Nasrullah, M. Effect of process parameters on bio-oil yield from lignocellulosic biomass through microwave-assisted pyrolysis technology for sustainable energy resources: Current status. *J. Anal. Appl. Pyrolysis* **2023**, *171*, 105958. [CrossRef]
2. Rajpoot, L.; Tagade, A.; Deshpande, G.; Verma, K.; Geed, S.R.; Patle, D.S.; Sawarkar, A.N. An overview of pyrolysis of de-oiled cakes for the production of biochar, bio-oil, and pyro-gas: Current status, challenges, and future perspective. *Bioresour. Technol. Rep.* **2022**, *19*, 101205. [CrossRef]
3. Dong, F.; Wang, Y.; Su, B.; Hua, Y.; Zhang, Y. The process of peak CO_2 emissions in developed economies: A perspective of industrialization and urbanization. *Resour. Conserv. Recycl.* **2019**, *141*, 61–75. [CrossRef]
4. Gadkari, S.; Fidalgo, B.; Gu, S. Numerical investigation of microwave-assisted pyrolysis of lignin. *Fuel Process. Technol.* **2017**, *156*, 473–484. [CrossRef]
5. Kamari, M.L.; Maleki, A.; Daneshpour, R.; Rosen, M.A.; Pourfayaz, F.; Nazari, M.A. Exergy, energy and environmental evaluation of a biomass-assisted integrated plant for multigeneration fed by various biomass sources. *Energy* **2023**, *263*, 125649. [CrossRef]
6. Contescu, C.I.; Adhikari, S.P.; Gallego, N.C.; Evans, N.D.; Biss, B.E. Activated carbons derived from high-temperature pyrolysis of lignocellulosic biomass. *C-J. Carbon Res.* **2018**, *4*, 51. [CrossRef]
7. *3060 Blue Book on Zero-Carbon Biomass Energy Development Potential*; Biomass Energy Industry Branch of China Industrial Development Promotion Association: Beijing, China, 2021.
8. Sun, J.; Luo, J.; Lin, J.; Ma, R.; Sun, S.; Fang, L.; Li, H. Study of co-pyrolysis endpoint and product conversion of plastic and biomass using microwave thermogravimetric technology. *Energy* **2022**, *247*, 123547. [CrossRef]

9. Wu, G.; Cheng, D.; Li, L.; Li, C.; Jiang, G.; Zheng, Y. Biomass energy utilization and soil carbon sequestration in rural China: A case study based on circular agriculture. *J. Renew. Sustain. Energy* **2018**, *10*, 13107–13111. [CrossRef]
10. Shi, C.; Shi, H.; Li, H.; Liu, H.; Mostafa, E.; Zhao, W.; Zhang, Y. Efficient heating of activated carbon in microwave field. *C-J. Carbon Res.* **2023**, *9*, 48. [CrossRef]
11. Guo, R.M.; Cheng, T.C.; Ng, J.H.; William, W.F.C.; Su, S.L.; Hwai, C.O.; Farid, N.A. Microwave pyrolysis for valorisation of horse manure biowaste. *Energy Convers. Manag.* **2020**, *220*, 113074.
12. Asif, F.C.; Saha, G.C. Graphene-like carbon structure synthesis from biomass pyrolysis: A critical review on feedstock-process-properties relationship. *C-J. Carbon Res.* **2023**, *9*, 31. [CrossRef]
13. Qiu, T.; Liu, C.; Cui, L.; Liu, H.; Kashif, M.; Zhang, Y. Comparison of corn straw biochars from electrical pyrolysis and microwave pyrolysis. *Energy Sources Part A Recovery Util. Environ. Eff.* **2023**, *45*, 636–649. [CrossRef]
14. Shvets, A.; Vershinina, K.; Vinogrodskiy, K.; Geniy, K. Microwave pyrolysis of woody biomass: Influence of radiation power on the composition of conversion products. *Appl. Sci.* **2023**, *13*, 7926. [CrossRef]
15. Zhang, Y.; Fu, W.; Cui, L.; Tariq, M.; Li, B. Experimental microwave-assisted air gasification of biomass for syngas production. *Fuel* **2023**, *339*, 126954. [CrossRef]
16. Fu, W.; Zhang, Y.; Cui, L.; Liu, H.; Tarique, M. Experimental microwave-assisted air gasification of biomass in fluidized bed reactor. *Bioresour. Technol.* **2023**, *369*, 128378. [CrossRef] [PubMed]
17. Mao, X.; Kang, Q.H.; Liu, Y.; Asif, A.S.; Ao, W.; Ran, C.; Fu, J.; Deng, Z.; Song, Y.; Dai, J. Microwave-assisted pyrolysis of furfural residue in a continuously operated auger reactor: Characterization and analyses of condensates and non-condensable gases. *Energy* **2019**, *187*, 583–584. [CrossRef]
18. Shi, K.; Yan, J.; Luo, X.; Luo, X.; Yang, G.; Chen, Y.; Edward, L.; Wu, T. Production of H_2-rich syngas from lignocellulosic biomass using microwave-assisted pyrolysis coupled with activated carbon enabled reforming. *Front. Chem.* **2020**, *8*, 3. [CrossRef]
19. Zhou, N.; Zhou, J.; Dai, L.; Guo, F.; Wang, Y.; Li, H.; Deng, W.; Lei, H.; Paul, C.; Liu, Y.; et al. Syngas production from biomass pyrolysis in a continuous microwave assisted pyrolysis system. *Bioresour. Technol.* **2020**, *314*, 123756. [CrossRef]
20. Huang, F.; Tahmasebi, A.; Maliutina, K; Yu, J. Formation of nitrogen-containing compounds during microwave pyrolysis of microalgae: Product distribution and reaction pathways. *Bioresour. Technol.* **2017**, *245*, 1067–1074. [CrossRef]
21. Li, Z.; Zhong, Z.; Zhang, B.; Wang, W.; Gabriel, V.; Fernando, L. Effect of alkalitreated HZSM-5 zeolite on the production of aromatic hydrocarbons from microwave assisted catalytic fast pyrolysis (MACFP) of rice husk. *Sci. Total Environ.* **2020**, *703*, 134605. [CrossRef]
22. Wan, Y.; Chen, P.; Zhang, B.; Yang, C.; Liu, Y.; Lin, X.; Ruan, R. Microwave-assisted pyrolysis of biomass: Catalysts to improve product selectivity. *J. Anal. Appl. Pyrolysis* **2009**, *86*, 161–167. [CrossRef]
23. Lu, Q.; Zhao, W.; Xia, Y.; Liu, J.; Meng, H.; Guo, X.; Hu, S.; Hu, B. Research on the migration and transformation mechanism of nitrogen during biomass pyrolysis. *J. Fuel Chem. Technol.* **2023**, *51*, 1047–1059+1025.
24. Yao, X.; Xu, K. Pyrolysis characteristics of corn cob and release rule of gas products. *J. Agric. Eng.* **2015**, *31*, 275–282.

Disclaimer/Publisher's Note: The statements, opinions and data contained in all publications are solely those of the individual author(s) and contributor(s) and not of MDPI and/or the editor(s). MDPI and/or the editor(s) disclaim responsibility for any injury to people or property resulting from any ideas, methods, instructions or products referred to in the content.

MDPI
St. Alban-Anlage 66
4052 Basel
Switzerland
www.mdpi.com

C Editorial Office
E-mail: carbon@mdpi.com
www.mdpi.com/journal/carbon

Disclaimer/Publisher's Note: The statements, opinions and data contained in all publications are solely those of the individual author(s) and contributor(s) and not of MDPI and/or the editor(s). MDPI and/or the editor(s) disclaim responsibility for any injury to people or property resulting from any ideas, methods, instructions or products referred to in the content.

www.ingramcontent.com/pod-product-compliance
Lightning Source LLC
LaVergne TN
LVHW070713100526
838202LV00013B/1087